Federico Etro
Università Milano-Bicocca
Dipto. Economia Politica
Piazza dell' Ateneo Nuovo, 1
20126 Milano
Italy
federico.etro@unimib.it

Front cover picture:
The Fortune Teller by Michelangelo Merisi called Caravaggio
Courtesy of the Louvre Museum ©

ISBN 978-3-540-87426-3 e-ISBN 978-3-540-87427-0
DOI 10.1007/978-3-540-87427-0
Springer Dordrecht Heidelberg London New York

Library of Congress Control Number: 2009928781

Cover design: WMXDesign GmbH, Heidelberg, Germany

Printed on acid-free paper

Springer is part of Springer Science+Business Media (www.springer.com)

Federico Etro

Endogenous Market Structures and the Macroeconomy

Springer

Endogenous Market Structures and the Macroeconomy

A Vanessa

Preface

This is a key year for the evolution of international markets. The global economy is experiencing the most severe downturn since the thirties, it is temporarily leaving a path of sustained growth that characterized the last decades, and is facing an impressive decline of trade between countries. Banks are going bankrupt, the stock market has crashed, firms are going out of business or drastically reducing their production and exports, workers are being fired and investment in new business creation or innovation is shrinking. Meanwhile, consumers' confidence has dropped at its minimum, aggregate demand has been declining for months and expansionary policies and international coordination have failed to counteract the crisis until now. It is quite likely that all this will change sooner or later, but at the end of this crisis our understanding of the macroeconomy may change as well.

In front of these crucial events, this book is not an attempt at proposing a radically new way of interpreting macroeconomic phenomena, and, as a matter of fact, it is not even a book on macroeconomic theory. My more modest goal is to collect a number of insights derived from recent research on the role of competition and innovation in the analysis of three topics: business cycles, trade and growth through innovations. These topics are usually analyzed in different fields of research with limited communication, but they all have one common aspect: they study aggregate phenomena of the macroeconomy starting from the microeconomic analysis of markets. This book analyzes three main issues in a unified framework: the role of market structures in shaping the reaction of the economy to shocks and macroeconomic policies in the short and long run, the impact of globalization and trade policy on international market structures, and the role of investments in R&D and of innovation and competition policy in determining technological progress and growth. The novel aspect of our research is that we endogenize the market structures departing from two usual assumptions adopted in the modern theories, that is, on one side perfect competition (which leads to indeterminate market structures), and on the other side monopolistic behavior by an infinity of firms *à la* Dixit-Stiglitz (which leads to exogenous market structures).

The Endogenous Market Structures (EMSs) approach, as I will call it, is based on theories of imperfect competition with strategic interactions, as Cournot competition, Bertrand competition, Stackelberg competition, imperfect collusion or patent races based on investments in R&D, and introduces them in a macroeconomic framework where entry in the markets is endogenous and constrained by fixed costs of entry. This realistic characterization of the supply side allows us to explain a number of stylized facts that remain largely unexplained in the traditional approaches to business cycle, trade and growth, and to revisit a number of policy implications concerning macroeconomic policy, trade policy and innovation policy. In this sense, we hope that the EMSs approach will contribute also to the understanding of the current crisis, of the policies that we need to implement to solve such a crisis and avoid future ones, and of the scenarios for the destiny of globalization and growth.

From a research point of view, this book is a follow up of my earlier one, *Competition, Innovation, and Antitrust* (2007, Springer: New York and Berlin), which focused on microeconomic and industrial policy issues. This one is dedicated to the macroeconomic implications of the theory of EMSs, a topic on which economists as John Sutton (London School of Economics), Russell Cooper (University of Texas at Austin), Pietro Peretto (Duke University), Marc Melitz (Princeton University), Fabio Ghironi (Boston College) and many others have been working in the last years.[1]

The book is organized as follows. Chapter 1 reviews the standard neoclassical approach to macroeconomics. The aim is to summarize in the shortest space all the main results of the traditional approach to microfounded general equilibrium models of growth, trade and business cycle and to emphasize the advantages and the disadvantages of this approach. The neoclassical approach is based on the crucial assumption that all markets are perfectly competitive and that there are constant returns to scale, which leads to the indeterminacy of the market structures: nothing can be said about how many firms are active in any market, about their production or investment levels and about their (stock market) value, while mark ups and operative profits are zero. Recent developments have introduced the analysis of monopolistic firms in otherwise standard models, leading to important investigations in the fields of endogenous growth, intra-industry trade and New-Keynesian macroeconomics, but in most of these models the market structure remains exogenous, with constant mark ups, absence of strategic interactions and a continuum of monopolistic firms. These limitations motivate our approach,

[1] We should cite on the theoretical front at least Michael Devereux (University of British Columbia), Jean-Pascal Benassy (Cepremap, Paris), Olivier Blanchard (M.I.T.), Francesco Giavazzi (Bocconi University), Florin Bilbiie (HEC Paris Business School), Nir Jaimovich (Stanford University), James Markusen (University of Colorado at Boulder), Elhanan Helpman (Harvard University), Krešimir Žigić (CERGE-EI, Prague), Frank Stähler (University of Otago), Anthony Creane (Michigan State University), Toshihiro Matsumura (University of Tokyo), and on the empirical front Timothy Bresnahan (Stanford University), Steven Berry (Yale University), Mark Manuszak (Federal Reserve Board), Hugo Hopenhayn (U.C.L.A.), Jeffrey Campbell (Federal Reserve Bank of Chicago), Christian Broda (University of Chicago) and David Weinstein (Columbia University).

that is aimed at studying the determinants and the consequences of genuinely endogenous market structures.

Chapter 2 reviews the foundations of the theory of EMSs and applies it to simple macroeconomic models. I build the concept of EMSs gradually, starting from a general definition that applies to models of competition in and for the market and that derives from my earlier book. Then, I focus on a class of microfounded partial equilibrium models that can be used to study both competition in quantities and in prices. The third step is to specialize this framework to the case of isoelastic preferences leading to constant elasticity of substitution between goods and to derive Cournot, Bertrand and Stackelberg equilibria. Fourth, I extend this static equilibrium model to a dynamic situation with two periods and provide a first example of the dynamic effects due to the presence of EMSs. Fifth, I introduce general equilibrium considerations in the basic model. Last, I develop a fully dynamic general equilibrium model with Cournot competition between firms producing homogenous goods and EMSs, which will be the workhorse model of a large part of the book. I characterize the EMSs in the short run and in steady state. This set up without physical capital or any form intertemporal substitution (that is, without the traditional mechanisms of business cycle propagation) provides the first insights of the EMSs approach to macroeconomics. Consider a positive shock to such an economy (a reduction of variable costs, for instance due to lower energy prices, or the introduction of a general purpose technology which reduces entry costs, for instance cloud computing). The shock increases profits and the stock market value of the firms, which attracts entry of new firms and leads to stronger competition between them, lower mark ups, larger individual and aggregate production, and larger total consumption (while any additional effect would be absent in a neoclassical version of the same model with perfect competition). In conclusion, I discuss the role of trade between countries and growth due to technological progress in such a framework.

Chapter 3 starts with an empirical analysis of the U.S. aggregate data over the last sixty years, emphasizing a few stylized facts concerning the process of business creation. In particular, entry of firms and profits appear to be strongly procyclical while mark ups exhibit a countercyclical pattern, which is confirmed by a Vector Auto Regressive analysis. This motivates the development of a dynamic stochastic general equilibrium model with EMSs that extends the one of the previous chapter with the introduction of endogenous savings, endogenous labor supply and imperfect substitutability between goods. In this framework the equilibrium interest rate is not governed by the marginal productivity of capital as in the neoclassical approach, but by the dynamics of the stock market value, in particular by the stock market return in terms of capital gains and dividends, which depends on the entry process and on the endogenous level of competition. Notice that the stock market affects the real economy not only because it reflects productivity changes, but also because it reflects the strategic interactions between firms engaged in competition and the entry/exit process due to various shocks. The introduction of EMSs creates a competition effect associated with positive temporary shocks which enhances their propagation by reducing the mark ups and increasing the real wages (so as to magnify the impact on consumption and labor supply due to intertemporal substitution). Both supply and demand shocks induce impulse response functions that are largely in line with the evidence. Moreover, the analysis of the second moments of the basic model (and of its extension to the accumulation of physical capital) shows that the introduction of EMSs allows us to outperform Real Business Cycle models in explaining the cyclical variability of U.S. data, in particular of output, consumption, labor supply, profits and mark ups. The model can also be used to evaluate the impact of shocks to the fixed cost of entry, and, more important, the role of fiscal and monetary policy. One one side, I show that the optimal fiscal policy requires countercyclical tax rates on sales and labor income to optimize the process of business creation along the business cycle. One the other side, I emphasize the distortions induced by inflation on the mechanism of entry and discuss the impact of strategic interactions on the effectiveness of monetary policy. Finally, I discuss the impact of labor and credit market imperfections in the dynamic model with EMSs.

Chapter 4 begins with an empirical discussion on the impact of market size on the endogenous elements of a market structure (prices, production levels and number of firms) and provides some support for the implications of the EMSs approach through case studies and a panel data analysis of the German manufacturing sector. In particular, larger market sizes appear to increase less than proportionally the number of firms, while increasing their individual production and reducing their mark ups: these results are in contrast with the traditional Dixit-Stiglitz model of monopolistic competition and support the existence of strategic interactions in endogenous entry models. This empirical analysis motivates an extension of the Krugman model of trade to strategic interactions. In particular, I introduce Cournot and Bertrand competition in a standard model of intra-industry trade and show that the gains from trade derive from price reductions due to the strength of global competition rather than from an increase in the number of consumed varieties. Afterward, I extend our dynamic stochastic general equilibrium model to trade between two countries: the basic version of this model is due to Ghironi and Melitz, and I discuss its pathbreaking implications for the behavior of the real exchange rate and also its extension to EMSs. The rest of the chapter is dedicated to policy issues in a simple (but quite general) model of EMSs, and derives implications for trade policy, export promotion policy, R&D policy and exchange rate policy. The main result concerns the general optimality of strategic export promotion, in particular through positive export subsidies, in the presence of endogenous entry in global markets. This result is in contrast with the neoclassical theory, for which export taxes are optimal (to improve the terms of trade) and with the strategic trade policy approach with exogenous market structures, that recommends export taxes on price setting firms (for profit shifting reasons). In case of international EMSs, I show that it is always optimal to subsidize exports to induce the domestic firms to reduce their prices (or expand their production) so as to gain market shares abroad and limit entry of competitors. I characterize the optimal trade policy under alternative forms of competition and show that the optimal unilateral export subsidy is inversely related to the elasticity of foreign demand (actually, when goods are homogenous the optimal subsidy under EMSs is the exact opposite of the optimal export tax in the neoclassical trade theory). The same idea can be applied to competitive devaluations. Finally, I apply the same framework to show the general optimality of R&D subsidies and protection of domestic intellectual property rights to strengthen the incentives of the domestic firms to invest in R&D and lead the competition for international markets.

Chapter 5 starts with an empirical test of the EMSs approach to the competition for the market. The test is based on the different role played by incumbent leaders in investing in R&D, and is built around a unique dataset for the German manufacturing sector. In line with the theoretical predictions of the EMSs approach, Tobit regressions (augmented with instrumental variable analysis) show that incumbent leaders tend to invest more than the other firms if and only if they face an endogenous entry threat. Moreover, the analysis emphasizes a number of market specific and institutional factors (such as IPRs protection) that are correlated with investment in R&D, and suggests the need for a microeconomic analysis of the drivers of technological progress. On this basis, first I analyze endogenous growth models with EMSs in the competition in the market, and then I extend the Schumpeterian growth model with patent races characterized by strategic interactions between firms investing in R&D and endogenous entry. In this context I emphasize the emergence of a form of dynamic inefficiency in the business creation process, which is characterized by too many small firms: a better allocation of resources could be achieved with larger firms investing more efficiently in R&D. Moreover, I show that incumbent leaders tend to invest more than the outsiders when pressured by entry, and they end up leading the growth process. As a consequence, strengthening the protection of IPRs is going to increase the value of being an incumbent leader, which in turn is going to enhance the incentives to invest for both the leaders and the outsiders, and therefore to augment the growth rate: this implies that patent protection and the neutrality of antitrust policy toward IPRs (of high-tech leaders) do promote investments in innovation and they do so especially in case of sequential innovations. I also use the model to study fiscal and monetary policy in a growing economy, showing the general optimality of R&D subsidies and of a monetary policy that minimizes the distortions on the business creation process. Finally, I investigate international growth issues, with particular attention to the endogenous size of markets, to the world tecnological leadership and the international coordination of R&D policy.

Chapter 6 provides a non-technical discussion of the main results of the EMSs approach and employs them to discuss a number of recently debated issues. This chapter is accessible to a non-specialized audience and tries to point out the novel contributions of the EMSs approach for the understanding of real world phenomena and for policymaking, starting of course from the current recession and its possible solutions. I pay close attention to three broad topics: the evolution of global markets of the New Economy and their EMSs (as for cloud computing, online advertising or the browsers market), the evolution of macroeconomic phenomena (as business creation, long run growth, globalization and innovation), and the prescriptions for policymakers (on macroeconomic policy, trade policy, innovation policy and competition policy). With this mix of applications I want to make a key point: there is no way to understand the macroeconomy that does not start from the structure of the markets that belong to it, especially the high-tech and global markets whose shocks, innovations and exchanges are at the basis of economic fluctuations, growth and trade.

Chapter 7 concludes my two books on EMSs in partial equilibrium microeconomics and general equilibrium macroeconomics summarizing in a non-technical way their results and their implications for industrial and macroeconomic policy.

This book is the fruit of fifteen years of studies and I want to express my gratitude to all the teachers and colleagues who have contributed to my understanding of the macroeconomy. The first category includes at Southampton University Morten Ravn, James Malcomson and Jacques Cremer, at the London School of Economics Nobuhiro Kiyotaki, at U.C.L.A. Gary Hansen, Michele Boldrin, Pietro Reichlin, Costas Azariadis, Carlos Vegh and Edward Leamer, at M.I.T. Daron Acemoglu, Guido Lorenzoni and Jeaume Ventura, and at Harvard University Philippe Aghion, Gary Chamberlain, Elhanan Helpman, David Laibson, Ariel Pakes, Kenneth Rogoff, Andrei Shleifer, Joseph Zeira, and, most of all, Alberto Alesina and Robert Barro. At the University of Milan, Bicocca, I benefited from discussions with many colleagues from my Department, especially with Patrizio Tirelli, Emilio Colombo, Silvia Marchesi, Michela Cella, Dario Pontiggia and, most of all, Andrea Colciago. Andrea has been coauthor of two papers (one of which at the third round of submission for The Economic Journal) that are largely reflected in Chapter 3: I am extremely grateful to him for substantial help in the quantitative analysis of this chapter. I would also like to thank Andy Snell, John Moore and Jozsef Sakovics from the University of Edinburgh. Finally, I am grateful to Alessandro Penati: working for his asset management company Epsilon many years ago, I learnt all the little I know about applied finance.

The ideas of this book were presented in many recent seminars (University of St. Andrews; University of Amsterdam; Catholic University of Leuven; IMT, Lucca; Tilburg University; Charles University, Prague) and conferences (the Round Table on Competition in a Period of Crisis and the Intertic

Conference on Recent Developments in Antitrust Policy in Rome, the Anglo-French-Italian Macroeconomic Workshop in Pavia, the OECD Conference on Innovation in the Software Sector in Cáceres, Spain, the ZEW Conference on the Economics of Innovation and Patenting in Mannheim, Germany, and the 2008 CRESSE Conference in Athens). I am thankful to Fabio Ghironi, Marc Melitz, Jean-Pascal Bénassy, Andrew Scott, Charles Horioka, Guido Ascari, Huw Dixon, Frode Steen, Liam Graham, Jan Boone, Michal Kejak, Evangelia Vourvachaki, Manfredi La Manna, Marco Faravelli, John Beath, Richard Schmalensee and many others for useful comments.

There have been three important events that largely helped the preparation of this book. The first one, which forced me to think in a systematic way about the material of this book, was the Bi-annual Lecture of the Review of Business and Economics, organized by the Faculty of Economics, Applied Economics and Business of the Katholieke Universiteit Leuven (Belgium). I gave this Lecture on December 12, 2007, and I am extremely grateful to the Faculty and the Dean for the kind invitation, to the audience, and, for insightful discussions, to Raimond De Bondt, Christophe Crombez, Dirk Czarnitzki and Kornelius Kraft. The last two became also two coauthors later on, and I am extremely indebted with one of them, Dirk, for extensive help with the empirical analysis of Chapters 4 and 5.

The second event was the 2008 Intertic Conference, held at the University of Milan, Bicocca on "Endogenous Market Structures and Industrial Policy" in June 5-6, 2008, a unique occasion to develop the academic debate on this emerging body of literature. I am grateful to all the participants, starting with Avinash Dixit, who gave the Intertic Lecture, John Sutton, who gave the Stackelberg Lecture, and the three Vice-Presidents of Intertic (the International Think-tank on Innovation and Competition) Simon Anderson, Dirk Czarnitzki, and Krešimir Žigić. I am particularly thankful to the interesting discussions with Pietro Peretto, Tommaso Mancini-Griffoli, Lilia Cavallari, Kevin Tsui, Arijit Mukherjee, Emek Basker, Toshihiro Matsumura, Irina Suleymanova, Axel Gautier, Lidia Tsyganok, Yannis Katsoulacos, Michele Polo, Chiara Fumagalli, Michael Ward, and the other participants.

The last important event is the Dynamic Competition Lecture that I have been invited to give at the International Workshop on IPRs and Competition Policy in Osaka (Japan) on November 27, 2009. The preparation of that lecture relies heavily on the thesis developed in this book. I am thankful to Noryiuki Doi, Testuya Shinkai and the JSPS group for the kind invitation to Japan.

I present a lot of unpublished material in the book. The rest derives from articles of mine, as "Endogenous Market Structures and Strategic Trade Policy" (2010, *International Economic Review*), "The Economic Impact of Cloud Computing on Business Creation, Employment and Output in the E.U." (2009, *Review of Business and Economics*), "Stackelberg Competition with Endogenous Entry" (2008, *The Economic Journal*), "Growth Leaders" (2008,

Journal of Macroeconomics), "Endogenous Market Structures and Macroeconomic Theory" (2007, *Review of Business and Economics*), "Political Geography" (2006, *Public Choice*), "International Unions" (2005, *The American Economic Review*, with A. Alesina and I. Angeloni) and "Innovation by Leaders" (2004, *The Economic Journal*).

I used early drafts of the first two chapters for teaching purposes. The first one (in a very preliminary version) for a course of undergraduate advanced macroeconomics held at Luiss University (Rome) in 2002-2003 and (in part) for a postgraduate course in Political Economy that I gave between 2003 and 2006 at the DEFAP (PhD in Public Economics) in Milan. I use the second chapter for my classes of Industrial Organizations for the Scottish Graduate Programme in Economics held since 2007 at the University of Edinburgh. I am thankful to all of my students for many smart comments. The book could be also used as a textbook for advanced macroeconomic courses with a special emphasis on microfoundations and international economics.

Most of this monograph and of my earlier one is based on theoretical research that I have developed alone. The hope is that the associated reduction of academic productivity has been partly compensated by a coherent view on these topics. My first contribution to the EMSs approach in macroeconomics appeared in an article of 2004, which ironically generated what may be the most extreme forms of reaction that an academic paper can generate. On one side, journalists from the Economist, the Sunday Times, Le Libre Belgique and other newpapers publicized it as a theoretical defense of the position of Microsoft in its famous antitrust case.[2] My research did not have that case in mind, and it was aimed at more general and largely unrelated conclusions. Nevertheless, my position radically against the traditional view of incumbent leaders that jeopardize innovation, and my analysis of the beneficial role of these leaders in driving sequential innovations and technological progress were correctly associated with a critical view of the positions of the EU antitrust authorities. I am glad to notice (without causal implications) that, since then, Europe has witnessed a substantial change in the terms of the debate on the role of Microsoft in the software market.

On the other side, the same article has attracted a radical epistemological critique in an essay self-defined "Towards Good Social Science"[3] which had nothing else to do than comparing the nature of "good natural science", exemplified by the heliocentric cosmology, the theory of relativity and the theory of evolution, with the nature of "bad social science", which the essay exemplified with my theory of innovation and growth in the field of economics (!) and with the rational choice theory of religion by Rodney Stark in the field of sociology. The critique argued that, contrary to the good science,

---

[2] For instance, see "Combattre les barrières, pas les leaders" (*Le Libre Belgique*, June 18, 2004).

[3] Moss, Scott and Bruce Edmonds (2005, *Journal of Artificial Societies and Social Simulation*, 8, 4, 1-15).

both the last two theories were not based on a proper empirical motivation. The reader will find here old and new empirical motivations for my theory of innovation and for the EMSs approach in general, including a wide investigation with Tobit regressions and IV analysis of the determinants of R&D investments and of the role of incumbent leaders and entry pressure, an analysis of the relation between market size and endogenous components of the market structure through panel and cross -sectional data, and a VAR analysis of the cyclical behavior of the market structures along the business cycle.

However, as a consequence of this critique, a methodological remark is in order. I believe that economic theory is not about absolute or even relative truth as natural science, but about reasonable descriptive and normative principles that derive from assumptions that must be realistic (or more realistic than in other theories) but parsimonious, that provide explanations for important empirical regularities, and that become a useful benchmark for policymakers over time. Only in this general sense, social science can be, hopefully, useful.

A last word on the cover of this book, which reproduces a masterpiece by Michelangelo Merisi called Caravaggio, painted probably in 1595 and currently at the Louvre Museum of Paris. The work of Caravaggio and this painting in particular may be seen as a turning point of Western paintings, the ultimate achievement in the evolution of realism and the ultimate source of the following major international artists, as Velasquez, Rubens, de La Tour, van Honthorst, Rembrandt and, of course, Vermeer. In this early genre painting, the Milanese artist depicted a gypsy girl reading the palm of a boy who expects a prediction of his future. Something else, however, is going on in the scene, and we leave the reader to realize what.

Macroeconomics cannot predict the future, but hopefully it can provide the tools to understand some endogenous and structural aspects of the way markets work.

Federico Etro
Department of Economics, University of Milan, Bicocca
Milan, March 2009

Contents

1. Neoclassical Macroeconomics 1
1.1 The Microfoundation of Macroeconomics 2
1.1.1 Consumption demand and labor supply 3
1.1.2 Public spending and fiscal policy 6
1.1.3 Money demand and monetary policy 9
1.1.4 Labor and capital demand by firms 11
1.1.5 Perfect competition and market structure indeterminacy 12
1.1.6 Market clearing and general equilibrium 14
1.2 Growth 16
1.2.1 The Solow model 16
1.2.2 The Diamond model 17
1.2.3 Dynamic fiscal policy 20
1.2.4 Dynamic monetary policy 22
1.2.5 Neoclassical endogenous growth 24
1.3 International Finance and Trade 26
1.3.1 International finance 27
1.3.2 International fiscal policy 29
1.3.3 International monetary and exchange rate policy 30
1.3.4 International trade theory 33
1.3.5 International trade policy 36
1.4 Business Cycles 37
1.4.1 The Ramsey model 37
1.4.2 Real Business Cycles 40
1.4.3 Monopolistic firms with exogenous entry 46
1.5 Conclusions 50

2. The EMSs Approach to Macroeconomics 51
2.1 EMSs in Partial Equilibrium 55
2.2 Microfounded EMSs 60
2.2.1 EMSs with competition in quantities 60
2.2.2 EMSs with competition in prices 63
2.2.3 Optimal EMSs 64
2.3 EMSs with Isoelastic Sub-utility 65
2.3.1 Cournot competition 67

2.3.2 Bertrand competition ... 69
2.3.3 Stackelberg competition ... 71
2.3.4 Collusion, endogenous entry costs and other extensions 73
2.4 EMSs in a Two Period Economy ... 75
2.5 EMSs in General Equilibrium ... 78
2.6 EMSs in an Infinite Periods General Equilibrium Economy .. 81
2.6.1 A model of business creation with Cournot competition 82
2.6.2 EMSs in the long run ... 87
2.7 Business Cycle ... 88
2.8 Trade ... 90
2.9 Growth ... 91
2.10 Conclusions ... 93

3. Endogenous Market Structures and Business Cycles ... 95
3.1 Empirical Evidence on EMSs over the Business Cycle ... 98
3.1.1 Business cycle fluctuations in the U.S. ... 100
3.1.2 VAR analysis ... 103
3.2 A Model of the Business Cycle with EMSs ... 105
3.2.1 Cournot competition ... 108
3.2.2 Bertrand competition ... 109
3.2.3 Stackelberg competition and imperfect collusion ... 110
3.2.4 Short run EMSs ... 110
3.3 Steady State EMSs ... 114
3.3.1 Dynamics and steady state under Cournot competition 115
3.3.2 Dynamics and steady state under Bertrand competition 118
3.4 Impulse Response Functions and Second Moments ... 120
3.4.1 Supply shocks ... 121
3.4.2 Demand shocks ... 125
3.4.3 Second moments ... 128
3.5 EMSs and Accumulation of Physical Capital ... 130
3.6 Fiscal Policy ... 133
3.7 Monetary Policy ... 137
3.8 Labor and Credit Market Imperfections ... 141
3.9 Conclusions ... 144

4. Endogenous Market Structures and International Trade and Finance ... 145
4.1 Empirical Evidence on EMSs and Market Size Effects ... 149
4.1.1 Panel data on the number of firms ... 150
4.1.2 Cross-sectional data on firms' production ... 153
4.2 A Model of Trade with EMSs ... 156
4.2.1 The Krugman model ... 157
4.2.2 Cournot competition ... 158
4.2.3 Bertrand competition ... 160
4.2.4 Gains from trade and business destruction ... 161

4.3 The Ghironi-Melitz Model . . . 163
4.3.1 International trade and business cycle . . . 164
4.3.2 International finance and macroeconomic policy . . . 167
4.4 Trade Policy and Import Tariffs . . . 169
4.5 Strategic Export Promotion . . . 172
4.5.1 A model of international competition . . . 173
4.5.2 Optimal trade policy *à la* Brander-Spencer . . . 174
4.5.3 Optimal trade policy with EMSs . . . 175
4.6 Trade Policy and Export Subsidies . . . 177
4.6.1 Optimal trade policy with Cournot competition . . . 177
4.6.2 Optimal trade policy with Bertrand competition . . . 181
4.7 Protectionism, Lobbying and Trade Wars . . . 183
4.7.1 Indirect export promotion and infant industries . . . 184
4.7.2 The political economy of export promotion . . . 184
4.7.3 Nash equilibrium trade policy . . . 186
4.8 Exchange Rate Policy . . . 187
4.8.1 Competitive devaluations with Cournot competition . . . 189
4.8.2 Competitive devaluations with Bertrand competition . . . 190
4.9 R&D Policy and IPRs Protection . . . 193
4.10 Conclusions . . . 195

5. Endogenous Market Structures and Growth . . . 197
5.1 Empirical Evidence on EMSs and Innovation by Leaders . . . 200
5.1.1 Data Sources on R&D, Entry and Leadership . . . 201
5.1.2 Tobit Regressions . . . 203
5.1.3 Instrumental variables for entry . . . 207
5.2 A Model of Endogenous Growth . . . 212
5.2.1 Cournot competition . . . 215
5.2.2 Bertrand competition . . . 216
5.2.3 The Romer model . . . 216
5.2.4 The Peretto model with EMSs . . . 217
5.3 A Model of Schumpeterian Growth . . . 219
5.3.1 Patent races . . . 220
5.3.2 The Aghion-Howitt model . . . 222
5.3.3 A Schumpeterian model with EMSs . . . 223
5.4 EMSs with Growth Leaders . . . 226
5.5 Fiscal Policy . . . 231
5.6 Monetary Policy . . . 236
5.7 Growth and International Trade and Finance . . . 240
5.7.1 Endogenous size of national markets and global growth 241
5.7.2 World technological leadership . . . 243
5.7.3 Optimal R&D policy in the open economy . . . 245
5.8 Conclusions . . . 246

6. Economic Principles and Policy for the New Economy 247
6.1 Short-run EMSs and the Competition Effect 248
6.1.1 The boom of the 90s 250
6.1.2 The current global recession 252
6.2 Steady-state EMSs and Entry 258
6.2.1 Entry and business creation in the U.S. 261
6.2.2 Cloud computing: the impact of a General Purpose Technology 263
6.3 Gains from Trade and the Effects of Globalization 267
6.4 Growth Driven by Leaders 271
6.4.1 Online advertising: market structure and innovation .. 273
6.4.2 Google economics 276
6.5 Dynamic Inefficiency 277
6.6 Fiscal Policy 279
6.6.1 Fiscal policy in the U.S. and the E.U. 281
6.6.2 A supply-based fiscal policy 284
6.7 Monetary Policy 285
6.7.1 Monetary policy in the U.S. and the E.U. 287
6.8 Trade Policy 290
6.9 Exchange Rate Policy 294
6.10 Innovation and Competition Policy 299
6.10.1 Innovation policy and coordination 300
6.10.2 Competition policy in the U.S. and the E.U. 301
6.10.3 Entry and competition in the browsers' market 305
6.11 Conclusions 308

7. Epilogue 309

8. References 313

Index 339

1. Neoclassical Macroeconomics

It's amazing that the amount of news that happens in the world every day always just exactly fits the newspaper. **Jerry Seinfeld**

In this chapter we review the foundations of neoclassical economics, which is the dominant framework for the analysis of growth, trade and business cycles. Our objective is to introduce the fundamental tools of macroeconomic analysis, most of which will be widely used in the rest of the book, and to emphasize advantages and disadvantages of the traditional approach to the analysis of aggregate phenomena.

The neoclassical approach relies on three main assumptions. The first one is about the rational behavior of agents, who take decisions to maximize utility under rational expectations, and about the rational behavior in the management of the firms, whose aim is profit maximization. This leads to decisions based on marginal calculus: agents set marginal utility equal to zero and firms equate marginal revenues to marginal costs.

The second assumption is about the technological conditions, which are characterized by constant returns to scale of the factors of production, and the associated absence of costs of entry or fixed costs of production for any firm. Since the neoclassical approach has been largely developed during the industrialization phase of the Western world, the factors of production have been usually assumed to be labor and physical capital, the first as the endowment of the working class, and the second as the reproducible factor augmented through investments of the capitalist class.[1]

The third assumption is that all markets are characterized by perfect competition, in the sense that each firm decides on entry and production levels taking the prices as given, and the equality between supply and demand holds in every market so as to determine all the equilibrium prices (actually, in the absence of price-making agents, an ideal auctioneer is assumed to determine these prices and reach the general equilibrium). This, together with the assumptions of constant returns to scale and zero entry costs, leads to

[1] The classical approach, associated with the early works of Smith (1776), Ricardo (1817) and Marx (1867), focused also on land as an important factor of production in fixed supply.

F. Etro., *Endogenous Market Structures and the Macroeconomy*,

DOI: 10.1007/978-3-540-87427-0_1,

equilibria with zero profits for each firm and prices equal to the marginal cost of production in every market. Moreover, as we will see, the equilibria are characterized by indeterminate market structures, in the sense that the neoclassical approach has nothing to say about how many firms should be active in each market, how much each one of them should produce and what their (stock market) value should be. For the same reason, the neoclassical analysis of shocks (leading to business cycles), of openness (leading to trade) and of innovation (leading to growth) has no relations with the market structures.

To anticipate the core differences between the endogenous market structures approach and the neoclassical one, notice that in the rest of the book we will depart from the above assumptions in the following way. First, we will assume imperfect competition, in the sense that firms will not take prices as given, but they will choose their strategies (prices or quantities) and they will interact strategically. Second, the technological conditions will be characterized by positive fixed costs of entry, and we will depart from the (often obsolete) conflict between labor and capital as the two factors of production, to embrace a (more modern) concept of investment that is needed to enter in the market with new products (or with better products). Finally, we will endogenize the entry decision of the firms as a rationale profit maximizing decision, which will lead to the full characterization of the endogenous market structures in models of business cycle, trade and growth.

In this chapter we review the neoclassical approach paying particular attention to the consequences of its basic assumptions for the characterization of macroeconomic equilibria, and we try to emphasize the merits and the limits of this approach. In Section 1.1 we study the behavior of consumers and firms in a two period setup, first in partial equilibrium (that is taking interest rate, wages and prices as exogenous), and then in general equilibrium (that is endogenizing these prices). In Section 1.2 we introduce the two period setup in a general equilibrium overlapping generations model to study growth. In Section 1.3 we open up the economy to trade with the rest of the world. In Section 1.4 we deal with the Ramsey model of growth, and we describe the results of the Real Business Cycle approach and of its main extensions. At each step we examine the role of macroeconomic policy, namely fiscal and monetary policy.

1.1 The Microfoundation of Macroeconomics

In this section we study the behavior of consumers and firms in the simplest dynamic framework: a two-period model. As we will see, this simple framework is sufficient to show some of the most important principles of economic theory.[2]

[2] For an excellent undergraduate introduction to economic theory see Mankiw and Taylor (2006). For a good textbook of macroeconomics see Burda and Wyplosz (2005).

1.1.1 Consumption demand and labor supply

Our analysis of the microfoundations of macroeconomics starts from the consumption side, to examine the choices of rational consumers. Economic analysis assumes that consumers are characterized by rational preferences over different goods, and that they maximize a well-behaved utility function under a budget constraint.[3] Here, we consider an agent with a utility function which is separable between consumption C_t and working time l_t over two periods $t = 1, 2$:

$$U = u(C_1) - v(l_1) + \beta[u(C_2) - v(l_2)] \tag{1.1}$$

The function $u(\cdot)$ is an increasing and concave utility function, $\beta \in (0, 1)$ is the discount factor (smaller than one because current utility is preferred to future utility), and $v(\cdot)$ is an increasing and convex disutility function of labor. Current consumption equals (labor) income in the first period less savings S, and future consumption equals (labor) income in the second period plus savings and interests at the real rate r. Therefore, we have $C_1 = Y_1 - S$ and $C_2 = S(1 + r) + Y_2$, where $Y_t = w_t l_t$ and w_t is the real wage at time t.[4]

For now, let us focus on the allocation of consumption for a given income. Savings can be positive or negative (in the latter case the consumer is borrowing in the credit market), and they are chosen to maximize $u(Y_1 - S) + \beta u\left[S(1 + r) + Y_2\right]$, which implies the following optimality condition, also known as the *Euler condition*:

$$u'(C_1) = \beta(1 + r)u'(C_2) \tag{1.2}$$

This generates a savings function $S = S(Y_1, Y_2, r)$, which is increasing in current income, decreasing in future income, but with an ambiguous relation with the interest rate. If we define ρ as the time preference rate - such that $\rho = 1/\beta - 1$, we can characterize the:

OPTIMAL CONSUMPTION PATH. **Optimal consumption is increasing (decreasing) over time if the interest rate r is larger (smaller) than the time preference rate ρ. Consumption smoothing is optimal if they are equal.**

---

[3] Preferences over baskets of goods are rational if they satisfy completeness, transitivity and non-satiety (i.e.: baskets with more of each good are always preferred to baskets with less). When they can be characterized through continuous and differentiable utility functions they are usually referred to as "well-behaved".

[4] If P_t is the price level in t, W_t is the nominal wage and $1 + i$ is the nominal interest factor that is the relative price of current versus future consumption, we have $w_t = W_t/P_t$ and $1 + r = (1 + i)P_t/P_{t+1}$. Assuming that the price of the single consumption good is the same in both periods (which turns out to be true under perfect competition, constant technology and without nominal money growth), we can normalize it to one, which turns all nominal variables into real variables as well.

The result of constant consumption for $(1+r)\beta = 1$ corresponds to the outcome of the theory of consumption of Modigliani and Brumberg (1954) based on the *life-cycle hypothesis*, according to which people forecast their future earnings and smooth their consumption over the whole life, and to the *theory of permanent income* of Friedman (1957), according to which agents calculate the constant level of consumption which enables them to exhaust all of their accumulated wealth during lifetime. The main implication of these theories is that, in general, consumption should depend on wealth rather than current income, and should be smoother than income. A crucial consequence is that, since the typical lifetime income profile is shaped like an inverse U (with high earnings in the medium part of life), young people should consume more than their income by borrowing in the credit market and repaying their loans when older.

To derive more specific results, let us adopt an isoelastic utility function, namely one with a degree of relative risk aversion $\gamma(C) \equiv -u''(C)C/u'(C) > 0$ that is independent from the consumption level, say $\gamma \in [0, \infty)$ - this implies that also the intertemporal elasticity of substitution (between current and future consumption) is constant. A utility function which satisfies this requirement is $u(C) = C^{1-\gamma}/(1-\gamma)$, whose properties for $\gamma \to 1$ converge to those of a logarithmic utility: $u(C) = \log C$. With these preferences, the Euler condition becomes:

$$\frac{C_2}{C_1} = [\beta(1+r)]^{\frac{1}{\gamma}} \tag{1.3}$$

which shows that more risk averse agents decide to smooth consumption more. Using the budget constraint we can obtain the actual consumption functions and, in particular, the savings function:

$$S(Y_1, Y_2, r) = \frac{\beta^{\frac{1}{\gamma}}(1+r)^{\frac{1}{\gamma}} Y_1 - Y_2}{1 + r + \beta^{\frac{1}{\gamma}}(1+r)^{\frac{1}{\gamma}}} \tag{1.4}$$

Notice that higher current income (income growth) induces higher (lower) savings, but the relationship between savings and the interest rate is complicated by the presence of an income effect (higher r increases the value of current income, increasing consumption in both periods and therefore reducing savings), a substitution effect (higher r makes future consumption more attractive, increasing savings), and a wealth effect (higher r decreases the current value of future income, increasing savings). Under the assumption that future income is zero, the wealth effect disappears and we have $S_r \gtreqless 0$ if $\gamma \lesseqgtr 1$.

Let us move on to consider the choice of labor supply. The corresponding first order conditions are:

$$u'(C_t)w_t = v'(l_t) \quad t = 1, 2 \tag{1.5}$$

Consider the case of logarithmic preferences in consumption ($\gamma = 1$) and isoelastic disutility from labor $v(l) = \upsilon l^{1+1/\varphi}/(1 + 1/\varphi)$, where φ is the so-called Frish elasticity of labor supply. The optimality conditions imply $C_2 = C_1(1 + r)\beta$ and labor supply:

$$l_t = \left(\frac{w_t}{\upsilon C_t}\right)^{\varphi} \tag{1.6}$$

More important, the optimal ratio between current and future labor supply becomes $l_1/l_2 = [(w_1/w_2)(1 + r)\beta]^{\varphi}$. Relative labor supply depends on the ratio of wages according to the Frish elasticity φ. In particular, it is immediate to verify that a permanent increase in the wage does not affect current or future labor supply, while a temporary increase, say only in the current wage, increases current labor supply relative to the future one. This change is bigger when the elasticity of substitution is higher. Most of the modern theory of business cycles, starting with Lucas and Rapping (1969), relies on this mechanism of intertemporal substitution to explain the propagation mechanism of technological shocks.

The model can be easily extended to the case of uncertainty about a stochastic interest rate, say z. Introducing rational expectations of future utility, the Euler condition becomes:[5]

$$u'(C_1) = \beta E[(1 + z)u'(C_2)] \tag{1.7}$$

or, using the property of covariance between two random variables,

$$u'(C_1) = \beta \left\{E[1 + z]E[u'(C_2)] + Cov\left[z, u'(C_2)\right]\right\} \tag{1.8}$$

In general, this result is compatible with savings above or below their level in the absence of uncertainty, but in the case of isoelastic utility, higher (precautionary) savings emerge. Moreover, a perfect financial market should provide assets consistent with this condition, therefore we can use this model for asset pricing. The expected-return premium that an asset must offer relative to the risk free real rate r can be derived as $E(z) - r = -Cov[z, u'(C_2)]/E[u'(C_2)]$. A larger covariance between the return and the marginal utility (i.e. higher interest rates when consumption is low) is associated with a lower expected return. A positive expected return premium that an asset must offer relative to the risk free rate emerges in case of positive covariance of its return with consumption, otherwise known

[5] Rational expectations of a variable are its expected value on the basis of all the available information. They were introduced by Muth (1961) and popularized by Lucas (1972) as an alternative to adaptive expectations, which were dominant before. In the theory of savings, rational expectations imply that consumption follows a random walk (Hall, 1978). Intuitively, if consumption is expected to change, the agent can do a better job at smoothing consumption, therefore only unexpected shocks can randomly affect consumption.

as "consumption beta".[6] Mehra and Prescott (1985) have shown that it is difficult to reconcile observed asset returns with this relation, emphasizing that the empirical equity premium is larger than the theoretical prediction (*equity premium puzzle*).[7]

1.1.2 Public spending and fiscal policy

Governments collect taxes and spend the tax revenue to provide different public goods as infrastructures, national defense or enviromental protection, and to redistribute resources and correct market distortions: this is the purpose of fiscal policy. The simple two period model allows us to introduce public spending and to obtain some basic results about optimal fiscal policy. We exploit this occasion to generalize the model of the previous section with a population of L agents and time varying preferences. Let us consider the utility function:

$$U = u^1(C_1) - v^1(l_1) + \boldsymbol{\alpha} H(G_1) + \beta \left[u^2(C_2) - v^2(l_2) + \boldsymbol{\alpha} H(G_2)\right] \quad (1.9)$$

where $u^t(\cdot)$ and $v^t(\cdot)$ are the subutilities for period t, $H(\cdot)$ is an increasing and concave function of public expenditure G_t, which is now taken as given, and $\boldsymbol{\alpha}$ is a weight for the subutility from public goods. The budget constraints become:

$$C_1 = w_1 l_1 (1-\tau_1) - T_1 - S \qquad C_2 = S(1+r) + w_2 l_2 (1-\tau_2) - T_2$$

where we introduced labour income taxation at rates τ_t and lump sum taxation T_t at time $t = 1, 2$. Summing the former constraint to the latter divided by $(1+r)$ we obtain the intertemporal budget constraint:

$$C_1 + \frac{C_2}{1+r} = (1-\tau_1) w_1 l_1 - T_1 + \frac{(1-\tau_2) w_2 l_2 - T_2}{1+r} \quad (1.10)$$

which equates the present discounted values of consumption and disposable income. For given tax rates, the Euler condition and the optimality conditions $u^{t\prime}(C_t)(1-\tau_t) w_t = v^{t\prime}(l_t)$ for $t = 1, 2$ provide the labor supply functions depending on the tax rates.

The revenue constraints are $G_1/L = \tau_1 w_1 l_1 + T_1 + B$ in the first period, and $G_2/L + (1+r)B = \tau_2 w_2 l_2 + T_2$ in the second, where B is per capita public debt issued in the first period to finance public spending. Accordingly, we can obtain the intertemporal revenue constraint (IRC):

6 This pricing model is known as the *Consumption Capital-Asset Pricing Model*, introduced in finance by the studies of Marcovitz (1958), Sharpe (1964) and Merton (1973). See also Lucas (1978)

7 However, Barro (2009) has shown that taking into account the risk of rare disasters (heavy stock market crashes and recessions) with generalized preferences can solve the puzzle.

$$G_1 + \frac{G_2}{1+r} = \left(T_1 + \tau_1 w_1 l_1 + \frac{T_2 + \tau_2 w_2 l_2}{1+r}\right) L$$

Imagine first that only lump sum taxes are used. These are not distorsive and lead to the independence of savings and labor supply decisions from the debt level for a given profile of public spending. To see this, substitute the IRC in the intertemporal budget constraint: the resulting equation between the presented discounted value of consumption one side and of exogenous income net of public spending on the other side shows that the optimal choices do not depend on debt. This is a simple version of the so-called *Ricardian Equivalence*, formalized by Barro (1974), who rediscovered an old argument by Ricardo (1817):

BARRO-RICARDO THEOREM. **With lump sum taxes, a perfect credit market and a life horizon equal to the government horizon, the debt level does not affect real choices, and it is not "net wealth".**

The crucial consequence is that, given public spending, changes in the composition of taxes between the periods do not affect the present discounted value of wealth and the optimal allocation of consumption and labor. For instance, a temporary (debt-financed) tax cut which increases available income does not increase current consumption and is entirely saved: the reason is that consumers see through the veil of public finance and are aware that a future tax increase will be needed to pay back the debt (which, therefore, does not represent net wealth in itself).[8]

One can also choose the optimal public spending which maximizes the utility function, deriving the optimality conditions:

$$L\left[\frac{\alpha H'(G_t)}{u^{t\prime}(C_t)}\right] = 1 \text{ for } t = 1, 2 \tag{1.11}$$

which have been introduced by Samuelson (1954).[9] More generally, he proved the:

SAMUELSON RULE. **Optimal public spending must equalize the sum of the marginal rates of substitution between private and public goods for all the agents to the marginal cost of production of the public good.**

[8] One can verify that in case of a smaller interest rate for the government compared to the interest rate for the private sector, public bonds become net wealth and a tax cut is saved only in part.

[9] For instance, in case of logarithmic subutilities, the optimal public good provision is $G_t = \boldsymbol{\alpha} L C_t$. Notice that when the preference parameter $\boldsymbol{\alpha}$ is negatively related to the size of the country L, for instance because this increases the locational distance from the public good (a metaphor for the heterogeneity of preferences), we obtain an inverse U relation between size of countries and public spending/consumption (Etro, 2006).

Imagine now that lump sum taxes are not available. The government can choose the optimal fiscal policy to maximize the utility function subject to the IRC and the optimality conditions. For simplicity, assume constant and unitary wages in both periods. If we define the elasticity of the marginal disutility of labor as $\varphi^t(l) \equiv v^{t\prime}(l)/v^{t\prime\prime}(l)l$, optimal taxation must satisfy the *Ramsey rule* (Ramsey, 1927):[10]

$$\frac{\tau_1}{\tau_2} = \frac{(1-\tau_1)/\varphi^1(l_1) + \gamma^1(l_1 - G_1/L)}{(1-\tau_2)/\varphi^2(l_2) + \gamma^2(l_2 - G_2/L)} \tag{1.12}$$

This rule provides an inverse relation between the two tax rates and the elasticity of their tax bases. However, the optimal fiscal policy implies *tax smoothing* (Barro, 1979), that is $\tau_1 = \tau_2$, under reasonable conditions: if the subutilities in leisure and consumption are the same in both periods, tax smoothing is optimal in case the subutility in consumption is isoelastic (or linear in one period).

The consequences of tax smoothing are relevant: despite public spending and income may change a lot (even because of exogenous shocks), it is optimal to keep constant the tax rates over time to minimize tax distorsions, and consequently, it is optimal to finance deficits (due to increases in spending or recessions which erode the tax base) by issuing debt, and repay this in periods of lower public spending or boom. For different reasons, the neoclassical approach reaches the same implication of the theory of Keynes (1936): fiscal policy should be countercyclical (with deficit spending in recessions).

Finally, one can choose the optimal public spending through a modified Samuelson rule that takes tax distortions into account:

$$L\left[\frac{\alpha H'(G_t)}{u^{t\prime}(l_t - G_t/L)}\right] = 1 + \lambda\left[1 - \gamma^t(l_t - G_t/L)\right] \qquad t = 1,2 \tag{1.13}$$

---

[10] As long as $u^{t\prime} > 0$ for both periods, substituting the optimality conditions in the IRC, we have the following problem:

$$\max_{\{l_1,l_2,G_1,G_2\}} \sum_{t=1}^{2} \beta^{t-1}\left[u^t(l_t - G_t/L) - v^t(l_t) + \alpha H(G_t)\right]$$

$$s.v. : \frac{G_1}{L}u^{1\prime}\left(l_1 - \frac{G_1}{L}\right) + \frac{G_2}{L}u^{2\prime}\left(l_2 - \frac{G_2}{L}\right)\beta =$$

$$= \left[u^{1\prime}\left(l_1 - \frac{G_1}{L}\right) - v^{1\prime}(l_1)\right]l_1 + \left[u^{2\prime}\left(l_2 - \frac{G_2}{L}\right) - v^{2\prime}(l_2)\right]l_2\beta$$

The first order conditions:

$$\left[u^{t\prime}\left(l_t - \frac{G_t}{L}\right) - v^{t\prime}(l_t)\right](1+\lambda) + \lambda\left[u^{t\prime\prime}\left(l_t - \frac{G_t}{L}\right)\left(l_t - \frac{G_t}{L}\right) - v^{t\prime\prime}(l_t)l_t\right] = 0$$

with λ as Lagrange multiplier, can be combined to obtain the rule in the text.

Two remarks are in order. First, this optimal policy is dynamically inconsistent, in the sense that it cannot be credibly implemented without a commitment of the policymaker. It is easy to verify that, after choosing the optimal policy for the first period and creating a corresponding debt, the optimal taxation problem in the second period would not reproduce the same optimal policy which was ex ante optimal for the second period (except for the special case in which $\gamma^2(\cdot) = 0$). The reason is that the *ex post* tax elasticity of labor supply is smaller than the *ex ante* one, since intertemporal substitution is available before and not after.[11] Second, the optimal policy hardly matches the experience of many countries that tend to overissue debt in a systematic way. A political economy explanation for this bias has been advanced by Persson and Svensson (1989), and is based on the strategic role of debt in constraining future political choices.[12]

1.1.3 Money demand and monetary policy

Money is an asset employed as a medium of exchange to simplify transactions between agents. Holding money provides liquidity services at the cost of bearing a zero (nominal) interest.[13] The purpose of monetary policy is to control the supply of money, and with it the inflation rate and its impact on the real economy. The most common ways to microfound the demand of money assume that its real amount provides utility (Sidrauski, 1967), reduces waste of time and resources in the exchange process (*Baumol-Allais-Tobin approach*) or that it is *ex ante* necessary to buy goods (cash-in-advance constraint; see Lucas, 1980, and Svensson, 1985). In this book, we will adopt the first approach.[14] Let us return to the basic two period model, abstract from labor

[11] The time inconsistency problem is even stronger in the presence of capital income taxation (Kydland and Prescott, 1977) since savings are highly mobile *ex ante* and immobile *ex post*, inducing the temptation to expropriate them.

[12] Imagine that in the two period model (with $\gamma^2(\cdot) = 0$ to avoid time inconsistency) two politicians with different preference parameters $\boldsymbol{\alpha}_i$ choose fiscal policy in the two periods: B in the first period (assuming for simplicity absence of initial public spending) and g_2 in the second. It is easy to show that as long as $\boldsymbol{\alpha}^2 > \boldsymbol{\alpha}^1$, the first politician overissues debt (above the optimal level) as a strategic device to constraint the activity of the future politician. For related explanations see Tabellini and Alesina (1990) and Alesina and Drazen (1991). Persson and Tabellini (1990) and Drazen (1999) survey the literature.

[13] We are referring to physical currency in circulation and checking account deposits, which correspond to the common definition of the basic stock of money called M1. Wider definitions (M2, M3) include time deposits, savings deposits, and non-institutional money-market funds. The Central Bank directly controls the monetary base only (physical currency). Additional money supply is created by the commercial banks through their lending activity, which leads to new bank deposits.

[14] The first approach is equivalent to the second under certain conditions and the last one is a particular case of the first when there is zero substitutability between consumption and money in the utility function (Feenstra, 1986).

supply, and introduce money in the utility function:

$$U = u(C_1) + \beta u(C_2) + \chi z(m) \tag{1.14}$$

where $z(\cdot)$ is an increasing and concave function of real balances (that is the real value of money). Assuming that there is income Y only in the first period, the budget constraints are $C_1 = Y - S - M/P_1$ and $C_2 = S(1+r) + M/P_2$, with $m = M/P_1$. Money (M in per capita terms) provides its liquidity services when young, but it can be spent also when old. This implies the first order conditions for savings and real money demand:

$$u'(C_1) = (1+r)\beta u'(C_2) \qquad \frac{u'(C_1)}{P_1} = \beta\frac{u'(C_2)}{P_2} + \frac{\chi z'(m)}{P_1} \tag{1.15}$$

Let us define the inflation rate with the bold symbol $\boldsymbol{\pi} \equiv (P_2 - P_1)/P_1$ and its (rational) expectation with $\boldsymbol{\pi}^e$. The nominal interest rate is:

$$i = r + \boldsymbol{\pi}^e \tag{1.16}$$

a relation known as the Fisher equation. Using these definitions and combining the two optimality conditions, one can rewrite the second one as $\chi z'(m) = u'(C)i/(1+i)$, where C refers to the first period consumption. In the particular case where preferences in consumption are logarithmic and $z(m) = m^{1-\xi}/(1-\xi)$ we obtain:

$$m = \left[\frac{\chi\,(1+i)\,C}{i}\right]^{\frac{1}{\xi}} \tag{1.17}$$

It is now clear that the two optimality conditions imply standard consumption functions and a real money demand increasing in income and negatively related to the nominal interest rate, which is exactly the opportunity cost of holding money: therefore, we have a money demand function $m = m(\boldsymbol{\pi})$ with $m'(\boldsymbol{\pi}) < 0$ and a savings function that can be expressed as $S = S\left[Y - m(\boldsymbol{\pi}), m(\boldsymbol{\pi})/(1+\boldsymbol{\pi}), r\right]$ with $S_{\boldsymbol{\pi}} > 0$.

Let us imagine that the Central Bank controls inflation, either by choosing the money supply or by setting the nominal interest rate. If we look for the inflation rate that maximizes utility under the constraint that the nominal interest rate cannot be negative, we obtain a corner solution:[15] the optimal inflation rate is $\boldsymbol{\pi} = -r$, which corresponds to set a zero nominal interest rate. This result is due to Friedman (1968):

---

[15] Substituting for money demand and savings, we have the following problem:

$$\max_{\{\boldsymbol{\pi}\}} V(\boldsymbol{\pi}) = u[Y - S(\cdot) - m(\cdot)] + \beta u\left[(1+r)S(\cdot) + \frac{m(\cdot)}{1+\boldsymbol{\pi}}\right] + \chi z[m(\cdot)]$$

$$s.v.\ :\ \boldsymbol{\pi} \geq -r/(1+r)$$

Using the optimality conditions (or the envelope theorem) it is immediate to derive that $V'(\boldsymbol{\pi}) < 0$ therefore it is optimal to set $\boldsymbol{\pi} = -r/(1+r) \cong -r$.

FRIEDMAN RULE. **It is optimal to satiate people with money so as to reduce to zero their marginal utility of money by adopting a deflation rate equal to the real interest rate.**

After all, money is produced at no cost while it creates utility in real terms, so it is optimal to increase its real value as much as possible, which is possible if its real return is the same as the one of the productive asset. This important result is robust to different ways of modeling money adoption and other extensions. However, it should be noticed that the welfare gains derived from reasonable changes in the inflation rate are quite limited (this can be easily verified in the case of logarithmic subutilities).[16]

1.1.4 Labor and capital demand by firms

Let us finally turn to the production side of the economy, to examine the behavior of profit maximizing firms. These firms employ inputs as labor and capital to produce goods that are demanded by consumers. Their objective in the production process is to minimize the cost of production taking as given the prices of the inputs: the wage w and the rental rate of capital r. Assume that:

$$y_i = F(K_i, AL_i) \tag{1.18}$$

is a production function of firm i, which turns its capital K_i and labor L_i into output y_i. The term A represents the productivity of labor. It is reasonable to assume that this function is increasing and concave in both arguments,[17] because the marginal productivity of each input is positive but decreasing. In the neoclassical approach, the technology is characterized by constant returns to scale (CRS), in the sense that when the amount of all inputs is double the output is double as well, and there are no fixed costs to start production.[18] A typical example that we will use repeatedly is the so-called Cobb-Douglas production function:

$$y_i = K_i^{\alpha}(AL_i)^{1-\alpha} \tag{1.19}$$

with $\alpha \in [0, 1)$ representing the elasticity of production with respect to the capital input.

We can now characterize the optimal allocation of the inputs for a given level of production: this form of production efficiency is necessary for any

[16] For a recent generalization of the Friedman rule see Da Costa and Werning (2008). For an examination of the welfare costs of inflation see Lucas (2000).

[17] Namely $F_j(K, L) > 0$ and $F_{jj}(K, L) < 0$ for $j = K, L$.

[18] The rationale behind these assumptions is to avoid free lunches in the allocation of inputs: without them entrepreuners taking prices as given could always increase their profits just by splitting a firm into two firms (under decreasing returns to scale) or joining two firms into one (under increasing returns to scale).

profit maximizing firm (under any form of competition). Consider the production of a single good: the amounts of capital a_K and labor a_L that minimize the production cost of a single good $c = ra_K + wa_L$ under the constraint $F(a_K, Aa_L) = 1$ are always uniquely determined. For instance, in the Cobb-Douglas case they are:

$$a_K = \left(\frac{\alpha}{1-\alpha}\right)^{1-\alpha} \left(\frac{w}{rA}\right)^{1-\alpha}, \quad a_L = \left(\frac{1-\alpha}{\alpha}\right)^{\alpha} \left(\frac{rA}{w}\right)^{\alpha} \left(\frac{1}{A}\right) \tag{1.20}$$

which shows that the relative price of the inputs and the elasticity α determine the relative input requirements. This choice implies the following unitary cost of production:

$$c(r, w) = ra_K + wa_L = \frac{r^{\alpha} w^{1-\alpha}}{\alpha^{\alpha}(1-\alpha)^{1-\alpha} A^{1-\alpha}} \tag{1.21}$$

CRS implies that this is constant and always equal to the *marginal cost of production*, that is the additional cost of any further unit of production. Notice that when only labor is needed to produce goods ($\alpha = 0$), the marginal cost boils down to $c = w/A$. In general, a positive shock to the productivity parameter A reduces the marginal cost of production and allows the economy to produce more with the same inputs and/or to increase the remuneration of the inputs.

1.1.5 Perfect competition and market structure indeterminacy

The most important assumption of the neoclassical approach is perfect competition. This requires that the price is taken as given by all the firms (there are not strategic interactions), and entry occurs whenever there are instantaneous profitable opportunities. Since each good is sold at a price p taken as given, and the unitary cost of production is c for all firms adopting the same technology, any firm producing $y_i = F(K_i, AL_i)$ can obtain profits $\pi_i^{PC} = (p - c)y_i$, where PC stands for perfect competition. Of course, the firm is willing to supply any (zero) quantity of goods at a price above (below) the marginal cost and is indifferent at a price equal to the marginal cost. This implies that any equilibrium in which supply equates demand of goods must entail $\pi_i^{PC} = 0$ for any firm i independently from its production level, and a price equal to the marginal cost of production $p = c$. Nothing can be said on the division of the equilibrium production between the firms, a result on which we will return shortly. Finally, notice that, in the absence of changes to technology and input remuneration, the price remains constant and we can normalize it to $p = 1$.

We can also look at the problem of the firms in a different perspective. The real profits of firm i can be written as $\pi_i^{PC} = F(K_i, AL_i) - wL_i - rK_i$.

In every period, the labor input is chosen by this firm to maximize profits according to the first order condition:

$$AF_L(K_i, AL_i) = w \tag{1.22}$$

which equates the marginal productivity of labor to the real wage. This shows that labor demand is decreasing in the real wage and affected by the stock of capital in function of the substitutability between the two inputs. The capital input is chosen to maximize profits as well, according to the optimality condition:

$$F_K(K_i, AL_i) = r \tag{1.23}$$

which equates the marginal productivity of capital to the real interest rate on capital. The demand of capital is decreasing in the interest rate in a way that depends on the amount of labor available.[19] Of course, in case the firm has already a pre-existing stock of capital, its investment is the difference between the demand of capital and that level. In the Cobb-Douglas case, the demands of the two inputs can be derived as:

$$L_i = \frac{(1-\alpha)y_i}{w} \quad \text{and} \quad K_i = \frac{\alpha y_i}{r} \tag{1.24}$$

Plugging the optimality conditions in the profit function, it turns out that the entire revenue is spent to remunerate the inputs:[20] for instance, in the Cobb-Douglas case this occurs with a fraction α for capital income and a fraction $1-\alpha$ for labor income (so that the distribution of income never changes). This holds for any firm independently from its size, which again confirms that the market structure is indeterminate (for any CRS production function), in the sense that we have nothing to say about how many firms should be active and how much each one should produce in any market:

MARKET STRUCTURE INDETERMINACY. **Perfect competition, constant returns to scale and the absence of fixed costs of entry in the markets for goods imply that the size and the number of the firms are indeterminate.**

Notice that when the stock of capital depreciates at rate $\delta_K \in [0,1]$ between periods, it is optimal to equate the marginal productivity of capital net of the depreciation rate δ_K to the real return on capital r.[21] Finally, the

---

[19] Adjustement costs of capital accumulation and uncertainty about a stochastic interest rate can affect the investment decisions. See Dixit and Pyndick (1994) for a general theory of investment under uncertainty.

[20] CRS requires $F(\lambda K, \lambda AL) = \lambda F(K, AL)$ for any $\lambda > 0$ or, deriving with respect to λ and evaluating at $\lambda = 1$, $F(K, AL) = KF_K(K, AL) + ALF_L(K, AL) = rK + wL$.

[21] One can think of the firms as renting capital in each period. In case of depreciation at rate δ_K, renting K requires them to pay the rent rK and to replace the depreciated capital $\delta_K K$, therefore the opportunity cost of capital is $r + \delta_K$.

sum of labor demand and investment by all firms determines the aggregate demand of labor and the aggregate investment. If the aggregate amount of labor and capital are L and K, CRS allows us to express the aggregate production in each period simply as:

$$Y_t = F(K_t, AL) \tag{1.25}$$

and the indeterminacy of the market structure allows us to assume without loss of generality the existence of a single firm. In particular, in a two period framework with Cobb-Douglas technology, initial capital K_1 and positive depreciation rate, the following conditions must hold:

$$w_t = (1-\alpha)K_t^{\alpha}A^{1-\alpha}L^{-\alpha} \text{ for } t = 1,2, \text{ and } r = \alpha K_2^{\alpha-1}(AL)^{1-\alpha} - \delta_K \tag{1.26}$$

for given wages and interest rates.

After characterizing the behavior of the agents and of the firms, it is now time to put consumers and firms together in the markets and endogenize interest rate, wages (and prices), which requires a general equilibrium analysis.

1.1.6 Market clearing and general equilibrium

One of the most crucial assumptions of the neoclassical approach is about equilibrium prices. In the absence of price makers in every market for final goods and inputs, this approach assumes that an ideal auctioneer sets all the prices in such a way that demand is equal to supply in every market. In the general (or Walrasian) equilibrium, each market must clear in each period with appropriate price levels relative to the prices of the other markets. Actually, according to the so-called Walras' law, as long as all markets except one have reached the equilibrium price that equated demand and supply, also the remaining market must clear. Therefore, to avoid redundancy, it is convenient to fix one price at unity and express all the others in terms of that *numeraire*.

The market for goods is characterized by equality of consumption demand and production by firms in each period, and we normalize its equilibrium price to one. Labor market clearing determines the real wage that equates the labor demand of all the firms with the labor supply of all the agents in each period. A similar process takes place in the credit market, where the demand of capital net of the pre-existing capital determines the needed investment, whose equality with the savings determines the equilibrium real interest rate. Therefore, in a two period general equilibrium model with a given initial stock of capital we could derive the equilibrium real wage in the first period w_1, the one for the second period w_2 and the equilibrium real interest rate r.

In the two period model with money we can also characterize the equilibrium of the market for money (in such a case, one can normalize to one the price level of the first period). If the Central Bank controls the money supply, its equality with the money demand determines the inflation rate π. *Vice versa*, if the Central Bank controls the nominal interest rate, the inflation rate is the difference between the nominal and the real rate, while money supply is endogenous and must match money demand. Notice that full price flexibility is key for monetary policy not to have real effects. Keynes (1936) noticed that in the short run the nominal price level is often rigid, and the consequence of this rigidity is unemployment, the only way to keep output below the full employment level (with capital given as well in the short run). The important insight of the disequilibrium theory of Keynes was that both monetary and fiscal policy could be effective expansionary policies in such a short run situation.[22]

A wide literature has studied the general features of Walrasian equilibria, including its existence (Arrow and Debreu, 1954) and its welfare properties. As long as all the markets for all the goods are competitive, in the sense that they are characterized by perfect competition, CRS, and the absence of fixed costs of production, and if there are not externalities (basically goods without a market) or other market or tax distortions, one can show that the economy achieves a Pareto-efficient allocation, in the sense that there are not re-allocations that allow somebody to be better off without someone else to be worse off. Under stronger conditions, one can show the opposite as well. More precisely, we have the following fundamental results (see Mas Colell, Whinston and Green, 1995, for proofs and discussion):

WELFARE THEOREMS. **Consider a general equilibrium economy with a finite number of agents with rational preferences over goods, competitive markets for all these goods and no externalities.** I THEOREM: **Any equilibrium allocation is Pareto-efficient.** II THEOREM: **If all agents have a positive endowment and well-behaved preferences, any allocation of resources can be obtained as an equilibrium through appropriate lump sum transfers.**

---

[22] This can be easily seen in the atemporal IS-LM model with fixed prices (Hicks, 1937). Consider the goods market equilibrium where supply of goods Y equals demand given by private consumption, investment and public spending. Consumption is a fraction $1-s$ of income Y, investment is negatively related to the interest rate as $\bar{I}-br$, and public spending is fixed at G. The equilibrium production $Y=(G+\bar{I}-br)/s$ emphasizes the Keynesian multiplier of demand $1/s>1$, and provides a negative relation between income and the interest rate. If real money supply is M/P and liquidity demand is a linear function $kY-hr$, their equilibrium provides a positive relation between interest rate and income. In equilibrium $Y=\left[h(G+\bar{I})+bM/P\right]/(sh+kb)$, therefore an increase in G (M) is expansionary and increases (decreases) the interest rate. For a given price level (and nominal wage), the labor market generates unemployment whenever output is below its potential level (contrary to what happens in the neoclassical world where the wage and the price level adjust to clear the labor market as well).

These fundamental results hold also in complex models with multiple periods and multiple markets as long as the basic neoclassical assumptions are respected. Therefore, we will often rely on them in the rest of this chapter. Moreover, whenever the I Welfare Theorem holds and the economy is populated by a representative agent, the decentralized equilibrium can be derived without loss of generality by finding the allocation of resources that would be chosen by a hypothetical social planner maximizing utility under the aggregate resource constraint.

1.2 Growth

The neoclassical approach can be useful to analyze the process of accumulation of capital, which is the main source of output growth in any industrialization phase. This allows us to investigate the main determinants of the growth of nations in the short and long run in the neoclassical perspective. The skeleton of modern dynamic macroeconomics is the model of growth which was independently introduced by Solow (1956) and Swan (1956). This model is not microfounded on the consumption side, therefore we need to extend it to endogenous savings. For this purpose we follow Diamond (1965), who developed the so-called OLG model, in which OverLapping-Generations of agents living for two periods populate the economy forever. This is particularly convenient because it allows us to entirely adopt the two period framework introduced in the previous section, and to apply it to an infinite horizon economy. This dynamic and microfounded model can be used to introduce fiscal and monetary policy and to study long run growth due to different sources.

1.2.1 The Solow model

Imagine a standard aggregate CRS production function $Y_t = F(K_t, A_t L)$, where L is the exogenous labor force to be interpreted in terms of the number of agents/workers. For convenience, defining $k \equiv K/L$ the capital-labor ratio, we have output per capita $y = Y/L = F(K/L, A)$ so that we can define:

$$y_t = F(k_t, A_t) \tag{1.27}$$

In each period investment I_t increases the stock of capital, net of depreciation at rate δ_K, according to the equation of motion $K_{t+1} = K_t(1 - \delta_K) + I_t$ or, in per capita terms, $k_{t+1} = k_t(1 - \delta_K) + I_t/L$.

The simple but genial contribution of Solow (1956) was to use this capital accumulation equation with the fact that in a closed economy investment has to equal savings in each period and savings are some function of income.[23]

[23] Solow was building on previous contributions by Harrod (1939) and Domar (1946) whose models correspond to the Solow one with a production function

In the case of a constant rate of savings out of total income $s \in (0,1)$, one obtains the following equilibrium equation of motion for the stock of capital:

$$k_{t+1} = k_t(1-\delta_K) + sF(k_t, A_t) \tag{1.28}$$

which establishes a simple relation between current and future capital. When total factor productivity is constant at a level A, capital accumulation is gradual and depends on the savings rate, the depreciation rate and the marginal productivity of capital (which is affected by A).[24] In the long run, the stock of capital converges to a constant *steady state* level which satisfies $\tilde{k}\delta_K = sF(\tilde{k}, A)$. Long run savings have to match total depreciation to keep the capital-labor ratio and output per capita constant. Notice that augmenting the model with a constant growth rate of the labor force would reduce the steady state output per capita (because it is more costly to maintain the capital-labor ratio constant when labor is increasing over time), while augmenting it with a constant growth rate of labor productivity would lead to a constant rate of long run growth equal to this same rate of rate of exogenous technological progress.

1.2.2 The Diamond model

To microfound the savings function Diamond (1965) assumed that in each period two generations of agents are alive, the young and the old. Every agent born in any period t lives for two periods, being young in t and old in $t+1$, choosing consumption in both periods to maximize a bi-periodal utility function as (1.1). For simplicity we assume that each agent works when young only, so that the budget constraints are $C_{1t} = w_t l_t - S_t$ and $C_{2t+1} = S_t(1+r_{t+1})$. This implies a savings decision $S_t = S(w_t l_t, 0, r_{t+1})$ which is increasing in the wage and ambiguously depending on the interest rate. The equation of motion for the stock of capital becomes:

$$k_{t+1} = k_t(1-\delta_K) + S\left(w_t l_t, 0, r_{t+1}\right) \tag{1.29}$$

We will now assume that the labor supply is exogenous and normalized to one. To close the model we need to derive the wage and the interest rate in

---

with zero substitutability between inputs $Y_t = \min[bK_t, (1-b)A_t L]$, which is associated with Leontief (1941). The constant elasticity of substitution production function (Arrow *et al.*, 1961):

$$Y_t = \left[K_t^{\frac{\theta-1}{\theta}} + (A_t L)^{\frac{\theta-1}{\theta}}\right]^{\frac{\theta}{\theta-1}}$$

with $\theta > 0$ elasticity of substitution between capital and labor, nests the Cobb-Douglas case for $\theta = 1$, the linear case for $\theta \to \infty$ and the Leontief case for $\theta \to 0$.

[24] Introducing differences in the propensity to save out of labor and capital income would not affect the results (Pasinetti, 1962).

equilibrium. Profit maximization by firms implies a stock of capital which equates its marginal productivity net of the depreciation rate to the interest rate, which is therefore:

$$r_t = F_K(k_t, A_t) - \delta_K \tag{1.30}$$

while the zero profit condition $F(k_t, A_t) = k_t(r_t + \delta_K) + w_t$ implies the equilibrium wage:

$$w_t = F(k_t, A_t) - k_t F_K(k_t, A_t) \tag{1.31}$$

To abstract from technological progress, let us assume that A is constant. The equilibrium equation for the accumulation of capital is an implicit expression, since the future stock appears on both sides. Moreover, savings may be increasing or decreasing in the interest rate, which in turn is decreasing in the stock of future capital. Therefore the relationship between present and future stock of capital may be complicated and even non-monotonic.[25] In Figure 1.1 we reproduce in the space (k_t, k_{t+1}) a monotonic process of capital accumulation. The steady state must satisfy:

$$S[F(\tilde{k}, A) - \tilde{k}F_K(\tilde{k}, A), 0, F_K(\tilde{k}, A) - \delta_K] = \tilde{k}\delta_K$$

and, graphically, it corresponds to the crossing of the capital accumulation equation with the 45° line satisfying $k_{t+1} = k_t$. Whenever the slope of the former is between 0 and 1 at the steady state (as in Figure 1.1) convergence is monotonic: starting with a low stock of capital, investment gradually increases capital (and output and wages with it) until the stationary state is reached. In such a case, the growth rates of capital and output per capita are decreasing toward a zero steady state value. Augmenting the model with a constant growth rate of A (exogenous technological progress), they would tend to grow at the same growth rate of productivity in the long run.

The main testable implication of the neoclassical growth model is the so called *convergence*: poorer countries should grow at a higher rate and catch up with richer countries. Notice that Barro and Sala-i-Martin (1992, 1995) find a negative correlation between growth and GDP in a panel of countries, but only after taking into account structural factors which differentiate production functions.

A simple example of the Diamond model emerges with a Cobb-Douglas production function $y_t = k_t^\alpha A_t^{1-\alpha}$ and logarithmic utility so that from (1.4) and (1.24) we obtain $S_t = \beta w_t/(1+\beta)$ and $w_t = (1-\alpha)A_t^{1-\alpha}k_t^\alpha$. Substituting, we have:

[25] However, this relationship is always monotonic when $S_r \geq 0$ since its slope is:

$$\frac{dk_{t+1}}{dk_t} = \frac{1 - \delta_K - S_w k_t F_{KK}(k_t, A)}{1 - S_r F_{KK}(k_{t+1}, A)}$$

See Galor and Ryder (1989) and Azariadis (1990) for further details.

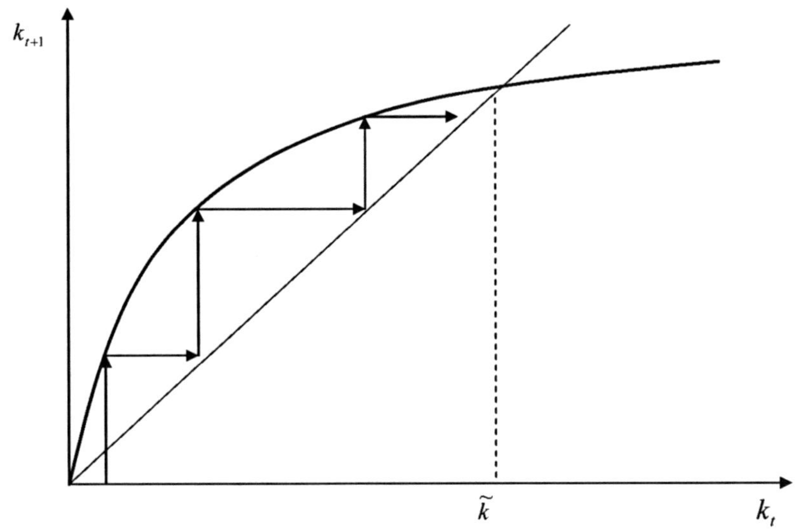

Fig. 1.1. Dynamic Accumulation of Capital in (k_t, k_{t+1}).

$$k_{t+1} = k_t(1 - \delta_K) + \frac{(1-\alpha)\beta}{1+\beta} k_t^{\alpha} A_t^{1-\alpha} \tag{1.32}$$

With constant productivity, the stock of capital converges to the following level:

$$\tilde{k} = A \left[\frac{(1-\alpha)\beta}{\delta_K (1+\beta)} \right]^{\frac{1}{1-\alpha}} \tag{1.33}$$

with a speed of convergence α.[26] The steady state stock of capital (and output) per capita are increasing in A and in the discount factor β, while they are decreasing in the depreciation rate δ_K.

The Diamond model can be extended to endogenous labor supply adopting the approach of Section 1.1.1. The simple case of logarithmic preferences leads to minor changes in the dynamic behavior of the economy,[27] but departing from this case, labor supply depends on the current wage and on the future interest rate and the evolution of the economy becomes bidimensional, possibly leading to complex dynamics (see Azariadis, 1990). More interesting

[26] Only when $\alpha = 1$, the model lacks convergence and the growth rate tends to a constant. We will return to this possibility in Section 1.2.3.

[27] A separable utility function as $U = \log C_{1t} - v l_t^{1+1/\varphi}/(1+1/\varphi) + \beta \log C_{2t+1}$ would lead to a constant labor supply $l_t = ((1+\beta)/v)^{\varphi/(1+\varphi)}$ and savings $S_t = \beta(1+\beta)^{1/(1+\varphi)} w_t / v^{\varphi/(1+\varphi)}$.

is to extend the model to take fiscal and monetary policy into account by adopting the approaches introduced in Sections 1.1.2-3.

1.2.3 Dynamic fiscal policy

Fiscal policy can be introduced as we did in the simple two-period economy of Section 1.1.2, but its role is more interesting now because the Diamond model is characterized by inefficiencies due to the fact that the current generations do not take properly into account the well being of the future ones in their savings decisions.[28] Fiscal policy can solve these inefficiencies.

Phelps (1961) has noticed that in steady state consumption per capita is equal to output per capita net of the steady state savings, $F(\tilde{k}, A) - \delta_K \tilde{k}$, which is maximized when:

$$F_K(\tilde{k}^{GR}, A) = \delta_K \tag{1.34}$$

where GR stands for the "golden rule" that maximizes long run consumption. In the Cobb-Douglas example this implies $\tilde{k}^{GR} = A\left(\alpha/\delta_K\right)^{1/(1-\alpha)}$ which is clearly different from the equilibrium steady state. In particular, when the equilibrium stock is higher than the golden rule stock, the equilibrium is dynamically inefficient because a lower savings rate would allow more consumption in each period. Dynamic inefficiency can be solved and the optimal allocation of resources (which maximizes a weighted sum of the utilities of all generations) can be reached through different forms of taxation that affect the incentives to save.

To introduce fiscal policy one can assume that in each period public spending G_t provides utility as in (1.9) and is financed with lump sum taxes on young generations T_{1t} and old generations T_{2t} or by issuing new debt per capita B_t. The usual optimality condition generates the savings function $S_t = S\left(w_t - T_{1t}, -T_{2t+1}, r_{t+1}\right)$ that determines the accumulation of capital. Moreover, new debt must be issued to finance the difference between public spending and taxes (i.e. the public deficit) which implies an accumulation equation as:

$$B_{t+1} - B_t = r_{t+1} B_t + d_t \tag{1.35}$$

where $d_t = G_t/L - T_{1t} - T_{2t}$ is the primary deficit per capita in period t. This, together with the equation determining the accumulation of capital and debt, allows one to study the impact of fiscal policy, or a social welfare system (Samuelson, 1975), in a growing economy and to characterize the optimal fiscal policy.[29]

[28] The I welfare theorem does not apply in the Diamond model because the number of agents is not finite.

[29] For instance, imagine that public spending is financed by the working generation only under budget balance, so that $T_{1t} = G/L$, and adopt the utility function

A first crucial result is that in this framework Ricardian Equivalence holds only for changes in taxes concerning a single generation and satisfying the government IRC, but not otherwise. The reason is that the horizon of the government is now longer than the life horizon: basically current debt may be paid back by the future generations, therefore it can be net wealth for its owners.

A second crucial result derives from pure accounting. Solving the government revenue constraint (1.35) for the current debt and iterating forward we get the IRC:

$$B_t = \frac{B_{t+1} - d_t}{1 + r_{t+1}} = \left(\frac{1}{1 + r_{t+1}}\right)\left(\frac{B_{t+2} - d_{t+1}}{1 + r_{t+2}} - d_t\right) = ... =$$

$$= \lim_{T\to\infty}\left(\frac{B_{t+T}}{\prod_{j=1}^{T}(1 + r_{t+j})}\right) - \sum_{k=0}^{\infty}\left(\frac{d_{t+k}}{\prod_{j=0}^{k}(1 + r_{t+1+j})}\right) \tag{1.36}$$

whose first term on the right hand side approaches zero. This result implies that outstanding debt has always to equalize the present discounted value of all future primary surpluses (of course in a steady state with a constant primary deficit we need to have $\tilde{B} = -d/r$). In practical terms, if the debt is higher than this (given the expected public spending), a permanent tax increase (or a corresponding spending cut) is necessary to sustain the level of outstanding public debt.[30]

Finally, we could introduce distorsive taxes to analyze the optimal fiscal policy in full fledged dynamic model. Each taxation would raise a revenue

---

$U = u(C_{1t}) + \beta u(C_{2t+1}) + \alpha H(G_t)$. Assume $\delta_K = 1$ for simplicity. In steady state we must have a negative relationship between steady state capital and public expenditure, $k = k(G)$ with $k'(G) < 0$: public expenditure crowds out private capital. The steady state indirect utility is maximized by an optimal public expenditure satisfying a modified Samuelson rule:

$$L\left[\frac{\alpha H'(G)}{u'(C_1)}\right] = 1 + Lk'(G)\left[1 - F_K\left(k(G), A\right)\right]\varepsilon$$

where $\varepsilon \equiv -F_{KK}(k, A)k/F_K(k, A)$ is the elasticity of the interest rate to the stock or capital. Notice that the marginal benefit of public expenditure is increased as long as the steady state stock of capital is above the golden rule level, since the crowding out of capital is dynamically efficient (the reverse happens otherwise).

30 For a calculation of the *fiscal gap* for OECD countries see Etro (2003,b). We need to remark that fiscal sustainability is easier when the economy grows at rate g: if $g > r$ the debt-output ratio can be stabilized even running a constant primary deficit, with $\tilde{B} = d/(g - r)$ growing at the same rate as output in the long run. Finally, notice that the IRC can be seen as an equilibrium equation expressing the price level. According to this *fiscal theory of the price level*, an increase in future real deficits induces an increase in the price level that keeps the real value of the outstanding debt equal to the discounted value of the future primary surpluses in real terms.

which is typically an inverted U function of the tax rate (remember that zero and full taxation induce zero revenue), something known as the *Laffer curve.* Given this, the usual principles of optimal taxation of Section 1.1.2 would govern the maximization of a weighted sum of the utilities of all generations.

1.2.4 Dynamic monetary policy

Monetary policy can be introduced in the Diamond model as we did in the simple two-period economy of Section 1.1.3, where we characterized jointly the demand of savings and real money balances of the young agents. However, a general equilibrium dynamic economy allows us to study the relation between monetary policy, the real economy and fiscal policy as well.

Consider a utility function as (1.14) with the budget constraints $C_{1t} = w_t - S_t - M_t/P_t$ in the first period and $C_{2t+1} = S_t(1+r_{t+1}) + (M_t + H_t)/P_{t+1}$ in the second period, where P_t is the price level in t and H_t is an additional amount of money that the government distributes in every period to the old generation. Money provides its liquidity services when young, but it can be also traded and spent when old.[31] This implies the same optimality conditions for savings S_t and money balances $m_t = M_t^D/P_t$ as in (1.15) for each young generation.

Defining B_t as the debt level and M_t^S as the supply of money, the new IRC in per capita terms can be written as:[32]

$$B_{t+1} - B_t = r_{t+1}B_t - \frac{M_t^S + H_{t+1}}{P_{t+1}} + \frac{M_t^S}{P_{t+1}} + d_t = (1+r_{t+1})B_t - m_{t+1} + \frac{m_t}{1+\pi_{t+1}} + d_t \tag{1.37}$$

where we used the definition of inflation and the fact that $M_{t+1}^S = M_t^S + H_{t+1}$, and we imposed market clearing in the money market. Notice that, assuming that the supply of money grows at the constant rate $\sigma = M_{t+1}^S/M_t^S - 1$, simple manipulations allow us to rewrite $1 + \pi_{t+1} = P_{t+1}/P_t$ as $m_t(1+\sigma)/m_{t+1}$.

For now, let us assume that there is no debt. This implies that the deficit is entirely financed with the real revenue obtained by the government issuing nominal money, i.e. seigniorage:

$$Se(\pi_{t+1}) = d_t = m_{t+1} - \frac{m_t}{1+\pi_{t+1}} \tag{1.38}$$

which shows that holding money raises fiscal revenues and has a cost for the agents (often referred to as inflation tax). Now, let us focus on the steady state

[31] For this reason, Samuelson (1958) introduced the OLG model without capital as a way to explain the emergence of money in a general equilibrium economy. Further classical investigations of this equilibrium under rational expectations are in Lucas (1972) and Grandmont (1985).

[32] Notice that we are implicitly treating the Central Bank as an arm of the government and consolidating their budget constraints.

with a constant primary deficit. Our previous relationships establish that in steady state we must have $\pi_t = \sigma$ and $m_t = d(1+\sigma)/\sigma$ for any t. The first result shows that ultimately, inflation is a monetary phenomenon. Moreover, the optimality conditions in steady state provide a standard savings function and a money demand $m(\pi)$ with $m'(\pi) < 0$. Consequently, seigniorage in steady state becomes $Se(\sigma) = m(\sigma)\sigma/(1+\sigma)$, which is an inverted U function of the growth rate of money: the so-called *inflation tax Laffer curve.* On the left hand side of this inverted U relation, the government can increase revenues by increasing the inflation rate until the revenue maximizing rate is achieved. In emergency situations maximization of seigniorage revenue has been an objective of monetary policy, often generating hyperinflations.

Moreover, Calvo (1978) has noticed that under uncertainty, money demand is actually a function of the expected inflation rate (because agents take economic decisions taking into account expected inflation), therefore the Central Bank can choose the inflation rate taking as given money demand. Under rational expectations, the choice of the Central Bank must be consistent with expectations and this implies a bias toward even higher inflation rates than the revenue maximizing one (see also Kydland and Prescott, 1977, and Barro and Gordon, 1983,a,b). The problem is avoided if the Central Bank is limited by a reputational constraint to keep inflation law (Backus and Driffill, 1985), by a precommitment, for instance because a conservative central banker has been appointed on purpose (Rogoff, 1985), or by contractual incentives (Walsh, 1995). Unfortunately, even if this is the case, agents may not believe such commitments, especially if there is a political influence on the Central Bank.[33]

Since a proper objective of a Central Bank should be to maximize long run welfare and not short run revenue, one can find the money supply growth rate and the associated inflation rate which maximize the utility function in the steady state. The Friedman rule of Section 1.1.3 does not hold anymore because of the emergence of the Tobin effect (Tobin, 1965): inflation reduces money demand and increases savings and hence investment in productive capital, which is growth enhancing. For this same reason, the optimal monetary policy typically implies a positive rate of inflation.[34]

[33] Imagine that the future Central Bank will be politically appointed after the elections: if a conservative (liberal) party wins, the Central Bank will implement an inflation rate π_L ($\pi_H > \pi_L$). (Rationally) expected inflation is $\pi^e = p\pi_L + (1-p)\pi_H$ where p is the probability that the conservative party wins. Clearly $m(\pi_L) > m(E(\pi)) > m(\pi_H)$, but ex post there is a surprise inflation and a greater seignorage if the liberal party wins and the opposite if the conservative party wins. As long as these surprises on inflation have real effects, elections induce political shocks on the real economy. Building on Nordhaus (1975), Alesina (1988) has developed on this idea a theory of the political business cycles. See Persson and Tabellini (1990) for a survey and Lossani, Natale and Tirelli (2000) for an evaluation.

[34] Assume $\delta_K = 1$ for simplicity. In steady state, the optimality conditions and the equilibrium condition $S = k$ provide capital $k = k(\pi)$ with $k'(\pi) > 0$

Finally, let us reintroduce debt. Solving (1.37) for current debt and iterating forward we obtain that outstanding debt has always to equalize the present discounted value of all future primary surpluses and seigniorage revenues, which sets a joint constraint on fiscal and monetary policy.[35]

1.2.5 Neoclassical endogenous growth

The basic neoclassical model is able to explain long run growth only as an exogenous phenomenon due to a constant growth of productivity. Since the pathbreaking work of Romer (1986),[36] however, a wide research program has been focused on the analysis of the long run determinants of growth. A basic explanation for endogenous growth relies on externalities in capital accumulation. For instance, if productivity is affected by the stock of capital of the economy so that $A_t = Bk_t$, the Diamond model with logarithmic preferences and Cobb-Douglas production function summarized in (1.32) leads to a constant long run growth rate:

$$\tilde{g} = \frac{(1-\alpha)\beta B^{1-\alpha}}{1+\beta} - \delta_K \tag{1.39}$$

A deeper explanation for endogenous growth can be found in the accumulation of other forms of capital, as human capital (Uzawa, 1964; Lucas, 1988)[37] or public capital (Barro, 1990), which enhance productivity. Let us

---

and money $m = m(\boldsymbol{\pi})$, and imply the indirect steady state utility $V(\boldsymbol{\pi}) = u[F(k,A) - kF_K(k,A) - k - m] + \beta u[F_K(k,A)k + m] + \chi z(m)$. Using the optimality conditions we can derive that $V'(0) > 0$, hence not only the optimal inflation rate is higher than the Friedman rule's prescription, but it is even positive (Phelps, 1973).

[35] In steady state we have:

$$\tilde{B} = \frac{m(\boldsymbol{\pi})\boldsymbol{\pi}}{(1+\boldsymbol{\pi})r} - \frac{d}{r}$$

Assuming that the Central Bank keeps the ratio money/debt constant at the level $\varpi = m/\tilde{B}$ with open market operations, we obtain that $\varpi = m(\boldsymbol{\pi})r/\left[Se(\boldsymbol{\pi}) - d\right]$. The numerator is decreasing in $\boldsymbol{\pi}$ while the denominator is increasing on the upward sloping side of the inflation tax Laffer curve, therefore an increase in $\boldsymbol{\pi}$ must be associated with a reduction in ϖ and vice versa. The consequence has been suggestively named *unpleasant monetaristic arithmetic* by Sargent and Wallace (1981): assuming that the Central Bank keeps the money-debt ratio constant, an increase of the debt level (or of the deficit) increases the equilibrium inflation rate (see Sargent, 1987b).

[36] See also Boldrin and Rustichini (1994) who emphasize the possibility of indeterminacy and, in a OLG context, Boldrin (1992) and Azariadis and Reichlin (1996).

[37] See also Rebelo (1991), Mankiw, Romer and Weil (1992) and, in a OLG context, Azariadis and Drazen (1991). On education and growth see the survery by Aghion and Howitt (2009, Ch. 13).

interpret A_t as a generic form of "social capital" that grows through investments and depreciates at the rate $\delta_A \in (0,1]$, and assume that a fraction $\tau \in (0,1)$ of labor income is invested in social capital, which accumulates according to:

$$A_{t+1} = A_t(1-\delta_A) + \tau w_t \tag{1.40}$$

with wage (1.31). Let us consider logarithmic preferences, which generate a constant savings rate out of net labor income. Then, private capital accumulates according to:

$$k_{t+1} = k_t(1-\delta_K) + \frac{(1-\tau)\beta w_t}{1+\beta} \tag{1.41}$$

and the two equations fully determine the behavior of the economy. One can think of τ as the rate of investment in education which generates accumulation of human capital A_t and enhances productivity, or as the tax rate which generates revenue used to build new public infrastructures which augment the stock of public capital A_t and enhance productivity.

The accumulation of private and social capital under perfect competition can lead to perpetual growth. To see this, notice that, after defining $x_t \equiv k_t/A_t$ as the ratio between private and social capital we can reduce the bidimensional equilibrium system to the following unidimensional one:[38]

$$x_{t+1} = \frac{(1-\delta_K)\,x_t + (1-\tau)\beta\left[f(x_t) - x_t f'(x_t)\right]/(1+\beta)}{1-\delta_A + \tau\left[f(x_t) - x_t f'(x_t)\right]} \tag{1.42}$$

where $f(x) = F(k/A, 1)$. This map determines the evolution of the ratio between private and social capital, and as a consequence the evolution of all the other variables. In case of endogenous growth, both forms of capital must grow at the same rate, so that their ratio remains constant. Intuitively, the accumulation of social capital allows the workers to be always more productive, which allows them to increase their savings and accumulate private capital on one side and to increase social capital on the other side: this creates perpetual growth. For instance, with a Cobb-Douglas production function, the equilibrium system becomes:

[38] Since the function $F(K, AL)$ exhibited CRS in K and L, the function $F(k, A)$ exhibits CRS in k and A, so that we can define a new variable $x \equiv k/A$ as the ratio between private and social capital and a new function $f(x) = F(x, 1)$ with $w/A = f(x) - xf'(x)$. If we solve the equilibrium system for the growth rates of the two forms of capital we have:

$$\begin{aligned} k_{t+1}/k_t &= 1-\delta_K + (1-\tau)\beta\left[f(x_t) - x_t f'(x_t)\right]/(1+\beta)x_t \\ A_{t+1}/A_t &= 1-\delta_A + \tau\left[f(x_t) - x_t f'(x_t)\right] \end{aligned}$$

Dividing the first expression for the second we obtain the unidimensional equilibrium map.

$$x_{t+1} = \frac{(1-\delta_K)x_t + (1-\tau)\beta(1-\alpha)x_t^{\alpha}/(1+\beta)}{1-\delta_A+\tau(1-\alpha)x_t^{\alpha}}$$

which generates a monotonic convergence to a steady state with endogenous growth. More in general, the equilibrium map can have an non-monotonic shape and generate a cycling convergence to the steady state.[39] This is what happens for low values of the elasticity of substitution between private capital and labor (and therefore social capital as well) and high depreciation of the social capital.[40]

The neoclassical model of growth and the last models augmented with new sources of sustained growth have provided useful insights on the industrialization process and on macroeconomic policy. However, on one side they have entirely neglected the role of innovation and endogenous business creation in driving growth and the impact of the growth process on the market structures, and on the other side they have been unable to provide useful prescriptions for competition and innovation policies.

1.3 International Finance and Trade

In this section we extend the neoclassical framework to an open economy to analyze the implications and the gains that derive from globalization. By definition this implies that the country trades with the rest of the world and either faces exogenous international prices (the case of a small open economy) or can affect international prices (the case of a large open economy). Trade can concern factor inputs like capital (capital flows) or labor (migration), or goods (pure trade). Initially, we follow the international finance literature adopting the standard assumption that there is just one consumption good (whose trade is a residual result of the different consumption/production choices of different countries) to focus on trade in factor inputs (capital in particular, since it is more mobile than labor). Later, we follow the international trade

---

39 See Devaney (1989) for an introduction to complex dynamical systems.

40 With a constant elasticity of substitution production function, the equilibrium map becomes:

$$x_{t+1} = \frac{(1-\delta_K)x_t + \frac{(1-\tau)\beta}{1+\beta}\left[1+x_t^{\frac{\theta-1}{\theta}}\right]^{\frac{1}{\theta-1}}}{1-\delta_A+\tau\left[1+x_t^{\frac{\theta-1}{\theta}}\right]^{\frac{1}{\theta-1}}}$$

This implies monotonic convergence for any elasticity of substitution $\theta \geq 1$ and non-monotonic for low enough values of θ. See Etro (2007d,e) for details and an empirical application to public capital. The model can also be used to study the political economy equilibrium for the choice of τ and consequently of the growth rate. A similar analyis has been introduced by Bertola (1993). See also Benabou (2000).

literature to consider two different consumption goods and discuss trade in goods without trade in inputs.

1.3.1 International finance

Imagine that a closed economy opens up to international capital flows facing the international interest rate r^*. Foreign capital starts flowing in and out of the country, with a net difference $\mathbf{B}_t$ representing net foreign assets owned by domestic agents. In such an environment, the current account (CA) is defined as the sum of the return on the net foreign assets ($r^*\mathbf{B}_t$) and the trade balance, which is the difference between exports EXP_t and imports Q_tIMP_t, where the real exchange rate Q_t is defined as the ratio between foreign and domestic price levels. The accounting identity $Y_t+Q_tIMP_t = C_t+I_t+EXP_t$ allows us to rewrite the trade balance as $Y_t - C_t - I_t$. In equilibrium of the balance of payments, the CA must correspond to the opposite of the capital account, which records all transactions between domestic and foreign residents that involve a change of ownership of assets (mainly foreign direct investments and portfolio investments): accordingly, the CA can be seen also as the change in net foreign assets:

$$CA_t \equiv \mathbf{B}_{t+1} - \mathbf{B}_t = r^*\mathbf{B}_t + (Y_t - C_t - I_t) = S_t - I_t \tag{1.43}$$

where we used $S_t = (r^*\mathbf{B}_t + Y_t) - C_t$. After opening up, a CA surplus tells us that the country is buying foreign assets thanks to a surplus in the trade balance, while a deficit tells us that the country is importing more than exporting and is selling domestic assets to do that. Solving (1.43) for net foreign assets and iterating forward we obtain that CA sustainability requires negative net foreign assets to be equalized by the present discounted value of all future primary trade surpluses.

A more interesting way to interpret the CA suggested by (1.43) is as the difference between savings, which derive from the consumption plan of the agents, and investments, which derive from the profit maximizing choices of the firms (and eventually from a public deficit). While in a closed economy savings and investments are equalized through adjustments in the interest rate, they can differ in an open economy because the country as a whole can lend or borrow from the rest of the world.[41] The so-called *intertemporal approach to the current account* has been introduced by Sachs (1981),[42] who analyzed a small open economy through a simple representative agent model with two periods as the one of Section 1.1 with the international interest rate r^* taken as given. As we know, the savings function derives from utility

[41] Notice that the interest rate r^* is not exogenous from a global perspective, but has to equalize global investment and savings. This creates interdependence between countries which is particularly relevant for large open economies, whose shocks and policies can affect the global equilibrium.

[42] See also Dornbusch (1983) and, for a wide survey, Obstfeld and Rogoff (1996).

maximization as $S = S(Y_1, Y_2, r^*)$. Let us adopt a Cobb-Douglas technology $y_t = k_t^\alpha A_t^{1-\alpha}$ in each period. Assuming zero depreciation, investment must be the difference between capital demand for the second period k_2, which equates the marginal productivity of capital to the world interest rate, $k_2 = A_2(\alpha/r^*)^{1/(1-\alpha)}$ and the initial stock of capital k_1, which is given. Accordingly, we can express the first period CA as:

$$CA_1 = S\left[k_1^\alpha A_1^{1-\alpha}, A_2(\alpha/r^*)^{\alpha/(1-\alpha)}, r^*\right] - \left[A_2(\alpha/r^*)^{1/(1-\alpha)} - k_1\right] \quad (1.44)$$

which shows that a current positive technology shock (a higher A_1) or an increase in the international interest rate tend to increase the CA surplus (or reduce the deficit), while a future positive technology shock (a higher A_2) has the opposite effect (since it reduces savings and increases investments). The impact of a permanent productivity shock (which affects both A_1 and A_2) is smaller, since both savings and investments go up. Sachs (1981) used this model to explain the opposite reactions to the oil shocks for the CAs of oil importing and exporting countries.

In general, savings and investments do not need to be highly correlated (as they should be in a closed economy), while consumption could be highly correlated across countries as long as preferences are not too different. Empirically, not only the correlation between savings and investment is very high (Feldstein and Horioka, 1980), but there is also a strong home bias - the savings of a country are destined for a great majority to finance investments in the same country - and the correlation between consumption across countries is very low. Imperfections in the international capital markets, sovereign risk and debt overhang problems may explain the constraints on international debt (and risk sharing),[43] and together with the existence of non-tradable goods and transaction costs may explain (in part) the above anomalies. Nevertheless, in the rest of this section we keep following the neoclassical approach and assume perfect capital mobility and perfect capital markets.

Let us consider the growth process of a small open economy facing the international interest rate r^*. Any convergence property of the growth model analyzed in the previous section becomes instantaneous. The equality of the net marginal product of capital with the international interest rate immediately fixes the stock of capital at the level:

$$\tilde{k} = A\left(\frac{\alpha}{r^*}\right)^{\frac{1}{1-\alpha}} \quad (1.45)$$

If the international interest rate is small enough, we have an immediate inflow of capital from the rest of the world (a CA deficit), and *viceversa* (a CA surplus) if the international interest rate is high enough (gradual convergence emerges only in case of imperfect capital mobility or adjustment costs in the

[43] See Bulow and Rogoff (1991) on a classic work on sovereign debt.

accumulation of capital). A general implication of this open economy model is that capital should fly to poor countries very easily looking for higher returns (unless there are unrealistic differences in productivity between countries). The fact that this does not happen can be seen as a puzzle (Lucas, 1990).[44]

Trade in the labor input, that is migration, can be studied in an analogous way to capital mobility, since it implies a tendency toward equalization of wages and flows of workers from poorer to richer countries. Historically, these processes have been quite evident, but the recent literature has emphasized more complex mechanisms explaining the migration phenomenon.[45]

1.3.2 International fiscal policy

Fiscal policy can be easily introduced in an open economy framework.[46] In such a context, fiscal policy is often regarded as a prosper-thy-neighbor policy, in the sense that public spending in a country often has a positive impact abroad (think of trans-national infrastructures, investments to reduce pollution or protect the environment and defence for neighboring countries). The simplest way to formalize this interdependence is to assume that the representative agent of country i is characterized by a subutility $\boldsymbol{\alpha}_i H(G_i)$ from public goods, depending on domestic and foreign public spending according to $G_i = \mathbf{g}_i + \boldsymbol{\beta} \sum_{j=1, j\neq i}^{\mathbf{N}} \mathbf{g}_j$, where the parameter $\boldsymbol{\beta} \in (0,1)$ represents the degree of spillovers between $\mathbf{N}$ countries. In such a case, the optimal global fiscal policy would require different spending for each country i according to its preference parameter $\boldsymbol{\alpha}_i$ and to the size of the spillovers $\boldsymbol{\beta}$. For instance, under linear subutility from consumption and lump sum taxation, the optimal provision of public good in each period would satisfy a system of modified Samuelson rules:

$$\boldsymbol{\alpha}_i H' \left(\mathbf{g}_i + \boldsymbol{\beta} \sum_{j=1, j\neq i}^{\mathbf{N}} \mathbf{g}_j \right) = \frac{1}{1 + \boldsymbol{\beta}(\mathbf{N} - 1)} \qquad i = 1, 2, ..., \mathbf{N} \tag{1.46}$$

which equates for each country the marginal utility of public spending to the social marginal cost, reduced because of the spillovers between countries. Nevertheless, sovereign countries would choose their fiscal policy equating the former to the unitary marginal cost of production of the public goods, leading to underprovision. This is a typical example in which a reputational commitment to the optimal policy can be sustained if policymakers are patient

---

[44] For a wide discussion of the applications of the open economy model to development macroeconomics see Agénor and Montiel (1999).

[45] For theoretical analysis see Wong (1995), for a recent empirical investigation see Mendola (2008).

[46] Augmenting the model with public spending G_t financed through lump sum taxes T_t and primary deficit $d_t = G_t - T_t$, the CA becomes $CA_t = r^* \mathbf{B}_t + (Y_t - C_t - I_t - G_t) = S_t - I_t - d_t$, which shows the twin-deficits argument: if the private sector is in balance, a CA deficit must be due to a public deficit.

enough (Etro, 2002). However, as long as preference-dependent policies cannot be decided at the central level, international unions of countries deciding by majority voting can only implement a uniform provision of public goods, which internalizes the spillovers according to the median country preferences, but ignores the heterogeneity: this leads again to limited participation in the international coordination and to inefficient outcomes.[47] Alesina, Angeloni and Etro (2005) have studied institutional solutions for this problem that include: 1) subsidiarity, which allows countries to choose their fiscal policy first and the union to decide *ex post* by majority voting on an additional common level of public spending to be provided by all the members, 2) federal mandates, which require all members to choose public provision above a minimum decided *ex ante* in the union by majority voting, and 3) matching grants, that subsidize decentralized public spending to eliminate the gap between its social and private marginal cost. These solutions are widely employed in federal states and international unions, and in particular in the European Union,[48] but they could be employed also at the global level, for instance to implement fiscal intervention in front of a global economic crises or to enhance investments in environmental policies.

1.3.3 International monetary and exchange rate policy

In a multicountry framework, a new nominal variable emerges, the nominal exchange rate. In a flexible exchange rate regime the equilibrium between demand and supply of each currency determines the nominal exchange rates between all the currencies. In the neoclassical approach this is irrelevant for the real economy. To see this, notice that under perfect competition, free trade in goods generates *Purchasing Power Parity* (PPP): the same goods have the same prices everywhere to avoid arbitrage opportunities according to the so-called law of one price, and the nominal exchange rates adjust accordingly. In particular, let us define the nominal exchange rate between two currencies as E_t (units of domestic currency per units of foreign currency), so that its increase corresponds to a depreciation of the domestic currency. If the domestic and foreign prices of the consumption good are P_t and P_t^*, it must be that:

[47] If an international union creates extra spillovers, the trade-off between the benefits of coordination and the loss of independent policymaking determines size, composition, and scope of the union. Policy uniformity reduces the size of the union, may block the entry of new members, and induce excessive centralization. The formalization of this idea is due to Alesina, Angeloni and Etro (2001b, 2005). Extensions of this approach include Lorz and Willmann (2005), Dur and Roelfsema (2005), Bordignon and Brusco (2006), Hartsad (2007), Kothenburger (2008), Lockwood (2008) and Acemoglu, Egorov and Sonin (2008).

[48] For a related empirical work see Alesina, Angeloni and Schuknecht (2005). The political economy of fiscal coordination in the presence of income heterogeneity has been investigated by Alesina, Angeloni and Etro (2001,a), Etro and Giarda (2007) and Hafer and Landa (2007).

$$E_t P_t^* = P_t \tag{1.47}$$

In other words, the real exchange rate (units of domestic consumption for units of foreign consumption), defined as $Q_t = E_t P_t^*/P_t$, must be always unitary. Under this extreme form of perfect price flexibility, monetary policy remains neutral as in the closed economy framework.

Empirical evidence shows systematic deviations from the absolute version of the PPP hypothesis. The typical explanations for these deviations rely on the presence of non-tradable goods (for which the law of one price does not hold) and imperfectly substitutable tradables,[49] and on nominal rigidities. Since the ratio between foreign and domestic prices affects (positively) the trade balance, these deviations from the PPP induce real effects of shocks and can lead to non-neutralities of monetary policy as well.[50]

There is another implication of the PPP. If this holds in each period, it implies a relation between the domestic inflation rate π_t and the foreign one, π_t^*, which can be approximated as:

$$\pi_t = \pi_t^* + e_t$$

where $e_t = E_t/E_{t-1} - 1$ is the rate of depreciation of the exchange rate. Domestic inflation must be matched by foreign inflation or a devaluation of the domestic currency (weak evidence for this form of relative PPP is available). Moreover, notice that perfect capital mobility leads to equalize the real interest rates: $r_t = r_t^*$. Using the Fisher equation (1.16), we have $i_t - \pi_{t+1}^e = i_t^* - \pi_{t+1}^{e*}$, which implies another arbitrage condition known as the *uncovered interest parity*, $i_t = i_t^* + e_{t+1}^e$. The differential between nominal interest rates must match the expected depreciation of the domestic currency. For this reason, changes in the monetary policy should be immediately reflected in the exchange rate.[51]

---

[49] If $C_t = C_{Tt}^a C_{Nt}^{1-a}$ and $C_{Tt} = C_{Ht}^b C_{Ft}^{1-b}$, where $j = T$, N, C,F denote tradables, non-tradables, and tradables produced at home and abroad, and a foreign country has identical preferences, the real exchange rate can be derived as $Q_t = (EP_{Ht}^*/P_{Ht})^{a(2b-1)} (EP_{Nt}^*/P_{Nt})^{1-a}$, which depends on the relative prices of tradables and non-tradables.

[50] The Keynesian analysis of a small open economy, due to Fleming (1962) and Mundell (1968), leads to radically different results from the neoclassical one. The new element of aggregate demand is the trade balance, which can be approximated with $TB = vQ - mY$, where v measures the sensitivity of exports to the real exchange rate and m is the marginal propensity to import. Under flexible exchange rates, but with a fixed international interest rate, the Mundell-Fleming model determines output - as $Y = (M/P - hr^*)/k$ - and the nominal exchange rate. In such a case monetary policy is expansionary (and induces an endogenous depreciation), while fiscal policy is not effective (and only appreciates the currency).

[51] Once again, price rigidities affect this result. In particular, Dornbusch (1976) has shown that price rigidities in the short run lead to exchange rate overshooting. See Tirelli (1993) for further discussion.

Until now we have assumed that the nominal exchange rate fluctuates freely. While this is true for many currencies, a number of alternative systems have been implemented, including a policy of fixed exchange rates. This requires that the Central Bank is ready to change domestic currency at fixed rates, so that any positive (negative) change in its reserves matches the surplus (deficit) of the balance of payments (which is defined as the sum of the current and capital accounts). Loosely speaking, a deficit in the trade balance implies positive net imports and hence net demand of foreign currency against domestic currency, while a deficit in the capital account implies a capital outflow and again net demand of foreign currency against the domestic one. To match demand and supply of domestic currency (and therefore defend the exchange rate), the Central Bank has to supply foreign currency reducing its reserves and demand domestic currency reducing the monetary base.[52] In other words, money supply becomes endogenous and the Central Bank looses control of monetary policy. As emphasized by Krugman (1979,b), when the nominal exchange rate becomes inconsistent with a long run equilibrium of the balance of payments (reserves are destined to end), a speculative attack leads to a currency crises and a devaluation.[53]

In the presence of price rigidities, monetary authorities can still affect the economy through changes of the fixed exchange rate. At least in the short run, exchange rate policy can affect the real economy through its impact on the trade balance.[54] In conclusion, notice that monetary independence and perfect capital mobility cannot be achieved at the same time as exchange rate stability.[55] Moreover, the interdependence between nominal variables can lead to difficulties in the implementation of the desired policy of each country. International monetary coordination can avoid these difficulties.[56]

[52] This is due to the fact that in the balance sheet of the Central Bank the monetary base must be equal to the sum of reserves in foreign currency and net domestic activities, that is credits toward domestic banks and companies.

[53] See Calvo (1987) and Lahiri and Vegh (2003) for a microfounded model of currency crises, Krugman (1991,b) on target zones and Carmignani, Colombo and Tirelli (2008) for an empirical investigation of exchange regimes.

[54] In the Mundell-Fleming model with fixed exchange rates (and fixed international prices and interest rate), output $Y = (G+\bar{I}-br^*+vEP^*/P)/(s+m)$ and money supply M become endogenous. Then, monetary policy is not effective, while fiscal policy and devaluations of the exchange rate are expansionary (and they both lead to an endogenous increase in money supply).

[55] See Calvo and Reinhart (2002) on credibility constraints on exchange rate policy.

[56] While fiscal policy is typically prosper-thy-neighbour, monetary policy (under flexible exchange rates) and exchange rate policy (under fixed exchange rates) tend to be beggar-thy-neighbour. A commitment to a fixed exchange rate or a monetary union can avoid competitive devaluations and limit time inconsistency problems in the implementation of monetary policy, but it implies giving up to policy independence. See Alesina and Barro (2002), Etro (2004,b) and Debrun, Masson and Pattillo (2005) on the political economy of monetary unions.

1.3.4 International trade theory

Let us now consider trade in goods in a model with two sectors and no factor mobility across countries. We will focus on the steady state properties of this model, which corresponds to the static neoclassical theory of international trade associated with Heckscher (1929) and Ohlin (1933).[57] Imagine that all countries have identical technologies and productivity levels, with two CRS production functions $y^1 = F(k^1, A)$ and $y^2 = F(k^2, A)$ for goods 1 and 2 with respective international prices $p^1$and p^2 (one of which could be normalized to one and defined as a numeraire). The domestic country has total endowments K and L of the two factors which must be shared between the two sectors.[58] This simple setup leads to a first immediate result. Under perfect competition, the real interest rates and wages must be the same in the two sectors and equal to their marginal productivities. This means that in terms of effective capital-labor ratios we have:

$$r = p^1 F_K(k^1, A) = p^2 F_K(k^2, A) \tag{1.48}$$
$$w = p^1 \left[F(k^1, A) - k^1 F_K(k^1, A)\right] = p^2 \left[F(k^2, A) - k^2 F_K(k^2, A)\right] \tag{1.49}$$

Given the prices of the two goods, this system of four equations provides unique capital/labor ratios for the two sectors and unique interest rates and wages for any endowment of factors, that is for any country facing the same international prices. In other words, according to the *factor price equalization theorem* (Samuelson, 1949, Lerner, 1952), under free trade and perfect competition, wages and interest rates are equalized across countries.

Now let us define with a_{Li} and a_{Ki} the amounts of labor and capital necessary to produce one unit of good $i = 1, 2$. Under efficient production, these parameters are uniquely determined for any production function - for instance, in the Cobb-Douglas case they are given by (1.20). Then, the zero profit conditions in the two sectors imply:

$$p^1 = wa_{L1} + ra_{K1} \qquad p^2 = wa_{L2} + ra_{K2} \tag{1.50}$$

Solving the system for the factor prices with $D \equiv (a_{K1}/a_{L1} - a_{K2}/a_{L2})$, we have:

$$r = \frac{p^1/a_{L1} - p^2/a_{L2}}{D} \qquad w = \frac{p^2/a_{K2} - p^1/a_{K1}}{D} \tag{1.51}$$

Noting that $D > (<) 0$ if and only if sector 1 is more capital (labor) intensive than sector 2, we can derive the Stolper-Samuelson (1941) theorem: under free trade and perfect competition, a relative increase in the price of one good increases the return on the factor which is intensively used in its production

[57] See Galor (1992) for an OLG model with two sectors.

[58] What follows assumes that the capital-labor ratio is not too high or too law (otherwise a country would specialize in one of the two sectors), and that there are not factor intensity reversals characterizing the production functions.

and decreases the return on the other factor (see also Jones, 1965). For instance, an increase in the price of a good produced with high intensity of capital increases the return on capital and reduces the wage.[59]

Finally, full employment of both factors in the total production of the two goods Y^1 and Y^2 implies:

$$a_{K1}Y^1 + a_{K2}Y^2 = K \qquad a_{L1}Y^1 + a_{L2}Y^2 = L \tag{1.52}$$

Solving the system for the production levels we have:

$$Y^1 = \frac{Ka_{L2} - La_{K2}}{a_{L1}a_{L2}D} \qquad Y^2 = \frac{La_{K1} - Ka_{L2}}{a_{L1}a_{L2}D} \tag{1.53}$$

which delivers the Rybczynski (1955) theorem: under free trade and perfect competition, an increase in the endowment of one factor increases the production of the sector which is intensive in that factor and decreases the production of the other sector.[60]

A crucial consequence of these results (under general equilibrium with endogenous relative prices) emerges under the assumption that two countries have identical production functions and identical and homothetic preferences (and therefore identical consumption functions which are linear in income for both goods). Imagine first that the domestic country has the same relative endowment of capital as the foreign country: $K/L = K^*/L^*$. This implies that the equilibrium relative prices of the goods will be the same under free trade as under autarchy and there will not be any actual trade. Assume now that domestic capital and/or foreign labor increase so that $K/L > K^*/L^*$. Then the Rybczynski theorem implies that the domestic country will produce more of the good produced with a more capital intensive technique; at the initial price this will induce excess supply of that good, which will require a reduction of its (relative) price. Consequently, the domestic country will export the more capital intensive good.

Notice that perfect competition implies always prices equal to the marginal costs and market structures that are indeterminate. Any change in the relative prices is uniquely due to a change in the relative marginal costs. For instance, the increase in capital in the domestic country (or in labor in the foreign country) have reduced the relative price of the capital intensive

59 Reading this result the other way around and redefining inputs as qualified labor and low-skilled labor, it tells us that, by reducing the prices of goods whose production is intensive in low-skilled labor, the effect of globalization in the developed world is to reduce the wages of low skilled labor and increase inequality between this and qualified labor. A similar analysis can be applied to derive the effects of sector- or factor-biased technological progress (see Leamer, 2000, and Krugman, 2000), a crucial explanatory element for the rising skill premium for educated workers.

60 Reading this result in a dynamic perspective, it tells us that capital accumulation or population growth lead to nonbalanced growth of the two sectors. For applications to structural change see Acemoglu (2009, Ch. 20).

good because they have reduced the return on capital compared to the wage, which has reduced the relative marginal cost of producing the capital intensive good. Finally, trade does not affect the structure of the markets (how many firms produce what), which remains indeterminate.

Another crucial implication is known as the Heckscher-Ohlin theorem: under free trade and perfect competition, each country exports the good whose production is more intensive in the factor of which that country is relatively richer. This is a version of the law of comparative advantage: here a country has a comparative advantage in the sector which is intensive in the factor with (relative) higher endowment.

These results crucially depend on the assumptions that there are two sectors and two factors that can be freely allocated between the two sectors, and that there are no technological differences between the countries. If there is just one factor of production, say labor, we obtain the situation associated with the Ricardian model, where there is no factor price equalization and each one of the two countries specializes in the production of the good on which has a technological comparative advantage (Ricardo, 1817).[61] If there are two factors, but one of them, say capital, is specific to a sector and cannot be allocated to the other one, we are in the so-called Ricardo-Viner model, where the mobile input is allocated to equalize its marginal productivity between sectors (Mussa, 1974); this model can be seen as a short run version of the neoclassical model. Finally, if there are more than two countries and more than two sectors, countries specialize in certain sectors and factor price equalization only holds between countries specializing in the same sectors (Leamer, 1987), but the comparative statics results do not hold in general.[62] The literature on the factor content of trade has been able to generalize some of the predictions for the impact of openness on production and trade. However, even the most basic prediction, the fact that more heterogeneous countries (in endowments) should trade more and in a direction related to their factor endowments, does not find a strong empirical support.[63] Differ-

[61] For instance, assume that the only input is labor. Then, good 1 is produced at home if $wa_{L1} < w^* a^*_{L1}$ and good 2 is produced by the foreign country if $w^* a^*_{L2} < wa_{L2}$. These conditions can be rewritten as $a^*_{L2}/a_{L2} < w/w^* < a^*_{L1}/a_{L1}$ which requires $a^*_{L2}/a^*_{L1} < a_{L2}/a_{L1}$, that is the foreign country is relatively (and not necessarily absolutely) more efficient at producing good 2, while the domestic country is relatively more efficient at producing good 1. The zero profit conditions imply the relative wages $w/w^* = (a_{L1}/a^*_{L2})(p^1/p^2)$ where relative prices depend on preferences and factor endowments. The result on full specialization survives even in extended versions of the model with more than two sectors (Dornbush, Fisher and Samuelson, 1977): here a country has comparative advantage in all the sectors with relatively more efficient technology.

[62] Recently, the basic model has been extended to account for endogenous effort (Leamer, 1999) and different distributions of talent (Grossman and Maggi, 2000).

[63] See Leamer (1980, 2000), Bowen, Leamer and Sveikauskas (1987), Trefler (1995) and Feenstra (2004) for a recent review.

ences in productivity across countries can contribute to solve the puzzle only partially.

1.3.5 International trade policy

In all these neoclassical models, the allocation of resources improves under free trade compared to autarchy, at least up to redistributive lump sum transfers. Loosely speaking, this is due to the fact that there are not inefficiencies and the I Welfare Theorem holds in the integrated economy, therefore if the allocation of resources under autarchy is still available but replaced by a new equilibrium allocation, it means that the latter is Pareto efficient (see Deardoff, 1980, and Dixit and Norman, 1980). Therefore, countries gain from opening up to factor inputs movements (as in Section 1.3.1) because they expand their investment possibilities, and they gain from opening up to trade in goods (as in Section 1.3.2) because they expand their consumption possibilities. For this reason a small country unable to affect international prices has no reason to adopt protectionist policies.

Nevertheless a large economy able to affect international prices finds it optimal to unilaterally adopt an import tariff, or equivalently an export tax, to improve its own terms of trade, that is to decrease the ratio between the price of imports and the price of exports (Lerner, 1934, 1936; Kaldor, 1940).[64] More precisely, one can calculate the specific tariff on import or the tax on exports that equate the social marginal cost and value of imported and exported goods. In case of homogenous goods, the optimal trade policy for a large open economy is given by the specific import tariff:

$$t = \frac{p^*}{\zeta} \tag{1.54}$$

where p^* is the price of imports and ζ the elasticity of foreign supply, or equivalently by the specific export tax:

$$\tau = \frac{p}{\varepsilon} \tag{1.55}$$

where p is the price of exports and ε the elasticity of foreign demand (see Krugman and Helpman, 1989). Notice that when we look at a small open economy for which both the foreign elasticities of imports' supply and exports' demand are infinite, free trade remains the optimal policy.

Until now we analyzed the optimal unilateral trade policy, but when multiple countries adopt their optimal protectionist policy inefficient outcomes emerge (Meade, 1952; Johnson, 1953-54). Once again, international policy coordination can improve the allocation of resources through a commitment to free trade.[65]

[64] The terms of trade is the same as the real exchange rate if there are no non-tradable goods.

[65] See Bagwell and Staiger (2004) and Facchini and Testa (2009) on the political economy of market integration.

In conclusion, the neoclassical approach to international finance and trade has provided interesting tools of analysis, but its results have been in contradiction with a number of stylized facts, including the limited flow of capital toward poor countries, the limited correlation between consumption across countries, the empirical failure of the PPP hypothesis, the limited correlation between factor endowments and factor content of trade, the existence of international market structures characterized by intra-industry trade and prices decreasing with the process of globalization. Moreover, from a normative point of view, the neoclassical approach has been unable to go beyond the neutrality of exchange rate policy and trade policy for a small open economy, and has been able to support the optimality of export taxes for a large open economy, a rather strange policy recommendation.

1.4 Business Cycles

Macroeconomic analysis aimed at studying the business cycle and the impact of shocks and policies requires a more realistic dynamic model than the two-period model or the Diamond model adopted until now. In particular, the need to match the timing of the theoretical framework with that of real world decisions taken by consumers and firms (and policymakers) requires a set up with infinitely living consumers and firms, whose maximizing choices are introduced in a dynamic stochastic general equilibrium (DSGE) framework. In this section we develop such a model, which is due to Ramsey (1928), and apply it to introduce the neoclassical theory of macroeconomic fluctuations, the Real Business Cycle theory. Finally, we extend the model with exogenous market structures characterized by positive and constant mark ups for an infinity of monopolistic firms and we employ this framework to review the results of the New-Keynesian approach to macroeconomics.

1.4.1 The Ramsey model

Consider a representative agent with utility from consumption in infinite periods:

$$U = \sum_{t=0}^{\infty} \beta^t u(C_t) \tag{1.56}$$

Assume CRS, perfect competition and the absence of fixed costs in the market for the production of the single final good. Since the market structure is indeterminate we can consider a representative firm producing $Y_t = F(K_t, AL)$. Labor force and productivity are assumed exogenous and constant. In each period, the agent allocates labor and capital income (summing up to total output because of CRS) between consumption and savings, which are entirely invested in capital accumulation. The decentralized equilibrium must be efficient by the I Welfare Theorem, and therefore it corresponds to the one chosen by a social planner to maximize the above utility under the following resource constraint (which equates investment and savings):

$$K_{t+1} = K_t(1 - \delta_K) + F(K_t, AL) - C_t \tag{1.57}$$

given the initial stock of capital K_0. The solution to this recursive problem can be found through dynamic programming (see Stockey and Lucas with Prescott, 1989), which amounts to find the consumption function $C(K)$ and the value function $V(K)$ satisfying the Bellman equation:

$$V(K) = \max_{C\in[0,K]} \{u(C) + \beta V[F(K, AL) - C + K(1 - \delta_K)]\} \tag{1.58}$$

This implies that the infinite horizon problem boils down to a sequence of problems that are equivalent to two period problems, characterized by the optimality of the usual Euler equation. A closed form solution for the equilibrium path of capital and consumption is not always available, but this path exists and it is unique and stable. Moreover, we can express it implicitly through the equilibrium system given by the resource constraint (1.57) and the optimality condition for the maximization of (1.56) s.v. (1.57) with respect to K_{t+1}:

$$K_{t+1}: \quad u'(C_{t+1}) = [\beta[1 - \delta_K + F_K(K_t, AL)]\}^{-1} u'(C_t) \tag{1.59}$$

The steady state is characterized by the following conditions:

$$\tilde{C} = F(\tilde{K}, AL) - \delta_K \tilde{K} \tag{1.60}$$

$$F_K(\tilde{K}, AL) = \delta_K + (1 - \beta)/\beta \tag{1.61}$$

Linearizing the equilibrium system around this steady state, we obtain:

$$\begin{bmatrix} K_{t+1} \\ C_{t+1} \end{bmatrix} = J \cdot \begin{bmatrix} K_t \\ C_t \end{bmatrix} \quad \text{with } J = \begin{bmatrix} \frac{dK_{t+1}}{dK_t} & \frac{dK_{t+1}}{dC_t} \\ \frac{dC_{t+1}}{dK_t} & \frac{dC_{t+1}}{dC_t} \end{bmatrix} = \begin{bmatrix} \beta^{-1} & -1 \\ \frac{F_{KK}}{\gamma} & 1 - \frac{\beta F_{KK}}{\gamma} \end{bmatrix}$$

where all functions are evaluated at the steady state. As the local stability of a unidimensional system required a slope smaller than one at the steady state, the (saddle path) stability of the equilibrium of a bidimensional system requires that the Jacobian J has two eigenvalues being one larger and one smaller than one, which can be easily verified here.[66]

[66] The eigenvalues are the two roots of $\lambda^2 - TrJ \cdot \lambda + \det J = 0$ where trace and determinant of the Jacobian are $TrJ = 1 + 1/\beta - \beta F_{KK}/\gamma > 2$ and $\det J = 1/\beta > 1$. Notice that the monotonic properties of the equilibrium are not guaranteed in case of multiple sectors and low discounting: in such a case Boldrin and Montrucchio (1986) have actually proved that any dynamic pattern can emerge, including *chaos*.

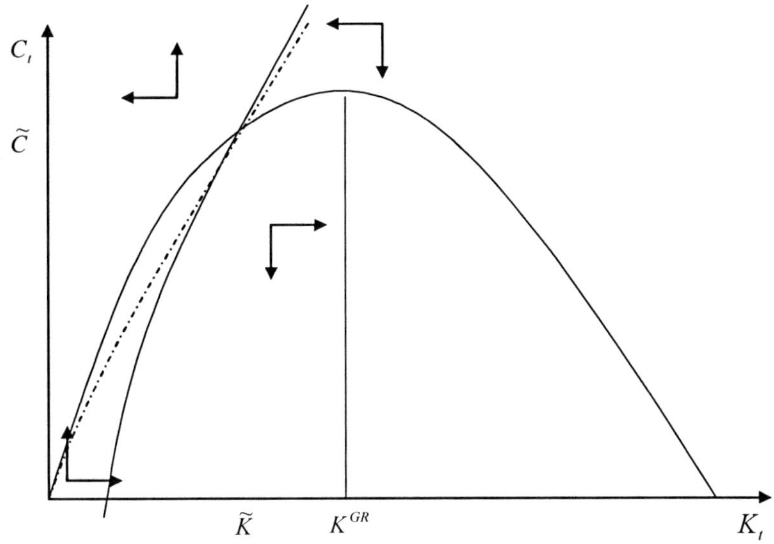

Fig. 1.2. Phase Diagram in (K_t, C_t).

In Figure 1.2 we show the saddle path in the phase diagram for capital and consumption, where the increasing relation represents the locus of the points where consumption is constant - more capital allows to produce and consume more, and the inverse U shaped curve the locus of the points where the stock of capital does not change - the steady state consumption level is maximized at the golden rule which satisfies $F_K(K^{GR}, AL) = \delta_K$. Notice that dynamic inefficiency never occurs in the neoclassical model: the steady state stock of capital is always below the golden rule level. This is not surprising, since the full equilibrium path is Pareto-efficient as well.

When the discrete interval of time between periods, which here was arbitrarily assumed unitary, converges to zero, we asymptotically obtain the Ramsey model in the continuous time. Defining $\dot{x}_t$ as the limit of $\Delta x_t \equiv x_{t+\Delta} - x_t$ when the interval between periods Δ goes to zero, that is the instantaneous rate of change of the variable x, we obtain the following system of accumulation equation and Euler equation:

$$\dot{K}_t = F(K_t, AL) - C_t - \delta_K K_t \tag{1.62}$$

$$\frac{\dot{C}_t}{C_t} = \frac{r_t - \rho}{\gamma(C_t)} \tag{1.63}$$

where $r_t = F_K(K_t, AL) - \delta_K$ and $\rho = 1/\beta - 1$. The steady state conditions are the same as before, (1.60) and (1.61).[67]

While the main focus of this section will be on business cycle fluctuations, the Ramsey model can be used to study growth, trade and policy issues, since it inherits many of the properties examined in the previous sections. Growth is characterized by convergence toward the steady state, and externalities in capital accumulation or the accumulation of different forms of capital (that enhances the marginal productivity of physical capital) can lead to endogenous growth. In case of a small economy open to capital flows at the given international interest rate r^* (and with the same preferences as the rest of the world), immediate convergence occurs unless there are adjustment costs in the accumulation of capital (see Turnovsky, 1997, 2000, for a wide discussion). Under isoelastic utility, the basic Euler equation shows that permanent growth takes place when the marginal productivity of capital (and therefore the interest rate), has a lower bound high enough.[68] Public spending and money will be introduced in the next section.

1.4.2 Real Business Cycles

The basic Ramsey model can be extended with endogenous labor supply and public spending financed with lump sum taxes, and it can be perturbed with supply shocks to productivity and demand shocks to public spending. The reaction to these shocks is at the basis of the neoclassical theory of the business cycle (Kydland and Prescott, 1982).[69] As we have seen earlier in the two period model, endogenous labor supply will be crucial to propagate small temporary shocks.

The standard RBC model is built on isoelastic separable subutilities and a Cobb-Douglas production function (King, Plosser and Rebelo, 1988,a,b). The utility function of the representative consumer derives from private consumption, leisure, public spending, and real money balances, and can be expressed as:

---

[67] Other extensions of the Ramsey model that can account for more realistic consumption dynamics include a stochastic death process (see Blanchard and Fisher, 1989), hyperbolic discounting (Laibson, 1997, Barro, 1999), and risk diversification through capital markets integration (Obstfeld, 1994).

[68] This is what happens with a production function linear in capital, or even with a CES production function with high enough elasticity of substitution between capital and labor, namely $\theta > 1$ (Jones and Manuelli, 1990). Permanent growth is also perceived (in the short run) by a small open economy facing an interest rate above its time preference rate or trading two intermediate goods used to produce a final non-tradable good (because factor price equalization fixes the interest rate), as shown by Ventura (1997). See Acemoglu (2009, Ch. 18 and 19) on trade and growth.

[69] See also Long and Plosser (1983), Prescott (1986) and Campbell (1994). An introduction to stochastic dynamic programming and growth models can be found in Ljungqvist and Sargent (2004) or Acemoglu (2009, Ch. 16 and 17).

$$U = E\sum_{t=0}^{\infty}\beta^t\left[\log C_t - \frac{\upsilon L_t^{1+1/\varphi}}{1+1/\varphi} + \boldsymbol{\alpha}\log G_t + \frac{\chi m_t^{1-\xi}}{1-\xi}\right] \tag{1.64}$$

The equilibrium optimality conditions are simple generalizations of the optimality conditions of the two period models of Section 1.1, that is (1.3), (1.6) and (1.17) with $\gamma = 1$:

$$\frac{C_{t+1}}{C_t} = \beta(1 + r_{t+1}) \tag{1.65}$$

$$L_t = \left(\frac{w_t}{\upsilon C_t}\right)^{\varphi} \tag{1.66}$$

$$m_t = \left[\frac{\chi\left(1+i_t\right)C_t}{i_t}\right]^{\frac{1}{\xi}} \tag{1.67}$$

We assume that the public spending is in terms of the same good consumed by the representative agent (even if it provides a separate utility). As long as taxation is lump sum Ricardian equivalence holds and we can assume budget balance in every period without loss of generality. Moreover, money is neutral in the sense that monetary shocks do not affect real variables, and the assumption of separability in the utility function allows us to ignore them: therefore we will assume $\chi \to 0$ and, for now, we will neglect real money balances without loss of generality. The model is closed with the resource constraint:

$$K_{t+1} = K_t^{\alpha}\left(A_t L_t\right)^{1-\alpha} + K_t(1-\delta_K) - C_t - G_t \tag{1.68}$$

and the equilibrium wage and interest rate, simple generalizations of the conditions obtained in Section 1.1. with (1.26):

$$w_t = (1-\alpha)K_t^{\alpha}A_t^{1-\alpha}L_t^{-\alpha} \text{ and } r_t = \alpha K_t^{\alpha-1}\left(A_t L_t\right)^{1-\alpha} - \delta_K \tag{1.69}$$

Given the initial stock of capital K_0 and a process for productivity and public spending, these equations fully characterize the behavior of C_t, L_t, K_t, w_t and r_t.[70]

To verify the qualitative properties of the model we need to simulate its behavior after shocks, and to compare it with the empirical reactions of the real economy (for a VAR analysis of these reactions see Chapter 3.1). As standard, we will assume that total factor productivity $\mathbf{A}_t \equiv A_t^{1-\alpha}$ follows a first order autoregressive process $\hat{\mathbf{A}}_{t+1} = \rho_A\hat{\mathbf{A}}_t + \varepsilon_{At}$, where $\rho_A \in (0,1)$ is the autocorrelation coefficient and ε_{At} is a white noise disturbance, with zero expected value and standard deviation σ_A. Government spending follows a similar process $\hat{G}_{t+1} = \rho_G\hat{G}_t + \varepsilon_{Gt}$ around a steady state value which is a constant fraction of output, with $\rho_G \in (0,1)$ and ε_{Gt} white noise.

[70] As in the basic Ramsey model, the I Welfare Theorem holds, and we could have obtained the same results by solving the social planner problem.

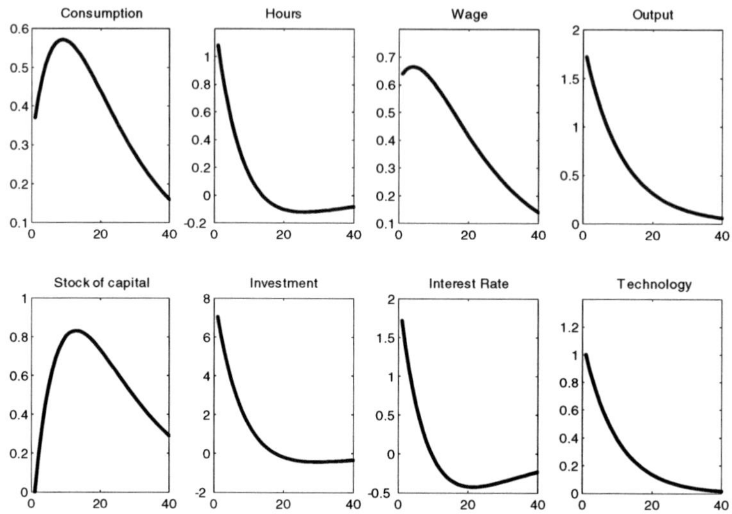

Fig. 1.3. Impulse response to a persistent technology shock. RBC with perfect competition. Percentage deviations/quarters.

We will calibrate the structural parameters as in King and Rebelo (2000). The time unit is meant to be a quarter. The discount factor, β, is set to 0.99, while the depreciation rate δ_K equals 0.025, implying an annual depreciation of 10 % of the capital stock. The labor share of capital is set to $\alpha = 1/3$. The value of υ is such that steady state labor supply is constant and normalized to unity. Public spending is assumed to be 20% of output, a reasonable value for public consumption. The Frish elasticity of labor supply φ is assumed equal to 4 as in King and Rebelo (2000). Cycles propagate from supply and demand shocks as shown in Figures 1.3 and 1.4.

Supply shocks. Consider a temporary but persistent supply shock first (a 1% increase in TFP with persistency $\rho_A = 0.9$). On impact, the shock increases the marginal productivity of both inputs and therefore their remuneration. Since the increase in the wage is only temporary, agents find it convenient to temporarily work more: this intertemporal substitution effect boosts output and amplifies the impact of the shock. Notice that, since the market for goods is perfectly competitive, there is not a mark up wedge between marginal costs and prices, therefore goods keep being priced at the marginal cost. At the same time, only part of the increase in income is consumed, since agents increase their savings to smooth consumption over time. The associated increase in savings turns into an increase in investment and leads to capital accumulation, which sustains the growth of wages and consumption, at least temporarily, but reduces the marginal productivity of capital. While

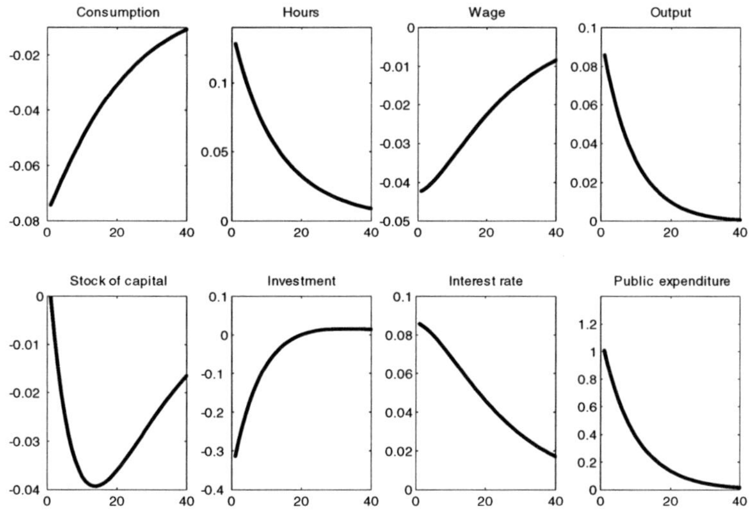

Fig. 1.4. Impulse response to a persistent government spending shock. RBC with perfect competition. Percentage deviations/quarters.

the shock gradually disappears, consumption reverts toward its initial steady state level.

Demand shocks. Consider a demand shock due to a 1 % increase in government spending (with persistency $\rho_G = 0.9$). This shock reduces available income (net of taxes), which leads to a reduction of private consumption, but lower than the increase in public consumption, so that total output goes up (the government spending multiplier is larger than one). The lower available income reduces not only private consumption, but also savings and leisure; the latter is a pure income effect due to the fact that taxes are non-distorsive (Barro and King, 1984). The associated increase in labor supply is strong enough to reduce wages, while the reduction of investment reduces the stock of capital below its steady state level and induces an upward pressure on the interest rate. After the initial reduction, consumption grows monotonically toward its initial level - a counterempirical consequence of an expansionary fiscal shock.

An alternative demand shock would be a preference shock associated with a disturbance Θ_t to the subutility from consumption $\log(C_t - \Theta_t)$. A positive shock to Θ_t would increase the marginal utility so as to increase consumption, but this would have the counterempirical implication of crowding out investment (Baxter and King, 1992)

Second moments. Finally, we can evaluate the unconditional moments of the basic RBC model assuming that the only source of random fluctu-

$Variable$	$\sigma(X)$	$\sigma(X)/\sigma(Y)$	$E(X_t, X_{t-1})$	$Corr(X,Y)$
Y	1.66, 1.39	1	0.84, 0.72	1
C	1.19, 0.60	0.75, 0.43	0.78, 0.78	0.76, 0.94
I	4.97, 4.09	2,99 2.59	0.87, 0.70	0.79, 0.98
L	1.82, 0.67	1.10, 0.48	0.90, 0.70	0.88, 0.97
Π	8.08, *n.a.*	4.87, *n.a.*	0.76, *n.a.*	0.67, *n.a.*
μ	0.99, *n.a.*	0.60, *n.a.*	0.79, *n.a.*	−0.28, *n.a.*

Table 1.1. Second moments. Left: US data. Right: RBC model

ations are technology shocks. We calibrate the productivity process as in King and Rebelo (2000), with persistence $\rho_A = 0.979$ and standard deviation $\sigma_A = 0.0072$. We report in Table 1 the performance of the standard RBC model[71] with respect to the statistics on U.S. data (1948-2008) for output Y, consumption C, investment I, labor force L, profits Π and mark up μ.[72] The measure of the mark up reported in Table 1.1 is the labor-share based measure introduced by Rotemberg and Woodford (1999) - see Chapter 3.

The main problems of the RBC framework are the limited variability of output and especially of consumption and labor force (while the large variability of investment is broadly matched by the model), and the lack of explanations for the cyclical movements of profits and mark ups. Its performance can be improved introducing adjustment costs in capital accumulation (Kydland and Prescott, 1982), labor market rigidities (Hansen, 1985, Rogerson, 1988), habits affecting consumption patterns (Boldrin, Christiano and Fisher, 2001) or other additional features.[73] However, a deeper problem of the neoclassical approach to the business cycle is given by the impossibility of explaining the cyclical variation (and the same existence) of price mark ups and profits, because the perfectly competitive paradigm excludes any interactions between profits, entry and pricing.

Fiscal Policy. Under lump sum taxation Ricardian equivalence holds and public spending can be chosen in each period according to a Samuelson rule which is equivalent to the one derived in the two period model (1.11) of Section 1.1. In particular, under our functional form we have:

$$G_t = \boldsymbol{\alpha} C_t \quad \text{for any } t \tag{1.70}$$

[71] The benchmark RBC model we consider is that by King and Rebelo (2000). Our utility function differs from theirs in the subutility from labor supply, but the second moments are equivalent under the same calibration and the same time period.

[72] Variables have been logged. We report theoretical moments of Hodrick-Prescott filtered variables with a smoothing parameter equal to 1600 (see Prescott, 1986). For a discussion of the data see Section 3.1.

[73] See Cooley (1995) for an early review of the main applications. Backus, Kehoe and Kydland (1992) have extended the model to an open economy framework with capital mobility. In such a case international risk sharing leads to a correlation of consumption between countries that is high and higher than the correlation between their output, which does not match empirical evidence.

Public spending should be a constant fraction of consumption which, by the way, is consistent with our steady state assumption of a constant public spending-output ratio. In the presence of distorsive taxation, however, Ricardian equivalence does not hold anymore,[74] and the effects of government spending shocks are changed. Baxter and King (1993) have shown that in such a case an increase in the tax rate on labor income needed to finance higher public spending induces a substitution effect in favor of higher labor supply: if this is strong enough, consumption can increase in reaction to a government spending shock.

Concerning the optimal fiscal policy, usual principles of the theory of optimal taxation studied in Section 1.1.2 apply here as well (see Lucas and Stockey, 1983). Public spending should follow a modified Samuelson rule and labor taxation should follow a Ramsey rule (possibly with tax smoothing) leading toward countercyclical fiscal policy. Moreover, capital taxation should tend toward zero while approaching the steady state to avoid long run distortions to the marginal productivity of capital (Chamley, 1986), a peculiar result of the neoclassical approach.

Monetary Policy. As we have already noticed, the allocation of resources in the neoclassical model is independent from nominal variables, therefore monetary policy can affect the nominal variables but not the real ones.[75] Imagine that the instrument of the Central Bank is the nominal interest rate i_t, related to the real rate r_t and to the future expected inflation rate π_{t+1}^e by the Fisher equation $i_t = \pi_{t+1}^e + r_t$. Imagine a policy rule that associates the interest rate to the current inflation rate:

$$i_t = \bar{\imath} + \phi\pi_t \tag{1.71}$$

where $\bar{\imath} > 0$ and $\phi > 0$. From the two relations, we obtain a difference expectation equation $\pi_t = \left(\pi_{t+1}^e + r_t - \bar{\imath}\right)/\phi$, whose solution depends on ϕ. If $\phi < 1$ (including the case of a constant interest rate rule), any inflationary process satisfying $\pi_{t+1} = \phi\pi_t - r_t + \bar{\imath} + \xi_{t+1}$ (where ξ_t are arbitrary shocks with zero expected value) solves the above equation: this implies indeterminacy of the price level and therefore of all the nominal variables (wages and interest rate). If $\phi > 1$, however, there is only one stationary solution, namely $\pi_t = \sum_{k=0}^{\infty} \phi^{-(1+k)} E\left[r_{t+k} - \bar{\imath}\right]$, therefore the full path of inflation, the price level and the other nominal variables are fully determined. This result is known as the *Taylor Principle*: the determinacy of the price level requires that the Central Banks adjust the nominal interest rate more than one for one in response to any change in inflation. Finally, notice that the optimality of the Friedman rule (zero nominal interest rate) still holds in this neoclassical environment, therefore an interest rate rule as $i_t = \phi\left(r_{t-1} + \pi_t\right)$ with $\phi > 1$

[74] As in the Diamond model, the evolution of the debt follows (1.35) and fiscal policy must respect the IRC (1.36).

[75] Also in this case, the evolution of debt and money must respect the same constraint of the Diamond model (1.37).

implements the Friedman rule as the only stationary solution (while money demand passively determines the needed supply of money).

1.4.3 Monopolistic firms with exogenous entry

Departures from perfect competition, in which prices are taken as given, can be easily introduced in the neoclassical framework. Most of the modern macroeconomic literature has adopted a model with an infinity of monopolistic firms having market power on their goods and taking as given the price index of the economy.[76] Assume that aggregate consumption summarizes consumption of a continuum of differentiated goods according to the following index:

$$C_t = \left[\int_0^1 C_t(j)^{\frac{\theta-1}{\theta}} dj \right]^{\frac{\theta}{\theta-1}} \tag{1.72}$$

where $\theta > 1$ is the elasticity of substitution between goods whose prices are $p_t(j)$. The price index P_t corresponding to the consumption of the index C_t, defined in such a way that $\int_0^1 p_t(j)C_t(j)dj = C_t P_t$, is:

$$P_t = \left[\int_0^1 p_t(j)^{1-\theta} dj \right]^{\frac{1}{1-\theta}}$$

Utility maximization generates a direct demand for good i given by $C_t(i) = [p_t(i)/P_t]^{-\theta} C_t$, therefore the profits of firm i at time t are:

$$\Pi_t(i) = [p_t(i) - c_t] C_t(i) = [p_t(i) - c_t] \left[\frac{p_t(i)}{P_t} \right]^{-\theta} C_t \tag{1.73}$$

where c_t is the (nominal) marginal cost of production. Since there is an infinity of monopolistic firms, each one acts independently in the choice of its price in every period, and has no impact on the price index or the consumption index. Accordingly, the profit maximizing price is $p_t(i) = \theta c_t/(\theta - 1)$ for each firm, which corresponds to a common and constant mark up for all goods:

$$\mu^M = \frac{\theta}{\theta - 1} \tag{1.74}$$

where M stands for monopolistic behavior. In the symmetric equilibrium we have $P_t = p_t(i)$ for any i, and unitary and aggregate consumption are the same, just like in the presence of a single monopolist. Therefore, the nominal

[76] The infinity of monopolistic firms is a clear oxymoron, but a standard assumption in the literature. Moreover, notice that an infinite number corresponds to an assumption of exogeneity on the number of firms.

profits are $\Pi_t^M = c_t C_t/(\theta - 1)$, and their real counterpart $\pi_t^M = \Pi_t^M/P_t$ becomes:

$$\pi_t^M = \frac{C_t}{\theta} \tag{1.75}$$

In this environment, the market structure is exogenous because strategic interactions do not take place and there is an exogenous number of firms (a continuum between 0 and 1). Monopolistic behavior creates a wedge between prices and marginal costs and the profits are directly distributed to the households, but the dynamic properties of the RBC model are not altered. Supply and demand shocks have similar effects to the case of perfect competition, with the only difference that they affect profits without altering the market structure. Moreover, as Blanchard and Kiyotaki (1987) have shown, monetary shocks remain inconsequential in such a model. However, things change in the presence of price rigidities, as we will see below.

The new trade literature has studied the opening up of such an economy to trade with an identical one producing a different set of goods: the consequence is that consumers enjoy a double number of goods available for consumption (Krugman, 1980), but mark ups, prices and output per firm do not change.

Similarly, the new growth literature has considered monopolistic firms selling differentiated inputs (rather than consumption goods) and has linked technological progress to the competition *for* the market, assuming for the innovation sector either perfect competition (Romer, 1990; Aghion and Howitt, 1992) or monopolistic behavior (Aghion and Howitt, 2009): both assumptions lead to endogenous growth, in the first case with indeterminate market structures and in the second one with exogenous market structures.

New-Keynesian macroeconomics. The above model of monopolistic behavior with an exogenous number of firms has been largely employed in the New-Keynesian literature, which has embraced the neoclassical microfounded approach augmenting it with monopolistic firms and different forms of price (and wage) stickiness (Rotemberg, 1982,a,b; Mankiw, 1985; Akerlof and Yellen, 1985).[77] This allows one to study the impact of technological and monetary shocks in the presence of nominal frictions. Consider the simpler case of an economy without capital ($\alpha = 0$) where, in each period, a fraction λ of the firms cannot adjust the nominal price and maintains the pre-determined price, and a fraction $1 - \lambda$ can reoptimize the nominal price to the optimal level p_t, which maximizes the discounted value of future profits. This form of price stickiness has been introduced by Calvo (1983) and generates the following equation for the inflation rate:[78]

[77] Classic studies on the consequences of nominal rigidities due to staggered price adjustments are due to Fisher (1977) and Taylor (1980).

[78] The price index must be $P_t = \left[\lambda P_{t-1}{}^{1-\theta} + (1-\lambda)p_t^{1-\theta}\right]^{1/(1-\theta)}$, whose log-linearization around a zero inflation steady state provides the inflation rate as $\boldsymbol{\pi}_t \equiv \hat{P}_t - \hat{P}_{t-1} = (1-\lambda)(\hat{p}_t - \hat{P}_{t-1})$. In every period t, each opti-

$$\boldsymbol{\pi}_t = \beta\boldsymbol{\pi}^e_{t+1} + \frac{(1-\lambda)(1-\beta\lambda)}{\lambda}(\hat{w}_t - \hat{A}_t) \tag{1.76}$$

which is usually defined as the *New-Keynesian Phillips curve*, and links current inflation to its expected value and to changes in the real marginal cost.[79] The reason is that the firms would like to keep their prices as a fixed mark up

---

mizing firm chooses the same new price p_t to maximize the expected profits until the next adjustment, $\sum_{k=0}^{\infty} \lambda^k E\{Q_{t+k}[p_t - W_{t+k}/A_{t+k}]\, C_{t+k}(i)\}$, where $Q_{t+k} = \beta^k (P_t C_t / P_{t+k} C_{t+k})$ is the discount factor (from the Euler equation) and demand at time $t+k$ is $C_{t+k}(i) = [p_t(i)/P_{t+k}]^{-\theta}\, C_{t+k}$. The first order condition:

$$\sum_{k=0}^{\infty} \lambda^k E\left\{Q_{t+k} C_{t+k}(i)\left[p_t - \frac{\theta W_{t+k}}{(\theta-1)A_{t+k}}\right]\right\} = 0$$

can be log-linearized around the steady state to obtain:

$$\begin{aligned}\hat{p}_t &= (1-\beta\lambda)\sum_{k=0}^{\infty}(\beta\lambda)^k E[\hat{w}_{t+k} - \hat{A}_{t+k} + \hat{P}_{t+k}] = \\ &= (1-\beta\lambda)\left(\hat{w}_t - \hat{A}_t + \hat{P}_t\right) + \beta\lambda E\left[\hat{p}_{t+1}\right]\end{aligned}$$

where we defined the real wage. Subtracting $\hat{P}_{t-1}$ from both sides and rearranging, one obtains:

$$\hat{p}_t - \hat{P}_{t-1} = (1-\beta\lambda)\left(\hat{w}_t - \hat{A}_t\right) + \boldsymbol{\pi}_t + \beta\lambda E[\hat{p}_{t+1} - \hat{P}_t]$$

which allows us to obtain (1.76) after substituting in the earlier expression for the inflation rate.

79 The original Phillips curve was an empirical relation between unemployment and (wage) inflation first verified by Phillips (1958) for the U.K. A simple rationale can be obtained when firms and unions bargain and set the expected real wage W_t/P^e_t as a mark up on a reservation wage $w(u_t)$, which is inversely related to unemployment. Then, the price rule becomes $P_t = \mu P^e_t w(u_t)/A_t$, where μ includes both price and wage mark ups. After dividing both sides for P_{t-1}, this can be approximated in logs as $\boldsymbol{\pi}_t = \boldsymbol{\pi}^e_t + \log \mu w(u_t)/A_t$. Assuming $w(u_t) = 1 - \gamma u_t$, this can be rearranged as:

$$\boldsymbol{\pi}_t = \boldsymbol{\pi}^e_t - \gamma(u_t - \tilde{u})$$

where $\tilde{u} = (\mu - A)/\gamma$ is the steady state (natural) rate of unemployment, or Non-Accelerating Inflation Rate of Unemployment (NAIRU). This relation can be translated in terms of output using the Okun's Law, an empirical relation between unemployment change and output gap x_t. Adaptive expectations $\boldsymbol{\pi}^e_t = \boldsymbol{\lambda}\boldsymbol{\pi}_{t-1}$ generate a trade-off between unemployment and the inflation rate for $\boldsymbol{\lambda} = 0$ or the change in inflation for $\boldsymbol{\lambda} = 1$. Rational expectations generate a relation between unemployment and unexpected inflation only (Lucas, 1972; Sargent and Wallace, 1975); the New Keynesian Phillips curve generalizes this insight through price rigidities and forward looking firms. Rotemberg (1982a,b) has derived a similar relation assuming price adjustment costs that are quadratic in the price change for all firms. See Fuhrer and Moore (1995) and Mankiw and Reis (2002) for alternative versions that can better explain the persistence of inflation.

over marginal cost, and a higher ratio of marginal cost to price sparks inflationary pressures because the firms that are re-setting prices will, on average, be raising them. It seems very likely that marginal costs are procyclical, and more so than prices (when production levels are high relative to potential output, there is more competition for the available factors of production, and this leads to an increase in real costs), therefore the New-Keynesian theory implies that we should expect higher inflation in booms than in recessions.

The real marginal cost can be directly linked to changes in the output gap x_t, the difference between current output and equilibrium output under flexible prices. In turn, the output gap can be derived by the equilibrium conditions augmented with a real interest rate depending on the nominal one according to the Fisher equation.[80] The model is closed by a monetary policy rule for the choice of the nominal interest rate, which is now necessary to derive the path of the real variables. The Taylor principle holds: a rule as (1.71) is consistent with a determinate equilibrium only if $\phi > 1$, that is if the Central Bank responds to deviations of the inflation rate from its target level with sufficient strength.[81] In such a case we can study the impact on the economy of real shocks and of a monetary shock as well. In particular, an increase in the nominal interest rate leads to a persistent decline in output and inflation, and to an increase in the real interest rate (see Galì, 2008).[82]

Optimal monetary policy differs from the neoclassical one. As long as the flexible price equilibrium is efficient and $\chi \to 0$, the optimal policy must implement zero inflation. This neutralizes price stickiness (all prices remain optimal) and contemporaneously stabilizes the output gap (a result known as the "divine coincidence"), therefore it reproduces the allocation of the flexible price equilibrium and avoids distortions on the steady state levels of consumption and utility. Nevertheless, when there are other real imperfections the optimal policy faces a trade-off between price and output stabilization, and one can derive the optimal state-contingent policy. Rotemberg and Woodford (1997) have shown that the log-linear approximation of (1.64) is quadratic in the inflation rate and in the output gap x_t:

---

80 In the absence of capital all output is consumed, and the simple log-linearization of the Euler equation around the steady state provides $x_t = x^e_{t+1} - (i_t - \pi^e_{t+1} - \rho)$ where the time preference rate ρ is the steady state real interest rate. This and the New-Keynesian Phillips curve $\pi_t = \beta\pi^e_{t+1} + \xi x_t$ are usually augmented with random shocks representing supply and demand shocks.

81 Intuitively, after a raise in inflation, an increase in the real interest rate induces substitution of future consumption for current consumption through the Euler equation, which reduces current output and therefore, current inflation through the New-Keynesian Phillip curve. A weaker response to inflation is sufficient for determinacy if the monetary rule implies a positive response to deviations of the output gap as well (see Benhabib, Schmitt-Grohe and Uribe, 2001).

82 There is a general consensus that nominal frictions can improve the performance of the RBC model (see Christiano, Eichenbaum and Evans, 1999, 2005), especially in reproducing the hump shaped response of output to supply and demand shocks.

$$U \approx const - E\sum_{t=0}^{\infty} \beta^t \left[\vartheta(x_t - \tilde{x})^2 + \pi_t^2\right]$$

where ϑ is the relative weight on the variance of output gap, and the target for the output gap, $\tilde{x}$, may differ from zero. The optimal policy maximizes this under a sequence of constraints from (1.76), but it is time inconsistent because of an inflation bias in the Barro and Gordon (1983,a) tradition. The optimal time-consistent interest rate policy is linear in the expected inflation rate, and its structure resembles the so-called Taylor rule proposed by Taylor (1993), for which the nominal interest rate should be increased by 1.5 % in front of a 1% increase in expected inflation, and by 0.5 % for a 1 % increase in the output gap.[83]

Even if the New-Keynesian literature has been successful in complementing the neoclassical approach with a deeper understanding of monetary policy, it is still based on weak foundations, namely on an exogenous characterization of the market structures: the monopolistic behavior leads to optimal mark ups independent from technological and policy parameters and eliminates any strategic interaction between the price-setters, and the exogenous number of monopolistic firms eliminates any causal relation between profitability, entry and price-setting behavior.

1.5 Conclusions

In this chapter we have presented the most important results of the neoclassical theory. Our aim was to emphasize merits and limits of the leading approach to macroeconomic issues. Most of the analysis was based on a simple characterization of the production side, based on perfect competition and the absence of fixed costs of entry, and it led to indeterminate market structures. For this reason, aggregate shocks leading to cycles, openness to trade and the growth process did not affect in any way the structure of the markets and *vice versa*. The mechanism of propagation of the shocks is due to intertemporal substitution decisions of the consumers on savings and labor, and not on decisions of the firms on entry and market strategies, the gains from trade are associated with differences in factor endowments between trading countries, but these do not affect their market structures, and finally the absence of profits excludes any endogenous incentive to invest in innovations and eliminates any link between market structures and growth.

In the rest of the book we will endogenize the market structure in the presence of imperfect competition with strategic interactions between the firms and endogenous entry. For this reason, the next chapter is dedicated to rebuild the microfoundation of macroeconomics on the supply side.

---

[83] See Clarida, Gali and Gertler (1999). See also Svensson (1997) and the surveys by Woodford (2003), Galì (2008) and Wickens (2008).

2. The EMSs Approach to Macroeconomics

Sometimes the road less traveled is less traveled for a reason?
Jerry Seinfeld

In this chapter we start our review of an approach to macroeconomics that is in part alternative and in part complementary to the neoclassical one. We depart from the perfectly competitive environment, in the sense that firms do not take prices as given, but they do choose their strategies and they interact strategically. We focus not only on the choice of prices as the strategic variables, but also on the choice of output levels, and on the choice of entry to produce new or better goods. In most of the analysis of this book we adopt either symmetric Cournot competition or symmetric Bertrand competition as the main models of static strategic interactions, but we will occasionally introduce other forms of competition, as Stackelberg competition or models of imperfect collusion, and we propose a general approach that can be employed with more sophisticated competitive structures borrowed from research in the field of industrial organization. As a matter of fact, one of the main aims of this book is exactly to build a solid bridge between macroeconomics and industrial organization.

The new ingredients of the endogenous market structures (EMSs) approach will be on the supply side of the economy. The technological conditions will be characterized by positive fixed costs of entry so as to move beyond the constant returns to scale hypothesis. To a large extent, we will also depart from the neoclassical assumption that investment (of final goods) builds the physical capital that is used as factor of production together with labor supplied by the working class. That was a good assumption to describe production in the industrialization phase, characterized by the dominance of the secondary (manufacturing) sector and by the social conflict between capital and labor, but not such a good one to describe production in the modern age, dominated by the tertiary (service) sector and by the New Economy, where ideas, innovations, intellectual property rights and creativity are the main inputs needed to create new products, and where the value of start ups without any capital can be high because of these intangible inputs. For this reason, we will embrace a concept of investment (in terms of labor or consumption

F. Etro., *Endogenous Market Structures and the Macroeconomy*,
DOI: 10.1007/978-3-540-87427-0_2, © Springer-Verlag Berlin Heidelberg 2009

units) needed to enter in the market with new products (or with better products) produced through labor.[1] This will establish a two-way link between investment and market structure: profitability in the market attracts investment to create new products, and the creation of new products by means of investment enhances competition and reduces profitability in the market.

Finally, we will endogenize the entry decision of the firms as a rational profit maximizing decision. As we have seen at the end of the previous chapter, the New-Keynesian literature has taken into consideration the rational behavior of monopolistic firms in the choice of their profit maximizing prices, but it has typically neglected the rational behavior of the same firms in the choice of entering in the market if and only if positive profits can be expected. As a consequence, there was no link between profit opportunities and production or any other aggregate variable. Our analysis of the entry process leads to the final characterization of the EMSs.

In a sense, our approach can be seen as a natural evolution of the neoclassical approach, which has been guided by the attempt of introducing rational behavior in all the aspects of decision making. The rational theories of consumption and labor supply and the theory of rational expectations (as opposed to adaptive expectations) have been the building blocks of the neoclassical approach. However, a rational theory of entry in markets in which there are profit-maximizing strategic firms has not been introduced until recently.[2] This is the additional contribution of the EMSs approach.

An EMS is defined as *an equilibrium organization of a market where each firm chooses its own strategy to maximize profits taking as given the demand conditions and the strategies of the other firms, and where the number of firms is such that all of them make non-negative profits and further entry cannot provide positive profits.* We will often refer to a simplified situation with a symmetric equilibrium in which all firms choose the same strategy and they obtain the same profits, and we will approximate the exact equilibrium assuming that the number of firms is a natural number. In such a case an

---

[1] This does not mean that we will ignore the accumulation of stock variables, but only that they will play a different role: we will focus on the development of the stock market value and on the accumulation of innovative ideas.

[2] There is an old partial equilibrium literature which investigates the endogenous entry process on the basis of an adaptive mechanism rather than a rational one. Suppose that gross profits in a market with N_t firms at time t are $\Pi(N_t)$, and that entry of N_t^e new firms depends on the excess profits compared to a fixed cost F according to:

$$N_{t+1} = (1-\delta_N)\left(N_t + N_t^e\right) \quad \text{with } N_t^e = \boldsymbol{\lambda}\left[\Pi(N_t) - F\right]$$

where $\delta_N \in (0,1)$ is a rate of exit from the market and $\boldsymbol{\lambda} > 0$ parametrizes the speed of entry. The evolution of this system can exhibit monotonic or cyclical convergence to the steady state, but complex dynamics can emerge as well. The exogenous and adaptive nature of this process is its limit, which will be avoided by the EMSs approach, where the number of entrants N_t^e derives from an endogenous and rational process.

EMS is defined by *a pair* (x, N) *where* x *is the strategy adopted by each firm and* N *is the number of firms, and the equilibrium satisfies the conditions for profit maximization and endogenous entry.* Notice that the strategy can be given by the production level of the firms or by their prices in case of competition in the market respectively *à la* Cournot or *à la* Bertrand, or by the investment in R&D in case of competition for the market.

In general, in the presence of multiple markets, each market k is characterized by an EMS with (x_k, N_k) and, in the presence of multiple periods, each period t is characterized by EMSs for each market (x_{kt}, N_{kt}) with associated dynamic paths for the equilibrium strategies and the equilibrium number of firms. These converge to steady state EMSs $(\tilde{x}_k, \tilde{N}_k)$ that depend on structural (technological, behavioral, strategic and policy) factors and can be interpreted as the long run EMSs. The crucial aspect of substituting perfect competition or exogenous market structures with EMSs in macroeconomics has to do with the link between demand and supply in general equilibrium. The demand functions perceived by the firms must be the result of the maximization of utility by rational consumers (or by a representative consumer), whose income includes both the remuneration of the factors of production and the eventual profits of the firms (that were zero in the neoclassical approach with perfect competition, or constant in models with an exogenous number of monopolistic firms). In a dynamic model, the discounted value of the firms'profits, represented by the stock market capitalization, reflects both the strategic interactions and the entry/exit process and it affects aggregate demand as well. Therefore, the EMSs approach creates a novel and complex channel that links competition, the stock market and the aggregate economy. In the book we will gradually introduce all these complex elements, but in this chapter we start sketching a simpler model with a single market and a single period to introduce the reader to the main aspects of the EMSs approach. Later on in the chapter we introduce a dynamic setup, and we provide preliminary discussions about the role of EMSs in explaining the determinants of the business cycle, the international trade between countries and the growth process.

In the analysis of industrial organization there are well developed studies on strategic interactions in the Cournotian tradition and on endogenous entry in the presence of fixed costs of production in the Marshallian tradition. The systematic adoption of both elements is more recent, but it is rapidly becoming the standard way to model market structures. One of the first characterizations of EMSs is due to Dasgupta and Stiglitz (1980), who studied competition in quantities and cost reducing strategies with homogenous goods and free entry. Only recently their results have been generalized to product differentiation and competition in prices by Vives (2008). However, the Dasgupta-Stiglitz model has largely inspired the investigations of Sutton (1991, 1998, 2008), who has analyzed markets with strategic interactions in the choice of production and quality, endogenous entry and endogenous sunk costs from both a theoretical and empirical point of view. The analysis of strategic investments and asymmetries in the presence of EMSs has been introduced only recently, with the first general characterization of Stackelberg equilibria with endogenous entry by Etro (2008,b).[3]

The modern empirical literature on EMSs started with the works of Bresnahan and Reiss (1987, 1990) and Berry (1992), which moved beyond the naive view for which lower mark ups are due to more competition associated with a larger number of firms. Such a mechanism definitely holds in the presence of exogenous market structures, but when entry is endogenous there is an opposite mechanism at work: lower mark ups attract a lower number of firms and higher mark ups attract a higher number of firms. In general, the empirical analysis of EMSs requires a different methodology. One possibility is an approach based on the effect that exogenous factors, as the size of demand or other technological conditions, have on the endogenous variables: mark ups, number of firms and production of these firms. Berry and Reiss (2007) review empirical studies within this approach, paying particular attention to equilibrium models that interpret cross-sectional variation in the number of firms or firm turnover rates, and to applications that analyze EMSs in airline, retail, professional, auction, lodging, and broadcasting markets. A more recent approach is based on the impact that entry conditions of different markets exert on the strategic behavior of some firms, and in particular on the leaders. When there are independent variables (or natural experiments) that can discriminate between markets with exogenous or endogenous entry, the predictions of the EMSs approach for the behavior of the leaders can be tested.[4]

The introduction of EMSs in macroeconomic analysis is very recent, even if the microeconomic tools have been available for a while. As we noticed in the previous chapter, the microfounded model of Dixit and Stiglitz (1977) has been widely used in the New-Keynesian macroeconomics assuming monopolistic behavior by an exogenous number of firms, therefore both strategic interactions and endogenous entry have been systematically neglected. The trade literature has mainly focused on one of the two aspects: endogenous entry of monopolistic firms in general equilibrium (Krugman, 1980) or strategic interactions between an exogenous number of firms in partial equilibrium (for instance Brander and Spencer, 1985). Growth theory has endogenized entry

---

3 For a comprehensive survey on the industrial organization literature on EMSs see Etro (2007,a). On recent advances of the theory of EMSs in partial equilibrium see Erkal and Piccinin (2007), Ino and Matsumura (2007), Mukherjee (2008), Ishida, Matsumura and Matsushima (2008), Tesoriere (2008,a,b), Žigić (2008), Anderson and de Palma (2008) and Creane and Konishi (2009).

4 For recent works within the first approach see Manuszak (2002), Mazzeo (2002), Campbell and Hopenhayn (2005), Czarnitzki and Etro (2009) and Chapter 4. For the second approach see Czarnitzki, Etro and Kraft (2008) and Chapter 5. See also Basker (2008) for an interesting analysis of entry in U.S. grocery distribution.

in the competition in the market neglecting strategic interactions (Romer, 1990) and has avoided any strategic consideration in the analysis of the competition for the market (Aghion and Howitt, 1992).

A recent class of models has augmented all these frameworks with the introduction of genuine EMSs, obtaining a number of new positive and normative predictions that we will examine in the next chapters. A few early works on the business cycle (summarized by Cooper, 1999) have introduced monopolistic behavior and endogenous entry in each period within otherwise standard neoclassical models. Other important works by Peretto (1996, 1999) have provided the first systematic attempt to introduce EMSs in the competition in the market in a dynamic general equilibrium model of endogenous growth, and to show the relevance of EMSs for the aggregate behavior of the economy. Etro (2004,a) has provided the first attempt to introduce EMSs in the competition for the market in a dynamic general equilibrium model of Schumpeterian growth. Ghironi and Melitz (2005) have nested trade models with monopolistic behavior and endogenous entry in a DSGE model of the open economy, and this important contribution has opened new research opportunities to study EMSs in macroeconomics. Etro (2007,b) has provided simple examples of the impact of EMSs on trade and business cycles: first, by analyzing strategic interactions and endogenous entry in trade theory and trade policy and second, by studying the impact of shocks on simple two periods models with EMSs. This chapter is based on some of the insights of that work.

The chapter is organized as follows. Section 2.1 introduces strategic interactions and endogenous entry in general models of competition in and for the market. Section 2.2 restricts the attention to microfounded profit functions in partial equilibrium focusing on competition in quantities and in prices with endogenous entry. Section 2.3 studies the particular case of isoelastic utility which will be adopted in multiple applications in the following chapters. Section 2.4 applies the EMSs approach to the simplest dynamic model, that is a two periods exchange economy with endogenous entry in each period. Section 2.5 extends the simple analysis to a general equilibrium context. Section 2.6 develops the first full fledged dynamic model with endogenous entry in the long run and characterizes the equilibrium and steady state EMSs. The analysis keeps savings and labor supply as exogenous, postponing their endogenous characterization to the next chapter. Nevertheless, this simple model allows us to derive in Sections 2.7-2.9 preliminary implications for the three main topics of the book. Section 2.10 concludes.

2.1 EMSs in Partial Equilibrium

When we want to analyze the endogenous structure of a market the first step is to characterize the profit functions of the firms active in this market and to understand how these firms interact strategically. The second step is to understand which firms are endogenously going to be active in this market and to study how demand and supply conditions affect entry and the strategies of the firms. The third step is to understand how the aggregate demand conditions have determined the profit functions of the firms under consideration, which allows us to introduce the market under investigation in a microfounded framework. The fourth step is to introduce this framework in a general equilibrium context.

In this section we focus on the first two steps and we briefly introduce a general class of models of the market structure (used already in the partial equilibrium analysis of Etro, 2007,a) where the profit functions are exogenously given and the EMSs can be characterized in a general way. In the next section, we will restrict our attention to a subset of this class of models where the profit functions are endogenously derived from the utility maximizing behavior of the consumers.

Consider N firms choosing a strategic variable $x(i) > 0$ with $i = 1, 2, ..., N$. These strategies deliver for each firm i the gross profit function:

$$\Pi(i) = \Pi\left[x(i), \beta_i\right] \tag{2.1}$$

where $x(i)$ is the strategy of firm i and we assume that gross profits have always a unique maximum in $x(i)$: $\Pi_1 \gtreqless 0$ for any $x \lesseqgtr \hat{x}$ for some profit maximizing strategy $\hat{x}$. The second argument represents the effects (or spillovers) induced by the strategies of the other firms on firm i's profits, summarized by $\beta_i = \sum_{j=1, j\neq i}^{N} h(x(j))$ for some function $h(x)$ which is assumed positive, differentiable and increasing; these spillovers exert a negative effect on profits, $\Pi_2 < 0$, and of course they affect the profit maximizing strategy. This general framework nests models of competition with strategic substitutability ($\Pi_{12} < 0$), and with strategic complementarity ($\Pi_{12} > 0$). In the former case, typical of Cournot competition, there may be multiple asymmetric equilibria (with firms choosing different strategies), and in the latter case, typical of Bertrand competition, there may be multiple symmetric equilibria. Cooper and John (1988) have emphasized the Keynesian implications of models with multiple equilibria derived from strategic complementarities,[5] but in this book we will not stress this issue, and we will focus on unique symmetric equilibria.

Finally, we assume that entry requires a fixed sunk cost F, so that the net profits of firm i are:

$$\pi_i = \Pi\left[x(i), \beta_i\right] - F$$

Given these profit functions, under the standard assumption of Nash competition between the firms, we can easily characterize the symmetric EMS with the pair (x, N) satisfying the profit maximizing condition:

---

[5] See also Diamond (1982), Hart (1982) and Murphy, Shleifer and Vishny (1989) for related Keynesian models.

$$\Pi_1 \left[x, (N-1)h(x) \right] = 0 \tag{2.2}$$

and the endogenous entry condition:

$$\Pi \left[x, (N-1)h(x) \right] = F \tag{2.3}$$

where we used the equilibrium condition $\beta = (N-1)h(x)$.

Such an EMS satisfies a number of properties that are widely discussed in Etro (2007,a). The main properties are the following. First, the number of firms N is always decreasing in the size of the fixed cost of entry (relative to the size of the market).[6] Second, the strategy of each firm x is increasing with the fixed cost of entry (relative to the size of the market) under strategic substitutability, i.e. the firm becomes more aggressive, and it is decreasing under strategic complementarity, i.e. the firm becomes more accommodating. Third, any firm would gain by committing, before entry occurs, to a more aggressive strategy than x, which would reduce the endogenous number of firms N. Fourth, any firm would also gain by committing to strategic investments that lead to a more aggressive behavior than x, which would reduce the endogenous number of firms N.

Most of the common models of competition in the market, that is in the choice of production or pricing for given products, are nested in our general specification. For instance, consider a market with competition in quantities such that the strategy $x(i)$ represents the quantity produced by firm i. The corresponding inverse demand for firm i is $p(i) = p\left[x(i), \sum_{j \neq i} x(j)\right]$ which is decreasing in both arguments (if goods are substitutes). With a generic cost function $c(x(i))$ with $c'(\cdot) > 0$, it follows that the gross profits for firm i are:

$$\Pi(i) = x(i)p\left[x(i), \beta_i\right] - c(x(i)) \tag{2.4}$$

with $\beta_i = \sum_{j \neq i} x(j)$. Examples include the case of linear demand $p(i) = a - \sum_{j=1}^{N} x(j)$ for any i, the class of isoelastic demand functions, and other common cases.

Consider now models of competition in prices where $p(i)$ is the price of firm i. Any model with a direct demand $D_i = D\left[p(i), \sum_{j \neq i} g(p(j))\right]$ such that $D_1 < 0$, $D_2 < 0$ and $g'(p) < 0$ is nested in our general framework after setting $x_i \equiv 1/p_i$ and $h(x(i)) = g(1/x(i))$. This specification guarantees that goods are substitutes in a standard way since $\partial D_i / \partial p(j) = D_2 g'(p(j)) > 0$. Examples include models of price competition with isoelastic demand, Logit demand, or any constant expenditure demand. Adopting, just for simplicity, a constant marginal cost c, we obtain the gross profits for firm i:

---

[6] Notice that the size of the fixed cost must be compared to the size of the market, which determines the profit opportunities, therefore we can think of F as a the fixed cost relative to the market size. In other words, if gross profits were $\Pi(i) = E\Pi\left[x(i), \beta_i\right]$ with E as a size parameter, the comparative statics of F would be the same as that of F/E.

$$\Pi(i) = \left(\frac{1}{x(i)} - c\right) D\left(\frac{1}{x(i)}, \beta_i\right) = (p(i) - c)\, D\left[p(i), \beta_i\right] \tag{2.5}$$

with $\beta_i = \sum_{j \neq i} g(1/x(j))$. This model is nested in our general framework as well.

Notice that in a dynamic framework where entry costs F are born once and the firm remains active over time, the gross value of the firm can be seen as the discounted sum of its profits, something that should reflect the stock market capitalization of the same firm. If r is the constant interest rate, this corresponds to:

$$V(i) = \frac{\Pi(i)}{r} \tag{2.6}$$

and the endogenous entry condition equates this to the fixed cost of entry, so that:

$$V(i) = F \quad \Longleftrightarrow \quad \Pi(i) = rF \tag{2.7}$$

This dynamic framework can be easily extended with an exogenous probability of exit from the market, for instance due to the introduction of a new and better product.[7]

Models of competition for the market focus exactly on the competition to innovate and associate the exit of the incumbent firm with the introduction of a new and better product. These models are also known as patent races because they represent contests to obtain profits from intellectual property rights associated with innovations (which typically provide a temporary monopolistic power). Assume that firms invest a flow of resources in the continuous time to obtain an innovation of exogenous value V^M according to a stochastic process *à la* Poisson. If $x(i)$ is the flow of investment of firm i determining an instantaneous probability of innovation $h(x(i))$, which is assumed to be positive, increasing and strictly concave, the gross value of firm i can be derived as:

$$V(i) = \frac{h(x(i))V^M - x(i)}{r + \delta_N} \tag{2.8}$$

[7] In a discrete environment where $\delta_N \in [0,1)$ is the exit rate in the presence of N firms, the gross value from being in the market at time is:

$$V(i) = \left(\frac{1-\delta_N}{1+r}\right)\Pi(i) + \left(\frac{1-\delta_N}{1+r}\right)^2 \Pi(i) + ... = \frac{(1-\delta_N)\Pi(i)}{r+\delta_N}$$

In the continuous time, with the instantaneous rate of exit $\delta_N \in [0, \infty)$, we have:

$$V(i) = \int_{t=0}^{\infty} \Pi(i) e^{-(r+\delta_N)t} dt = \frac{\Pi(i)}{r+\delta_N}$$

where $\delta_N = \sum_{j=1}^{N} h(x(j))$ is the instantaneous probability of innovation, which corresponds to the rate of exit of the incumbent firm (now endogenous). It is easy to verify that this case is nested in our general framework after decomposing the exit rate as $\delta_N = h(x(i)) + \beta_i$. Assuming again that the entry cost F is born once and the firms keep doing research until an innovation emerges, endogenous entry must satisfy:

$$V(i) = F \iff rV(i) = h(x(i))V^M - \delta_N V(i) - x(i) \tag{2.9}$$

whose second expression equates the return on the value of the firm $rV(i) = rF$ with the expected net return from the R&D investment. This takes into account the expected net gain from innovation $h(x(i)) \left[V^M - V(i)\right] - x(i)$, and the expected loss in case others innovate $\beta_i V(i)$.

In all these models, we can derive the EMSs and characterize the equilibrium pair (x, N) as a function of the exogenous variables, which is a starting point for comparative static analysis and for the study of the strategic behavior of firms in a realistic market environment. This class of models has proved to be quite useful to investigate a number of positive and normative issues at the microeconomic level. Etro (2007,a) reviews the applications of the EMSs approach to strategic investments in R&D, advertising, quality choices, product differentiation, debt financing and other financial decisions, dynamic forms of competition, issues related to network effects, bundling, vertical restraints, price discrimination, mergers, collusion and liberalizations, and discusses the main implications for antitrust policy.

In this book, however, we want to introduce EMSs in a macroeconomic framework, therefore all of the above mentioned exogenous variables are going to be endogenized sooner or later. For instance, in Chapter 3 we study competition in the market within dynamic stochastic general equilibrium models of the aggregate economy, therefore the demand functions derive from endogenous choices of utility maximizing agents (and policymakers as well), and the cost functions depend on the technology but also on the equilibrium in the market for inputs. In Chapter 4 we study open economies in which decisions taken by firms and consumers in the foreign markets (and by policymakers as well) affect the profit functions of the domestic firms. In Chapter 5 we study models of competition for the market where the value of innovations and the interest rate depend on the general equilibrium of the economy (and by the action of policymakers once again), and they affect accordingly the expected profits of the firms investing in R&D.

Of course, a preliminary investigation of the EMSs in partial equilibrium must be our next step, and we now proceed in this direction focusing on a restricted class of static models of competition in the market whose demand structure can be easily derived from consumers' behavior.

2.2 Microfounded EMSs

In this section we follow the industrial organization literature and analyze a single static market with multiple products characterized by a set of demand functions that are directly derived from the optimal choices of a representative agent with an exogenous endowment. All the firms face common technological conditions. Given these elements, we derive the EMSs in the case of competition in quantities and in prices in partial equilibrium. This framework will be introduced in dynamic and general equilibrium macroeconomic models in later sections.

Consider a representative agent with the following utility depending on the consumption of N goods:[8]

$$U = U\left[\sum_{j=1}^{N} u\left(C(j)\right)\right] \tag{2.10}$$

where $C(j)$ is consumption of good j, $u(C) > 0$, $u'(C) > 0$ with $u''(C) \leq 0$, and $U(\cdot)$ is a positive and increasing function.[9] Notice that these preferences exhibit "love for variety", in the sense that spreading consumption through a larger number of goods increases utility: this reflects complementarities in consumption. The above utility is maximized under the budget constraint:

$$\sum_{j=1}^{N} p(j)C(j) = E \tag{2.11}$$

where $p(j)$ is the price of good j and E is the exogenous endowment of the representative agent. In partial equilibrium this endowment is taken as given. Utility maximization provides the demand for each good and allows us to analyze competition in quantities or in prices. Here we analyze the general case, but at a first reading, one may want to skip the rest of this section and move to the traditional case of isoelastic sub-utilities considered in Section 2.3.

2.2.1 EMSs with competition in quantities

Let us derive the inverse demand functions for the different goods. Utility maximization for each good i implies $u'(C(i)) = \lambda p(i)$ with λ La-

[8] As well known, this specification is due to Dixit and Stiglitz (1977), whose results are commented below. However, the original Dixit-Stiglitz model did not take into account strategic interactions.

[9] Moreover, we assume the regularity condition $u'(C) + Cu''(C) > 0$.

grange multiplier of the budget constraint.[10] Multiplying each side by $C(i)$, summing up over all goods and using the budget constraint, one obtains $\lambda = \sum_j C(j)u'(C(j))/E$. Therefore, if we define with $x(i)$ the production of good i, its inverse demand can be written as:

$$p(i) = \frac{u'(x(i))E}{\sum_{j=1}^{N} x(j)u'(x(j))} \tag{2.12}$$

which is increasing in the endowment and decreasing in the production of each good.[11] Notice that with linear sub-utilities ($u''(C) \to 0$) we would obtain the particular case of homogenous goods as a limiting outcome; namely, the inverse demand would become hyperbolic:

$$p = \frac{E}{\sum_{j=1}^{N} x(j)}$$

for every firm.

If each firm produces at a constant marginal cost c, the gross profit function is:

$$\Pi(i) = \frac{x(i)u'(x(i))E}{\sum_{j=1}^{N} x(j)u'(x(j))} - cx(i) \tag{2.13}$$

which is nested in our general formulation (2.1) with $\beta_i = \sum_{j \neq i} x(j)u'(x(j))$.

In case of Cournot competition between N firms, each one chooses its own output $x(i)$ to maximize profits given the strategies of the other firms, and in the symmetric equilibrium one can derive the following output per firm:

$$x = \frac{(N-1)\left[u'(x) + xu''(x)\right]E}{N^2 u'(x)c}$$

which generates the following price:

$$p = \mu^Q(N,x)c \ \text{ with } \mu^Q(N,x) = \frac{Nu'(x)}{(N-1)\left[u'(x) + xu''(x)\right]} \tag{2.14}$$

where the index Q stands for competition in quantities. The mark up rule is decreasing in the number of firms, but in general it also depends on the

---

[10] The problem of maximization of (2.10) s.v. (2.11) is equivalent to the problem of maximization of $\sum_{j=1}^{N} u(C(j))$ under the same constraint, which generates the Lagrangian:

$$\mathcal{L} = \sum_{j=1}^{N} u(C(j)) + \lambda \left[E - \sum_{j=1}^{N} p(j)C(j)\right]$$

Its maximization with respect to the consumption of all goods and the Lagrange multiplier λ provides the optimal consumption plan.

[11] This holds under our restrictions on the sub-utilities.

individual production x, and we assume that it is non-decreasing in x.[12] Budget balance requires $x = E/Np$, or $\mu^Q(N) = E/cNx$, which together with (2.14) uniquely defines the mark up as a decreasing function of the number of firms $\mu^Q(N)$. Moreover, for a given number of firms, the mark up is non-increasing in the marginal cost c, and non-decreasing in the endowment E.

The equilibrium gross profits become the following decreasing function of the number of firms:

$$\Pi(N) = \frac{[u'(x) - (N-1)xu''(x)]\,E}{N^2 u'(x)} \tag{2.15}$$

Now, let us use the fact that there is a fixed cost of entry in the market F. Then, when entry is endogenous, the number of firms must be such that these profits are zero. One can solve for the equilibrium number as:

$$N^Q = \sqrt{\frac{E}{F}\left(1 + \frac{xu''(x)}{u'(x)}\right) - \frac{xu''(x)E}{2u'(x)F}} - \frac{xu''(x)E}{2u'(x)F} \tag{2.16}$$

The profit maximizing condition (2.14) and the endogenous entry condition (2.16) together provide the equilibrium value for the pair (x^Q, N^Q), and therefore fully characterize the EMS in partial equilibrium. In this general case the analysis of the comparative statics is complex, but a special case can help us to derive a few basic results.

Consider the case of homogenous goods, corresponding to the limiting case of linear sub-utilities ($u''(x) = 0$). Now the mark up boils down to $\mu^Q(N) = N/(N-1)$, which is decreasing in the number of firms (and independent from marginal costs and endowment), but always larger than one. This allow us to consider the effect of strategic interactions in an otherwise standard setup with perfectly substitutable goods (which has been traditionally studied only under perfect competition in the neoclassical tradition of macroeconomics). Under endogenous entry the number of firms becomes simply:

$$N^Q = \sqrt{\frac{E}{F}} \tag{2.17}$$

which sets the equilibrium mark up at:

$$\mu^Q = \frac{\sqrt{E}}{\sqrt{E} - \sqrt{F}} \tag{2.18}$$

This relations show the simple link between the endowment of the representative agent and the cost of entry on one side, and the EMS on the other side.[13] We can easily verify that increasing the size of a market the number

[12] This turns out to be true under weak conditions on the preferences.

[13] Only when the fixed costs of production tend to zero, the market structure approximates the perfectly competitive one, with infinite firms producing an infinitesimal amount of the uniform good at a price equal to the marginal cost.

of firms increases but less than proportionally, and the mark up decreases. More precisely, the entry of at least N firms requires an endowment above the minimum level N^2F: in other words, if we want to double the number of active firms, we need an endowment that is more than the double.[14]

Finally, when the endowment increases, each firm has to produce at a larger scale, according to:

$$x^Q = \frac{\sqrt{EF} - F}{c}$$

This happens because a larger expenditure opens space for a larger number of firms, but this strengthens competition and reduces the mark ups, which requires a larger scale of production for each firm to cover the fixed costs.[15] In conclusion, notice that one could study alternative models of competition in quantities, as the Stackelberg model, in which one firm is the leader and has a first mover advantage in the choice of its production level. We will analyze this case later on.

2.2.2 EMSs with competition in prices

The utility maximization problem can be used also to express the direct demand functions. This allows us to analyze the case of competition in prices. In particular, inverting the utility maximizing condition $u'(C(i)) = \lambda p(i)$ we have $C(i) = u'^{-1}[\lambda p(i)]$, which must be decreasing in $\lambda p(i)$ by the concavity of the subutility function. Using our expression for the Lagrange multiplier $\lambda = \sum_j C(j)u'(C(j))/E$, we obtain the following function for the direct demand of good i:

$$C(i) = u'^{-1}\left[\frac{p(i)}{E}\sum_{j=1}^{N}\left\{u'^{-1}\left[p(j)\right]\right\}\right] \tag{2.19}$$

which is increasing in the endowment, decreasing in $p(i)$ and increasing in the other prices. Gross profits become:

$$\Pi(i) = \left[p(i) - c\right]u'^{-1}\left[\frac{p(i)}{E}\sum_{j=1}^{N}\left\{u'^{-1}\left[p(j)\right]\right\}\right] \tag{2.20}$$

---

[14] Such a prediction can be generalized to models of competition in quantities with imperfect competition, but not to models of competition in prices. It can be tested in the presence of markets of different sizes, for instance professional or retail markets in different towns. A wide empirical literature (Breshnan and Reiss, 1987; Manuszak, 2002) has found encouraging support for this view (see Chapter 4).

[15] This prediction holds in more general models of competition in quantities and prices as well. Campbell and Hopenhayn (2005) provide convincing empirical evidence in its support (see Chapter 4).

which are nested in our general formulation (2.1) with $\beta_i = \sum_{j \neq i} u'^{-1}[p(j)]$.

In a symmetric Bertrand equilibrium, all firms choose a profit maximizing price:

$$p = \mu^P(N)c \quad \text{with } \mu^{P\prime}(N) < 0 \tag{2.21}$$

where $\mu^P(N)$ is an implicit expression for the mark up (which may depend on marginal cost and endowment). Notice that budget balance requires a demand equal to $E/Np = E/N\mu^P(N)c$ for each firm, therefore the equilibrium gross profits must be:

$$\Pi(N) = \frac{\left[\mu^P(N) - 1\right]E}{\mu^P(N)N} \tag{2.22}$$

Under endogenous entry, the number of firms must be such that these profits are zero. An implicit expression for the equilibrium number of firms can be derived as follows:

$$N^P = \frac{\left[\mu^P(N^P) - 1\right]E}{F\mu^P(N^P)} \tag{2.23}$$

The profit maximizing and endogenous entry conditions (2.21) and (2.23) provide together the equilibrium values for the pair (p, N^P), and therefore fully describe the EMS under symmetric competition in prices. Also in this case, generality does not allow us to obtain simple comparative statics results, but the example of the next section will clarify the relation between exogenous and endogenous variables. Finally, notice that also in case of competition in prices one could study the role of a leader within a Stackelberg model, as we will do in the example of the next section. However, before focusing on this example, we need to derive the optimal market structure in this general model.

2.2.3 Optimal EMSs

In their pathbreaking work on monopolistic pricing with product differentiation and free entry, Dixit and Stiglitz (1977) characterized the (constrained) optimal market structure.[16] This is given by a common production level x^O for each firm and by the number of firms N^O that maximize utility $U = U[Nu(x)]$ subject to the zero profit constraint $x(p-c) = F$ and to the resource constraint $pxN = E$, that is:

$$\max_x \frac{Eu(x)}{F + cx}$$

[16] The constraint refers to the zero profits of the firms. The unconstrained first best would adopt marginal cost pricing to maximize utility under the resource constraint.

The optimality condition can be written as:

$$x^O = \frac{F\rho(x^O)}{[1-\rho(x^O)]\,c}$$

where $\rho(x) \equiv u'(x)x/u(x)$ is the elasticity of the subutilities (notice that, contrary to the case of EMSs, the optimal production per firm is independent from the total endowment). Of course this is only an implicit expression unless the sub-utility is isoelastic. The corresponding optimal number of firms can be implicitly written as:

$$N^O = \frac{[1-\rho(x^O)]\,E}{F} \tag{2.24}$$

which is linear in the endowment. Comparing this with (2.23), one can notice that the optimal market structure is compatible with a mark up $\mu^O = 1/\rho(x^O)$. Therefore, the EMS under competition in prices is efficient if and only if the equilibrium mark up happens to coincide with the inverse of the elasticity of the utility function.

In the next section, we will explore a particular case of our model in which sub-utilities are isoelastic. In this case the equilibrium mark up in the short run (i.e. with exogenous entry) is higher than the optimal one under both forms of competition, and depends on the number of firms and on the degree of substitutability between goods, but not on the marginal cost and the endowment. This simplifies things at the cost of loosing (in the short run, but not in the long run) the impact of supply and demand conditions on mark ups. Future research should try to take into account more general preferences that deliver richer short-run interactions between supply and demand conditions and mark ups.[17]

2.3 EMSs with Isoelastic Sub-utility

Let us simplify our analysis by introducing isoelastic subutilities, which will be used in a large part of this book. Assume that preferences depend on the consumption of the N goods according to the following index:

[17] More complex utility functions have been usefully analyzed. Feenstra (2003) has introduced translog preferences which microfound an elasticity of substitution increasing in the number of goods. Bertoletti, Fumagalli and Poletti (2008) have introduced a new class of "increasing elasticity of substitution" preferences finding that, even under constant returns to scale, a rise in the number of firms can be price-increasing under both monopolistic and Cournot competition (notice that, despite the price increase, consumers benefit from a rise in the number of monopolistic competitors because of higher product diversity, therefore higher prices are associated with higher consumer welfare).

$$U = \left[\sum_{j=1}^{N} C(j)^{\frac{\theta-1}{\theta}} \right]^{\frac{\theta}{\theta-1}} \tag{2.25}$$

where $\theta > 1$ is the degree of substitutability between goods. When $\theta \to \infty$ the goods become perfect substitutes and generate a hyperbolic demand, when $\theta \to 1$ they tend to complete independence. Of course, intermediate values of θ are associated with imperfect substitutability.

Notice that the elasticity of the sub-utility is $\rho(x) = u'(x)x/u(x) = (\theta - 1)/\theta$, which is constant. Using the results of the previous section, this allows us to determine the (constrained) optimal market structure as characterized by a number of firms:

$$N^O = \frac{E}{\theta F} \tag{2.26}$$

The optimal number of firms can be obtained if the firms adopt a mark up $\mu^O = \theta/(\theta - 1)$ and produce

$$x^O = \frac{F(\theta - 1)}{\theta c} \tag{2.27}$$

which are both independent from the endowment. Incidentally, this is exactly what would emerge if firms were behaving as monopolistic price setters (as in Dixit and Stiglitz, 1977), ignoring the impact of their choices on the price index.[18] However, our interest here, is not on the monopolistic behavior of an infinity of firms, but on strategic interactions between a limited number of firms active in the market.

To microfound the profit function, notice that the representative consumer allocates its endowment E across the available goods with prices $p(i)$ according to the direct demand function:

$$C(i) = C\left(\frac{p(i)}{P}\right)^{-\theta} = \frac{p_t(i)^{-\theta}}{P^{1-\theta}} CP = \frac{p(i)^{-\theta} E}{P^{1-\theta}} \qquad i = 1, 2, ..., N \tag{2.28}$$

where P is a price index defined as:

$$P = \left[\sum_{j=1}^{N} p(j)^{1-\theta} \right]^{\frac{1}{1-\theta}} \tag{2.29}$$

such that total expenditure satisfies $E = \sum_{j=1}^{N} p(j)C(j) = CP$.

[18] Of course, this would be a reasonable assumption in the presence of an infinity (or a very high number) of firms selling different varieties of the same good - by the way, this is a situation at odds with the same concept of monopolistic behavior.

Inverting the direct demand functions, we can also derive the system of inverse demand functions:

$$p(i) = \frac{x(i)^{-\frac{1}{\theta}} E}{\sum_{j=1}^{N} x(j)^{\frac{\theta-1}{\theta}}} \qquad i = 1, 2, ..., N \tag{2.30}$$

where $x(i)$ is the consumption of good i.

In the following sections we analyze different forms of competition that can take place between the firms and derive the associated EMSs.

2.3.1 Cournot competition

First, let us consider competition in quantities. Using the inverse demand function, we can express the profit function of a firm i as a function of its output $x(i)$ and the output of all the other firms:

$$\begin{aligned} \Pi(i) &= [p(i) - c]\, x(i) = \\ &= \frac{x(i)^{\frac{\theta-1}{\theta}} E}{\sum_{j=1}^{N} x(j)^{\frac{\theta-1}{\theta}}} - cx(i) \end{aligned} \tag{2.31}$$

Assume now that each firm chooses its production $x(i)$ taking as given the production of the other firms. The first order conditions:

$$\left(\frac{\theta-1}{\theta}\right) \frac{x(i)^{-\frac{1}{\theta}} E}{\sum_j x(j)^{\frac{\theta-1}{\theta}}} - \left(\frac{\theta-1}{\theta}\right) \frac{x(i)^{\frac{\theta-2}{\theta}} E}{\left[\sum_j x(j)^{\frac{\theta-1}{\theta}}\right]^2} = c$$

for all firms $i = 1, 2, ..., N$ can be simplified imposing symmetry of the Cournot equilibrium. This generates the individual output:

$$x = \frac{(\theta-1)(N-1)E}{\theta N^2 c} \tag{2.32}$$

Substituting into the inverse price, one obtains the equilibrium price $p = c\theta N/(\theta-1)(N-1)$, which is associated with the equilibrium mark up:

$$\mu^Q(\theta, N) = \frac{\theta N}{(\theta-1)(N-1)} \tag{2.33}$$

which is a particular case of (2.14). Notice that the mark up is decreasing in the degree of substitutability between products θ, with an elasticity $\epsilon_\theta^Q = 1/(\theta-1)$. As long as the number of firms is finite, the markup remains positive for any degree of substitutability. Finally, the mark up is decreasing and convex in the number of firms and it tends to $\theta/(\theta-1) > 1$ for $N \to \infty$.

Its elasticity is $\epsilon_N^Q = 1/(N-1)$, which is decreasing in the number of firms (the mark up decreases with entry at an increasing rate) and independent from the degree of substitutability between goods.

Gross profits can be expressed as:

$$\Pi^Q(\theta, N) = \frac{(N+\theta-1)E}{\theta N^2} \tag{2.34}$$

If the fixed cost of entry is F, entry will take place and will reduce the individual profits as long as the gross profits are higher than this fixed cost. In equilibrium, the zero profit condition leads to the following number of firms:

$$N^Q = \frac{E}{2\theta F}\left[1 + \sqrt{1 + \frac{4\theta(\theta-1)F}{E}}\right] \tag{2.35}$$

which is larger than the optimal number (2.26). This excessive entry result generalizes to a wider context (Mankiw and Whinston, 1986) and has also found some empirical evidence.[19] Moreover, the equilibrium number of firms increases in a less than proportional way with the size of the market (E/F), contrary to what happens in the case of monopolistic behavior of each firm (or in the optimal market structure). Larger markets induce stronger competition, as can be verified from the equilibrium markup:

$$\mu^Q(\theta, N^Q) = \frac{\theta}{(\theta-1)}\left(1 - \frac{2\theta F}{E + \sqrt{E^2 + 4\theta(\theta-1)FE}}\right)^{-1} \tag{2.36}$$

which is decreasing in ratio E/F. This implies that the size of a market has to more than double to allow the entry of a double number of firms. Nevertheless, comparing this EMS with the (constrained) optimal market structure we can conclude that competition in quantities leads to an excessive mark up and to an excessive number of firms.

Finally, we can calculate the production of each firm as:

$$x^Q = \frac{F(\theta-1)}{\theta c}\Phi(1-\Phi) \quad \text{with } \Phi = \frac{2\theta F}{E + \sqrt{E^2 + 4\theta(\theta-1)FE}} \tag{2.37}$$

which is decreasing in the marginal cost of production and increasing and concave in the endowment. The former result shows that positive cost shocks induce a larger production by each firm. The latter shows that positive demand shocks (increasing the endowment of the consumers) increases the production of each firm as well: this happens because each firm has to produce more to cover the same fixed costs at a lower mark up.

[19] Berry and Waldfogel (1999) have investigated EMSs in radio broadcasting, providing evidence that entry is systematically above the optimal level.

2.3.2 Bertrand competition

Let us now consider competition in prices. In each period, the gross profits of firm i can be expressed as:

$$\Pi(i) = \frac{[p(i) - c]\, p(i)^{-\theta} E}{\left[\sum_{j=1}^{N} p(j)^{-(\theta-1)}\right]} \tag{2.38}$$

Firms compete by choosing their prices. Contrary to the traditional Dixit-Stiglitz (1977) approach which neglects strategic interactions between firms, we take these into consideration and derive the exact Bertrand equilibrium. Each firm i chooses the price $p(i)$ to maximize profits taking as given the price of the other firms. The first order condition for any firm i is:

$$\left\{p(i)^{-\theta} - \theta\,[p(i) - c]\, p(i)^{-\theta-1}\right\} + \frac{(\theta-1)p(i)^{-\theta}\,[p(i) - c]\, p(i)^{-\theta}}{\sum_{j=1}^{N} p(j)^{1-\theta}} = 0$$

Notice that the last term is the effect of the price strategy of a firm on the price index: higher prices reduce overall demand, therefore firms tend to internalize their impact on the price index and set higher mark ups compared to the case of monopolistic behavior. Imposing symmetry between the N firms, the equilibrium price p must satisfy:

$$\left[\theta\,(p-c)\, p^{-\theta-1} - p^{-\theta}\right] N p^{-(\theta-1)} = (\theta-1)p^{-\theta}\,(p-c)\, p^{-\theta}$$

Solving for the equilibrium we have $p = c(\theta N - \theta + 1)/(\theta - 1)(N - 1)$, which implies the following mark up:

$$\mu^P(\theta, N) = \frac{1 + \theta(N-1)}{(\theta-1)(N-1)}$$

The mark up under competition in prices is always smaller than the one obtained before under competition in quantities, as well known for models of product differentiation. As in the previous case, the mark up is decreasing in the degree of substitutability between products θ, with an elasticity $\epsilon_\theta^P = \theta N/(1-\theta+\theta N)(\theta-1)$ which is always higher than ϵ_θ^Q: higher substitutability reduces mark ups faster under competition in prices. Moreover, contrary to the case of competition in quantities, the mark up under competition in prices vanishes in case of homogenous goods $\lim_{\theta\to\infty} \mu^P(\theta, N) = 1$, a well known result in industrial organization. Finally, the mark up is again decreasing in the number of firms, with elasticity $\epsilon_N^P = N/\left[1 + \theta(N-1)\right](N-1)$, which is decreasing in the level of substitutability between goods, and approaching zero when the goods become homogenous.

In conclusion, with competition in prices the individual gross profits can be expressed as:

$$\Pi^P(\theta, N) = \frac{E}{1 + \theta(N-1)} \tag{2.39}$$

Given total expenditure, the number of firms and the degree of substitutability, it is easy to verify that the profits under competition in prices are smaller than those under competition in quantities.

If the fixed cost of entry is F, the endogenous entry condition that sets net profits equal to zero provides the following number of firms:

$$N^P = \frac{E}{\theta F} + \frac{\theta - 1}{\theta} \tag{2.40}$$

which is linearly increasing in the endowment and decreasing in the fixed cost of entry. The corresponding equilibrium markup is:

$$\mu^P(\theta, N^P) = \frac{\theta E}{(\theta - 1)(E - F)} \tag{2.41}$$

which is increasing in the fixed cost of entry and decreasing in the endowment. Notice that, given the total expenditure, the fixed costs and the degree of substitutability, competition in prices generates a smaller number of firms compared to competition in quantities. Moreover, if we take the integer constraint (on the number of firms) into account, we can verify that the equilibrium number of firms can be above the (constrained) optimal number by at most one firm.

Finally, one can easily verify that $\mu^Q(\theta, N^Q)$ is always bigger than $\mu^P(\theta, N^P)$, which means that the EMSs under competition in quantities are characterized by more firms but they preserve higher prices than competition in prices:

$$N^Q > N^P \quad \text{and} \quad \mu^Q(\theta, N^Q) > \mu^P(\theta, N^P)$$

This shows that the index of concentration is a poor measure of the market power as an expression of the ability of firms to price above the marginal cost. When entry is endogenous, low mark ups are consistent with high concentration and *vice versa*.

Under competition in prices, the production of each firm is:

$$x^P = \frac{F(\theta - 1)(E - F)}{[E + (\theta - 1)F]\, c} \tag{2.42}$$

which is decreasing in the marginal cost and increasing in the endowment, as it was under competition in quantities. Therefore, cost and demand shocks affect the production of each firm in similar ways.

In conclusion, these EMSs provide two main differences compared to the case of monopolistic firms *à la* Dixit-Stiglitz: mark ups are reduced and individual production is increased when the size of the market increases, while they are constant in case of monopolistic firms. Moreover, under competition in prices the endogenous number of firms increases linearly with the size of the market, as in the case of monopolistic firms *à la* Dixit-Stiglitz, but under competition in quantities it increases in a less than proportional way.

2.3.3 Stackelberg competition

The EMSs can be used to study more complex forms of competition. In this section we extend the symmetric models of competition in quantities and in prices with the introduction of market leaders. In the industrial organization jargon, these are firms able to commit to their own strategies before the so-called followers. Since many markets are characterized by the presence of incumbent firms which typically have larger market shares than their rivals, taking them into account allows us to obtain a more realistic picture of the EMSs. The model of Stackelberg competition with endogenous entry has been introduced by Etro (2008,b) in a static set up as the one considered until now.[20]

For the sake of simplicity, let us consider first the case of competition in quantities and homogeneous goods, that is $\theta \to \infty$. With one leader and N followers playing simultaneously, the equilibrium mark up can be derived as:

$$\mu^S(N) = \frac{N}{N - 1/2}$$

which is lower compared to the mark up under pure Cournot competition. The profits of the leader and the representative follower are respectively larger and smaller than the profits under Cournot competition, but the impact of a change in the number of firms on the equilibrium mark up and production is qualitatively analogous to the Cournot case. In Chapter 3 we will employ also this market structure in a dynamic macroeconomic model to examine the role of market leaders over the business cycle.

Contrary to the case of an exogenous number of firms, the static model of Stackelberg competition with endogenous entry is characterized by a radically different market structure with only one firm active: actually, whenever the goods are homogeneous and the marginal cost of production is constant, the leader produces enough to deter entry. In our example, the equilibrium output of the leader is $x^L = (\sqrt{E} - \sqrt{F})^2/c$ and the equilibrium mark up is:

$$\mu^S = \frac{1}{\left(1 - \sqrt{\frac{F}{E}}\right)^2} \tag{2.43}$$

[20] Further extensions can be found in Maci and Žigić (2008), Žigić (2008), Tesoriere (2008,a,b), De Bondt and Vandekerckhove (2008) and Kováč, Vinogradov and Žigić (2009).

which is higher than the one emerging in the absence of the leader. Even if the EMS is radically different from the case of Cournot competition, the endogeneity of entry leads to similar comparative statics: an increase in the endowment or a reduction of the fixed cost of entry force the leader to produce more and to keep the mark up lower. The main difference compared to the Cournot case is that here the leader obtains positive profits in spite of free entry. In a recent important work, Kováč, Vinogradov and Žigić (2009) have extended the analysis to a dynamic setup: they analyze a oligopoly model in which a leader invests in process innovations facing subsequent endogenous entry by followers, and identify conditions under which it is optimal for the leader in an initially oligopoly setup with endogenous entry to undertake preemptive R&D investment (strategic predation) that eventually leads to the exit of all followers.[21]

The radical result of entry deterrence disappears when we introduce imperfect substitutability between the goods, that is when θ is low enough. Consider the general case of quantity leadership in the presence of imperfect substitutability. The Stackelberg equilibrium with endogenous entry is characterized by a larger production for the leader compared to the followers, and entry of a lower number of firms compared to the Cournot equilibrium with endogenous entry. The characterization of the equilibrium is relatively simple, with the leader selling at the monopolistic mark up, and with an endogenous number of followers adopting the same production level as under symmetric competition in quantities and the same mark up (2.36). Therefore, the equilibrium is summarized by the following mark ups for the quantity leader (index L) and for the representative follower (index F):

$$\mu^{LQ} = \frac{\theta}{\theta - 1}, \quad \mu^{FQ} = \left(\frac{\theta}{\theta - 1}\right)\left(1 - \frac{2\theta F}{E + \sqrt{E^2 + 4\theta(\theta - 1)FE}}\right)^{-1} \tag{2.44}$$

[21] The technological leader adopts the accommodation strategy only when (roughly speaking) his R&D efficiency is low or/and the size of the market is relatively small. In all other cases, the leader opts for strategic predation aiming to achieve the monopoly position after certain time T. During the predation period (up to T), the leader might be willing even to incur losses in order to enjoy monopoly profit from time T onward. Thus, unlike a static game, in a fully dynamic model the costs of predation last only for a limited period and have to be contrasted to the infinite stream of monopoly profit afterwards. The time pattern of R&D investment crucially depends on the equilibrium strategy. If accommodation is the optimal strategy, then the leader chooses an R&D path which steadily increases over time towards the unique steady-state value. When, on the other hand, strategic predation is the optimal strategy, the leader first invests significantly in R&D in order to achieve the monopoly position in time T. After all rivals are eliminated, the leader may continue to increase his R&D investment as an unconstrained monopolist or to prevent the rivals from re-entering the market. Nevertheless, this investment level is still higher than in the case of accommodation. From a welfare point of view, the predation regime is optimal because it implies high R&D investments, but the target time T is usually suboptimal.

In spite of this asymmetric EMS due to the presence of a leader, the endogeneity of entry leads to similar conclusions as before: a larger endowment or a lower fixed cost attract further entry of followers, increase their individual outputs and reduce their mark ups, with a positive impact on total production.

Let us finally consider the case of price leadership with imperfect substitutability. Under competition in prices, the Stackelberg equilibrium with endogenous entry is characterized by the leader committing to a lower mark up compared to the followers. In particular, the leader adopts again the monopolistic price, and the followers adopt the same price as under symmetric competition in prices (2.41). Therefore, the respective mark ups for the price leader and the representative follower become:

$$\mu^{LP} = \frac{\theta}{\theta - 1}, \ \mu^{FP} = \left(\frac{\theta}{\theta - 1}\right)\left(1 - \frac{F}{E}\right)^{-1} \tag{2.45}$$

This result is in striking contrast with the usual outcome under price leadership and exogenous entry, for which leaders adopt higher prices than the followers to relax competition. When entry is endogenous, the only way for the leaders to obtain positive profits is to adopt an aggressive strategy. When the endowment increases or the cost of entry decreases, more followers are attracted in the market, and they reduce their mark ups and increase their production, while the leader maintains the lowest price.

Notice that these results on the behavior of the market leaders have substantial implications for industrial policy, since they show that large market shares by leading firms can be the result of strong entry pressure rather than of market power, and antitrust policy should be more concerned about verifying the entry conditions in a market rather than associating large market shares with dominant positions. A similar result, which we will discuss in Chapter 5, emerges in case of competition for the market, where incumbent leaders tend to invest more than their rivals only when entry is endogenous: this leads to the conclusion that also the persistence of leadership can be the consequence of strong entry pressure rather than of market power.[22]

2.3.4 Collusion, endogenous entry costs and other extensions

The framework that we adopted is tractable enough to take into account other forms of competition. We could adopt the conjectural variations approach to introduce imperfect collusion in a stylized way: in such a case, each firm adopts an exogenous conjecture on the reaction of the other firms to its

[22] For a wider discussion of the antitrust implications of this model see Etro (2008,c). On recent related advances of antitrust theory see Fumagalli, Motta and Rønde (2008), Katsoulacos (2008), Polo and Immordino (2008) and Fernández, Hashi and Jegers (2008).

strategy, and this conjecture can reproduce competitive and collusive equilibria or any intermediate case (including Cournot equilibria). We will follow this *ad hoc* model of imperfect collusion in the dynamic general equilibrium model of Chapter 3.

We could also analyze multiproduct firms which choose the production levels or the price levels of their goods to maximize the joint profits. All these and other models would lead to equilibria with mark ups $\mu(\theta, N)$ and profits $\Pi(\theta, N)$ decreasing in the number of firms, and therefore to well defined EMSs. Notice that from an empirical perspective, one could be interested in estimating these mark up and profit functions as depending on the number of firms in different markets (defined according to the degree of substitutability between products).

Vives (2008) has extended the model to endogenous costs assuming that the fixed cost of production is an investment in R&D aimed at reducing the marginal cost of production. In general, he finds that increasing the endowment increases the investment in cost reduction and the output of each firm, but with ambiguous consequences on the number of firms. For instance, consider our case of isoelastic preferences with an isoelastic cost function. In such a case a larger market size is associated with such a larger fixed investment in cost reductions that the endogenous number of firms remains the same. This result for the case of competition in quantities and homogenous goods is originally due to Dasgupta and Stiglitz (1980). They assume that demand is hyperbolic and that the marginal cost depends on the fixed R&D investment F as in $c = \varkappa F^{-\varrho}$ with $\varkappa, \varrho > 0$. The Nash equilibrium in the choice of output and R&D investment with endogenous entry implies the investment $F = E\varrho^2/(1+\varrho)^2$, the number of firms:

$$N^Q = 1 + 1/\varrho \tag{2.46}$$

the mark up (on the endogenous marginal cost):

$$\mu^Q = 1 + \varrho \tag{2.47}$$

and the production per firm:

$$x^Q = \frac{E^{1+\varrho}}{\varkappa\varrho}\left(\frac{\varrho}{1+\varrho}\right)^{2(1+\varrho)} \tag{2.48}$$

Notice that the number of firms and the mark up are now independent from the size of the market, but the individual production is still increasing in it. Similar results emerge in the case of product differentiation and also with competition in prices. Finally, an increase in the degree of product substitutability increases per-firm output and cost reduction expenditure, while reducing the number of firms as a consequence of the stronger competition.

Until now, we have limited our analysis to the case in which each firm is active for a single period only. A more realistic situation emerges when each firm is active in multiple periods, or has always a positive probability of being active in the future. In the absence of credible commitments to future strategies, we can assume that in each period the existing firms compete according to one of our static models. In such a case, the gross value of the firms would be the present discounted value of the future profits and endogenous entry would still require equalization of the initial fixed costs of entry to the gross stock market value of the same firm. This creates a dynamic behavior of the number of firms that is reflected on the equilibrium mark ups and, through them, on the aggregate behavior of the macroeconomy. Starting with Section 2.6 we will extend our analysis in this direction studying dynamic market structures. Finally, we need to notice that a multi-period framework would allow one to study dynamic models of market competition in which firms can commit to multi-period strategies or in which forms of imperfect collusion can be sustained as subgame perfect equilibria of supergames (Rotemberg and Saloner, 1986; Rotemberg and Woodford, 1992) or as Markow perfect equilibria (Maskin and Tirole, 1988).

However, before embarking in more complex analysis, we still need to extend our basic framework to account for two important aspects: intertemporal links between markets and general equilibrium considerations. Following the strategy of Chapter 1, the next section extends the static analysis of EMSs to the simplest dynamic situation, that is the one characterized by two periods only. This allows us to appreciate the potential role of EMSs in a dynamic framework.

2.4 EMSs in a Two Period Economy

In this section we follow an example by Etro (2007,a) of a two-period economy where an exogenous endowment is allocated between current and future consumption. Imperfect competition and endogenous entry in the goods market of both periods generates a novel link between exogenous shocks and real choices which works through the impact on the endogenous mark ups.

Consider a two period model of an exchange economy with logarithmic subutilities:

$$U = logC_1 + \beta logC_2 \tag{2.49}$$

where $\beta \in (0, 1)$ is the discount factor. The consumption good is homogenous and it is produced by multiple firms in each period, so that the consumption index boils down to $C_t = \sum_{j=1}^{N_t} C_t(j)$ for $t = 1, 2$. Firms compete in quantities. The interest rate r and the endowment of the agent E are assumed exogenous for simplicity. One can think of this as a small open economy facing a given international interest rate.

Given the price levels in the two periods p_1 and p_2, the corresponding budget constraints are:

$$C_1 = \frac{E - S}{p_1} \qquad C_2 = \frac{S(1+r)}{p_2}$$

Utility maximization requires the demand of consumption $C_1 = E/(1+\beta)p_1$ in the first period and $C_2 = \beta(1+r)E/(1+\beta)p_2$ in the second one, which imply the inverse demand functions:

$$p_1 = \frac{E}{(1+\beta)C_1} \qquad p_2 = \frac{\beta(1+r)E}{(1+\beta)C_2}$$

In each period, N_t firms compete in quantities producing at a marginal cost c_t. For simplicity, assume $(1+r)\beta = 1$ in what follows. Defining $x_t(i)$ as the production of firm i in period t, we have the gross profit functions:

$$\Pi_t(i) = \frac{Ex_t(i)}{(1+\beta)\sum_{j=1}^{N_t} x_t(j)} - c_t x_{it}$$

In Cournot equilibrium, each firm produces $x_t(i) = E(N_t - 1)/(1+\beta)N_t^2 c_t$, and the equilibrium price is $p_t = \mu_t(N_t)c_t$ where the mark up function is:

$$\mu_t(N_t) = \frac{N_t}{N_t - 1} \tag{2.50}$$

which is decreasing in the number of competitors. Therefore, we obtain the following modified Euler equation:

$$\frac{C_2}{C_1} = \frac{c_1\mu_1(N_1)}{c_2\mu_2(N_2)} \tag{2.51}$$

The traditional outcome of perfect competition emerges in case of constant returns to scale, here equivalent to the absence of fixed costs of production. In such a case, endogenous entry implies an infinite number of firms, prices are equal to the marginal cost in both periods, and relative consumption is linked to the ratio of marginal costs only: $C_2/C_1 = c_1/c_2$. Of course, under constant technology we have consumption smoothing ($C_2/C_1 = 1$). The neoclassical theory of the business cycle is largely based on this mechanism: a permanent increase in productivity does not affect the relative marginal cost and consumption, but a temporary increase in productivity (a reduction in c_1/c_2) induces an increase in relative consumption (a decline of C_2/C_1). Finally, notice that an exogenous change of the endowment does not affect prices and relative consumption.

When the markets are characterized by positive fixed costs of production, however, only few firms can be active and entry strongly affects relative prices and consumption. As a preliminary example, imagine that the fixed cost of entry in period t is F_t, and entry is endogenous. Then, in each period t we have a markup:

$$\mu_t = \frac{1}{1 - \sqrt{\frac{(1+\beta)F_t}{E}}} \tag{2.52}$$

and a number of firms:

$$N_t = \sqrt{\frac{E}{(1+\beta)F_t}}$$

This result shows that an increase in the endowment (or a reduction in the fixed cost of production) increases the number of firms and reduces the markups. Relative consumption can be calculated as:

$$\frac{C_2}{C_1} = \left(\frac{c_1}{c_2}\right)\left[\frac{\sqrt{E} - \sqrt{(1+\beta)F_2}}{\sqrt{E} - \sqrt{(1+\beta)F_1}}\right]$$

This shows two mechanisms due to the endogeneity of the market structures (and completely absent under perfect competition). The first is rather straightforward: an increase in the fixed cost of entry in one period increases the relative consumption in the other period, and *vice versa*. In particular, a reduction in the future costs of entry leads to consumption growth: for instance, the introduction of a general purpose technology that is going to reduce entry costs (say cloud computing) should exert a positive impact on growth.

The second mechanism is less intuitive: an exogenous increase in the endowment increases the relative consumption of the good produced by a lower number of firms. Suppose $F_1 > F_2$, which implies that more firms are active in the second period and $p_1/c_1 > p_2/c_2$: under these circumstances, an increase in E increases C_1 relative to C_2.

Assume now that the fixed cost of production is related to the marginal cost $F_t = \eta c_t$, as it typically happens when both fixed and variable costs require the same combination of inputs (for instance just labor). In such a case, we obtain a magnification effect of the technology shocks. Rewriting the optimality condition as:

$$\frac{C_2}{C_1} = \left(\frac{c_1}{c_2}\right)\left[\frac{\sqrt{E} - \sqrt{(1+\beta)\eta \cdot c_2}}{\sqrt{E} - \sqrt{(1+\beta)\eta \cdot c_1}}\right] \tag{2.53}$$

one can notice that a reduction in the marginal cost of the first period is going to increase relative consumption in the first period more than proportionally. This new propagation mechanism works through endogenous entry. A temporary shock reduces the marginal cost, which makes current consumption more attractive. Moreover, the reduction in the entry costs induces more firms to enter in the market, temporarily increasing competition. This induces a temporary reduction in the equilibrium mark up, which exhibits countercyclicality. Accordingly, the shock makes current consumption even more attractive.

In conclusion, in the presence of EMSs characterized by competition in quantities and endogenous entry, the impact of a temporary productivity shock on consumption is magnified through the impact of entry on the mark up. Notice that the result would be affected by changes in the degree of intertemporal substitution (assuming a utility function with a higher elasticity of substitution than the logarithmic one, the impact of the temporary shock on the relative consumption in the first period would be strengthened).

Moreover, if we introduce endogenous labor supply, a temporary productivity shock would generate a standard intertemporal substitution mechanism in the labor choice. This would be magnified through the competition effect: a temporary productivity shock would induce a temporary increase in the real wage, which would generate higher labor supply in the short run. In other words, the EMSs create an additional channel through which traditional intertemporal substitution mechanisms (in consumption and labor supply) work.

Finally, the model could be extended to imperfect substitutability between goods produced by different firms introducing a separate consumption index as (2.25) for each period:

$$U = log\left[\sum_{j=1}^{N_1} C_1(j)^{\frac{\theta-1}{\theta}}\right]^{\frac{\theta}{\theta-1}} + \beta log\left[\sum_{j=1}^{N_2} C_2(j)^{\frac{\theta-1}{\theta}}\right]^{\frac{\theta}{\theta-1}}$$

and examining competition in quantities and prices. While the mark up rules would change, the same logic of the results above would go through, because the modified Euler equation (2.51) still holds.

Summarizing, our outcome is dependent on two differences from the standard neoclassical set up. The first is the departure from the assumption of constant returns to scale: fixed costs of entry imply increasing returns to scale in the production function. The second difference relies on the form of competition: here we adopted standard competition in quantities, but more general models of strategic interaction as those examined in the previous sections would deliver analogous results.

Before turning to more complicated dynamic extensions, we still need to introduce our tractable static model in a general equilibrium framework to complete our overview of the EMSs approach. We will do this in the simplest possible way in the next section.

2.5 EMSs in General Equilibrium

General equilibrium analysis concerns multiple markets interacting between each other. Most of the literature on general equilibrium has been focused on the case of perfectly competitive markets with price-taking firms and without fixed costs of production or entry.[23] Only limited efforts have been dedicated to the analysis of markets with imperfect competition or with strategic interactions between few agents - see Bonanno (1990), Mas Colell, Whinston and Green (1995, Ch. 18) and Gabszewicz (2003). Even this literature has been mostly aimed at providing non-cooperative foundations of Walrasian equilibria studying the strategic behavior of the agents (rather than the firms), verifying the limit properties of the equilibria when the number of agents increases, and deriving conditions for existence, uniqueness and stability of the equilibria. Moreover, it has systematically neglected the role of fixed costs of productions or other technological non-convexities in endogenizing entry of firms in the markets.

In our view, general equilibrium theory should try to provide a deeper understanding of aggregate phenomena in the presence of strategic interactions (of different kinds) between firms and with endogenous entry in each market. The aim of this section is not to provide such a full-fledged analysis of EMSs in general equilibrium, but to introduce the simplest general equilibrium extension of our partial equilibrium static model. This example will be generalized in the next section and in future chapters.

Imagine that firms produce the goods employing labor only, which in turn is supplied in fixed quantity. One unit of labor produces A units of good. Moreover, the fixed cost of creation of a new firm corresponds to the cost of η/A units of labor, where $\eta > 0$. The nominal unit wage is W.

The representative agent provides L units of labor and maximizes the same utility as in (2.25) under the same budget constraint (2.11). However, the endowment E is now endogenous and it depends on labor income and, using the fact that the representative agent is the only shareholder of all the firms, it depends on net profits too. Summing up, the endogenous endowment becomes:

$$E = WL + \left[\sum_{j=1}^{N} \Pi(j) - N\frac{\eta W}{A} \right] \tag{2.54}$$

This allows us to derive the demand function for each good as a function of both labor income and the profits of the same firms. Therefore, the individual profits of each firm depend on the aggregate profits as well, and so on in a circular way. However, assuming that the firms take aggregate profits as given, competition takes place as before and the aggregate profits amount to zero under endogenous entry.[24] Accordingly, the endogenous endowment simplifies to $E = WL$.

[23] See Ellickson (1993) for a nice overview.

[24] This is not the case with asymmetric forms of competition. For instance, under Stackelberg competition, there are positive profits for the industry leaders even if there is endogenous entry of followers. These profits should be taken into account in the demand functions.

Using this, one can express the EMS in general equilibrium as a function of total labor supply and of the fixed cost parameter. For instance, in the case of homogenous goods and competition in quantities we have a general equilibrium number of firms given by:

$$N^Q = \sqrt{\frac{AL}{\eta}} \tag{2.55}$$

and an equilibrium markup:

$$\mu^Q = \frac{\sqrt{AL}}{\sqrt{AL} - \sqrt{\eta}} \tag{2.56}$$

Both the two markets of this economy, the one for the goods and the one for labor, are in equilibrium, and this allows us to examine the new general equilibrium effect that EMSs create in the labor market. Adopting the price of the goods as the *numeraire*, we can derive an expression for the (real) wage:

$$w^Q = \left(1 - \sqrt{\frac{\eta}{AL}}\right) A \tag{2.57}$$

This shows that the real wage is increasing in the aggregate productivity, decreasing in the size of the fixed cost, and increasing in the total labor supply. The first comparative static result is standard, and the wage is lower than the marginal productivity because of market power. The last two results derive from the impact of endogenous entry on competition in general equilibrium: larger markets or lower fixed costs attract entry, which in turn strengthens competition, reduces the mark ups and shifts resources from extra profits toward labor remuneration. These general equilibrium EMSs can be easily extended to the case of product differentiation with competition in quantities with similar implications.

In case of competition in prices the general equilibrium number of firms becomes:

$$N^P = \frac{AL}{\theta\eta} + \frac{\theta - 1}{\theta} \tag{2.58}$$

and the equilibrium markup is:

$$\mu^P = \frac{\theta AL}{(\theta - 1)(AL - \eta)} \tag{2.59}$$

Again, we cannot derive equilibrium prices without a normalization. However, in case of product differentiation, it is convenient to express the real wage as the ratio between the nominal wage and the price index, which in the symmetric equilibrium is $P = pN^{1/(1-\theta)}$. Therefore, the real wage becomes:

$$w^P = \frac{(\theta-1)(AL-\eta)}{\theta AL}\left(\frac{AL}{\theta\eta}+\frac{\theta-1}{\theta}\right)^{\frac{1}{\theta-1}} A \tag{2.60}$$

which is again a fraction of the aggregate productivity, and is decreasing in the size of the entry cost, but increasing in the size of the labor force. Larger and more accessible markets attract entry, which reduces mark ups on one side and increases the number of different varieties on the other side. Now, both effects increase the real wage: the former because it leads to a reduction in the average price, the latter because it increases the purchasing power for a given income (because of the love for variety effect).

The competition effect of market size and entry costs on mark ups and real wages appears extremely simple and possibly trivial in such a static model with a single market, but it will be at the source of a number of crucial results in the presence of multiple markets or multiple periods, in particular when intertemporal substitution mechanisms are available (as in Chapter 3) or when intra-industry trade between countries occurs (as in Chapter 4). Loosely speaking, the fundamental reason is that, contrary to what happens in the neoclassical approach, here prices depend on both the marginal cost of production and the mark up, and the latter is affected by shocks through the entry mechanism. In turn, the impact on the mark ups is transmitted to the real wages through the general equilibrium mechanims shown in this section, and any change in mark ups and wages has an impact on consumption and labor supply choices. In turn, this feeds back on profits and affects the entry decisions and with them the mark ups and the real wages. In such a way, the EMSs create a new mechanism of propagation of shocks in general equilibrium.

As of now we have largely discussed microfounded EMSs in a static framework in partial and general equilibrium and in a simple dynamic framework with two periods only. It is time to approach a more ambitious task and to build a fully dynamic general equilibrium model, which should give to the reader the ultimate flavor of the EMSs approach to macroeconomics.

2.6 EMSs in an Infinite Periods General Equilibrium Economy

In this section we provide an application of the EMSs approach to a dynamic production economy with an infinite horizon both for the representative agent and the firms. These have to pay an initial fixed cost to enter in the market, and subsequently they compete *à la* Cournot in the production of a homogenous good. Production occurs with a single input, labor, which is inelastically provided by the agent, and business creation is driven by savings, that are inelastically provided as well.[25]

[25] In Chapter 3 we extend this same model to endogenous savings, endogenous labor supply, imperfectly substitutable goods and competition in prices.

We adopt the simplifying assumption used by Solow (1956) for which savings are a constant fraction of income. However, in our model income includes both the remuneration of inputs and the profits. This allows us to obtain a dynamic model in which it is not investment in physical capital to generate the accumulation of the reproducible input over time, as in the neoclassical Solow model, but it is entry of new firms to generate the creation of new productive business.

Since entry strengthens competition, it also induces a sort of decreasing marginal productivity of business creation, just like capital accumulation reduces the marginal productivity of capital in the neoclassical model. Here, however, it is entry that strengthens competition and reduces the marginal profitability of subsequent entry. Therefore, both models generate a gradual convergence toward a steady state: in the Solow case through a decreasing growth rate of the capital stock, in our case through a decreasing rate of business creation.

2.6.1 A model of business creation with Cournot competition

Consider a representative market for a homogenous good with N_t firms active in each period t.[26] Each firm i produces $x_t(i)$ according to a linear production function:

$$x_t(i) = A_t l_i \tag{2.61}$$

where A_t is the exogenous productivity of labor (or the total factor productivity in this case without other inputs), which is common to all firms, and l_i is the labor input used by firm i. Given the nominal wage W_t, the constant marginal cost of production is $c_t = W_t/A_t$. Total expenditure in the sector is:

$$E_t = p_t C_t = p_t \sum_{j=1}^{N_t} x_t(j)$$

where p_t is the equilibrium price equating consumption demand C_t, for the moment taken as given, and supply by all the firms in period t. Nominal profits for firm i are:

$$\begin{aligned} \Pi_t(i) &= \left[p_t(i) - \frac{W_t}{A_t}\right] x_t(i) = \\ &= \frac{x_t(i) E_t}{\sum\limits_{j=1}^{N_t} x_t(j)} - \frac{W_t x_t(i)}{A_t} \end{aligned} \tag{2.62}$$

[26] In the next chapter we will explicitly introduce multiple sectors of this kind, which adds realism to the description without changing the main insights of the representative sector model.

We assume that the firms cannot credibly commit to future production strategies, therefore they play Cournot competition in each period. If at time t firm i chooses its production $x_t(i)$ to maximize its profits taking as given the production of the other firms, the equilibrium generates individual output $x_t = (N_t - 1)E_t A_t / W_t N_t^2$. Substituting, one obtains the equilibrium price at time t:

$$p_t = \frac{N_t}{N_t - 1}\left(\frac{W_t}{A_t}\right) \tag{2.63}$$

which is associated with the usual equilibrium mark up $\mu(N_t) = N_t/(N_t - 1)$. This equilibrium generates individual profits $\Pi_t(N_t) = E_t/N_t^2$ in nominal terms. Since the equilibrium price of the consumption good is p_t, it is convenient to express all the variables in units of consumption, that is in real terms (alternatively one can use the consumption good as the *numeraire*). Then, the real profits $\pi_t(N_t) \equiv \Pi_t(N_t)/p_t$ become:

$$\pi_t(N_t) = \frac{C_t}{N_t^2} \tag{2.64}$$

and the real wage $w_t = W_t/p_t$ can be derived from the equilibrium pricing relation as:

$$w_t = \frac{N_t - 1}{N_t} A_t$$

This implies that each firm produces $x_t = (N_t - 1)C_t A_t / w_t N_t^2 = C_t/N_t$.

When the number of firms increases, the equilibrium price goes down and the wage goes up, with the former approaching the marginal cost and the latter approaching the productivity of labor for $N_t \to \infty$. However, here we do not want to approach the neoclassical paradigma with infinite firms, but we want to endogenize the number of firms. One way to do it is to assume as usual that there is a fixed cost of production in each period and that free entry occurs at all times. Such an assumption, however, would exclude any interesting dynamics because profits would be zero at any time. Another way to endogenize entry, which is more realistic and interesting for dynamic models, is to assume that entry is constrained by the expectations on future profitability and by a one-shot fixed cost of entry. This is the approach that we will adopt from now on.

In every period N_t^e new firms enter in the market, and a fraction $\delta_N \in (0, 1)$ of the (old and new) firms exits from the market for exogenous reasons. Therefore, the number of firms follows the equation of motion:

$$N_{t+1} = (1 - \delta_N)(N_t + N_t^e) \tag{2.65}$$

which is analogous to the equation of motion of capital in the Solow model.[27]

[27] This equation of motion for the number of firms is borrowed from Ghironi and Melitz (2005). Analogous results would emerge with a more traditional version as $N_{t+1} = (1 - \delta_N)N_t + N_t^e$, in which new firms are always active for at least one period.

The real gross value of a new firm V_t is the present discounted value of its future expected profits, which, using the expectations operator $E[\cdot]$, and taking into account the exit probability in each period, becomes:

$$V_t = (1-\delta_N)E\left[\frac{\pi_{t+1}(N_{t+1})}{1+r_{t+1}}\right] + (1-\delta_N)^2 E\left[\frac{\pi_{t+2}(N_{t+2})}{(1+r_{t+1})(1+r_{t+2})}\right] + ... =$$
$$= (1-\delta_N)E\left[\frac{\pi_{t+1}(N_{t+1}) + V_{t+1}}{1+r_{t+1}}\right]$$

where r_{t+k} is the real interest rate at time $t+k$, whose expectation is taken as given by the firms, and the second line rearranges the first one in a recursive form. In each period entry occurs until the real value of the representative firm equates the fixed cost of entry.

Since all firms produce the same homogenous good, it is reasonable to assume that entry of a new firm requires only an extra labor activity to prepare production (rather than a specific monetary investment in R&D to create a new or better product), therefore we assume that the fixed cost of entry F_t is equal to η/A_t units of labor, where $\eta > 0$. Given the wage $w_t = (N_t - 1)A_t/N_t$, the endogenous entry condition $V_t = F_t$ amounts to:

$$V_t = F_t = \eta\frac{(N_t - 1)}{N_t} \tag{2.66}$$

Notice that this endogenous entry condition determining the investment in business creation can be re-interpreted in terms of the Tobin (1969) approach, for which additional investment takes place if the (stock) market value of a unit of capital is higher than its replacement cost, or in other words if their ratio, known as the *Tobin's q*, is larger than one. In our framework, additional investment in business creation takes place when the stock market value of a firm is larger than the entry cost, that is when the *Tobin's q* defined as:

$$q_t \equiv \frac{V_t}{F_t} \tag{2.67}$$

is larger than one - augmenting the model with adjustment costs would generate a gradual (dis)investment for $q_t > (<)1$.

Investment is destined to the creation of new firms. Given the fixed costs of entry F_t and the number of entrants N_t^e, total investment is:

$$I_t = N_t^e F_t = \frac{\eta(N_t - 1)N_t^e}{N_t} \tag{2.68}$$

where we used the endogenous entry condition.

Assume that the number of workers is given by L_t and each one supplies a unit of labor in each period. Real income in each period must be the sum of profits and labor income in real terms:

$$Y_t = N_t\pi_t(N_t) + w_t L = \frac{C_t}{N_t} + w_t L_t$$

This income must be allocated between consumption C_t and savings S_t in each period.

The market clearing condition that equates savings and investments in every period links the equilibrium number of active firms to the equilibrium interest rate in each period. Therefore, the interest rate depends on the stock market evaluation of the return on the investment in business creation, which depends on the strategic interactions between firms and on the entry/exit process. Finally, total labor demand equates the exogenous labor supply in each period.

To close the model we need to introduce a consumption function. Following the standard approach of Solow (1956) we assume that savings are an exogenous fraction $s \in (0,1)$ of income, $S_t = sY_t$. From the aggregate resource constraint derived above, this implies:

$$\begin{aligned} Y_t &= \frac{(1-s)Y_t}{N_t} + w_t L_t = \\ &= \frac{(1-s)Y_t}{N_t} + \frac{(N_t-1)}{N_t} A_t L_t \end{aligned}$$

where we used the equilibrium expression for the wage. Solving for income we obtain:

$$Y_t = \frac{(N_t-1)A_t L_t}{N_t-(1-s)} \tag{2.69}$$

which is an increasing function of productivity and labor force, but also of the number of firms and of the propensity to consume $(1-s)$. The last effects have a Keynesian flavor, even if they operate on the supply side of the economy (rather than on the demand side as for the traditional Keynesian multiplier). Given the number of active firms, a stronger propensity to consume increases aggregate demand and total profits,[28] which in turn increases total output. Of course, an increase in the number of firms strengthens competition and reduces the profits while increasing labor income. However, as long as part of income is saved and not consumed, the reduction in total profits is more than compensated by the increase in labor income, so that more firms lead to higher output.

Applying the equality of savings:

$$S_t = sY_t = \frac{s(N_t-1)A_t L_t}{N_t-(1-s)}$$

[28] Total profits can be derived as:

$$N_t \pi_t(N_t) = \frac{(1-s)(N_t-1)A_t L_t}{N_t\left[N_t-(1-s)\right]}$$

which is increasing in productivity but decreasing in the number of firms and in the savings rate. One can notice similarities with the "big push" story by Murphy, Shleifer and Vishny (1989).

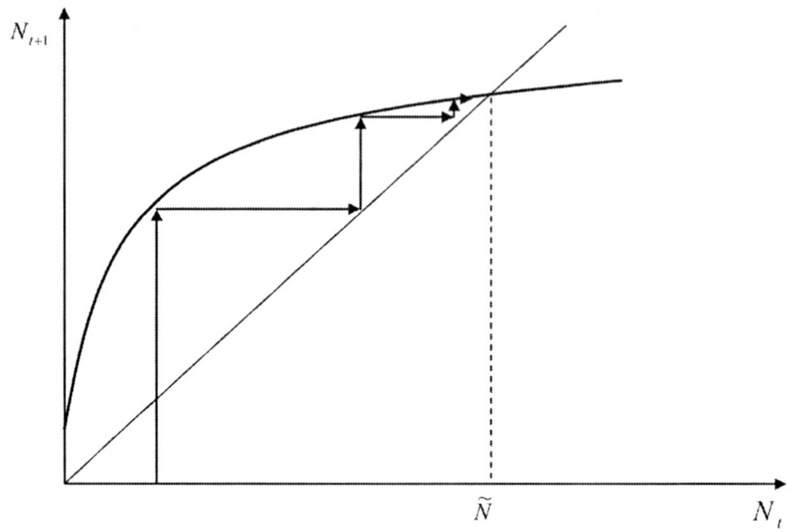

Fig. 2.1. Dynamic Creation of Firms in (N_t, N_{t+1}).

with investments I_t as defined in (2.68), we can solve for the equilibrium number of new firms:

$$N_t^e = \frac{sN_tA_tL_t}{\eta\left(N_t - 1 + s\right)}$$

Plugging the above expression in the equation of motion for N_t we have our final result for the evolution of the number of firms:

$$N_{t+1} = N_t\left(1 - \delta_N\right) + \frac{s(1-\delta_N)A_tL_t}{\eta - \frac{\eta(1-s)}{N_t}} \tag{2.70}$$

Assume that the labor force is constant at the level L at each point in time, and that the aggregate productivity is fixed at $A_t = A$. Then, the dynamic adjustment of the number of firms toward its steady state value is shown in Figure 2.1, which clearly resembles the dynamic adjustment of capital in the Solow model toward zero growth.

From the dynamics of the number of firms one can reconstruct the path of all the other variables. Two remarks are in order. First, the value of the stock market can be expressed as the value of all the firms $N_tV_t = \eta(N_t - 1)$, which follows the same dynamic path of N_t. For this reason, the aggregate behavior of the economy (consumption and output) is strictly related to the behavior of the stock market. Second, the model provides a dynamic path for income distribution, because the labor share $1 - \alpha_t$ is procyclical. This can be derived from:

$$1-\alpha_t=\frac{w_tL}{Y_t}=1-\frac{1-s}{N_t}$$

which is increasing in the savings rate and in the number of firms (i.e.: the fraction of income distributed as labor income is procyclical). Contrary to the neoclassical approach, in which the labor share was constant (at least under a standard Cobb-Douglas technology), the EMSs approach is able to generate more complex dynamics for the distribution of income between capital remuneration (in the form of dividends) and labor income.

2.6.2 EMSs in the long run

Let us consider the stationary situation to characterize the long run EMSs. Since the right hand side of (2.70) is increasing in the current number of firms but with a declining slope (smaller than one for a number of firms large enough), we can conclude that the dynamic path of the economy is stable around its unique steady state. When the initial number of firms is low, savings contribute to create new firms, but new firms strengthen competition reducing the profits and the incentives to enter. The steady state number of firms can be derived as:

$$\tilde{N}=1+s\left[\frac{(1-\delta_N)\,AL-\eta\delta_N}{\eta\delta_N}\right] \tag{2.71}$$

which is increasing in the savings rate s, in the productivity level A and in the labor force L, and decreasing in the exit rate δ_N and in the relative size of the fixed costs η. The equilibrium endogenously generates imperfect competition between a positive but limited number of firms producing the homogenous good, with a steady state mark up:

$$\tilde{\mu}=\frac{s\,(1-\delta_N)\,AL+(1-s)\eta\delta_N}{s\,(1-\delta_N)\,AL-s\eta\delta_N} \tag{2.72}$$

which is characterized by the opposite comparative statics of the number of firms.

Notice that dynamic inefficiency holds, since a better allocation of resources could be achieved through a reduction of the number of firms and an increase in the production of each firm (so as to reduce the waste in fixed costs of production). As we will see in the next chapters, this inefficiency result is a particular case of a much more general result that holds under different market conditions and also when firms produce differentiated goods.

Of course, the dynamic path of output and consumption (and of the real wage and the interest rate) can be determined residually from the evolution of the number of firms. When the latter increases toward its steady state value, output increases as well toward its steady state value:

$$\tilde{Y}=AL-\frac{\eta\delta_N}{(1-\delta_N)} \tag{2.73}$$

This does not depend on the savings rate: a larger propensity to save increases entry and the number of firms, which enhances competition and wages, but decreases consumption which reduces the profits, and the two effects balance each other.

In its simplicity, this model can be used for multiple purposes, and in the next sections we will provide a short overview of those that will be at the core of the following chapters.

2.7 Business Cycle

The EMSs approach can be used to study business cycles in an environment where, contrary to the neoclassical approach (Lucas and Rapping, 1969; Kydland and Prescott, 1982) competition in the market is not perfect, and, contrary to the New-Keynesian approach (Blanchard and Kiyotaki, 1987), the market structure is not exogenous. Our characterization of the market structure and of the incentives to create new firms gives raise to a new mechanism of propagation of the shocks that has nothing to do with the process of capital accumulation, with phenomena of intertemporal substitution (of consumption or labor supply) or with price rigidities, all elements that are absent here. The new mechanism is entirely driven by the relation between profits, firm's value, entry and mark ups.

To see the mechanisms at work in the simple model of the previous section, let us re-introduce a variable aggregate productivity A_t to study the reaction of the EMSs and of the aggregate variables to exogenous shocks and verify the business cycle properties of the model. We are mainly interested in temporary shocks, because permanent ones would simply lead to monotonic convergence to a new steady state. Therefore, consider a temporary positive shock to A_t. This would suddenly increase the productivity and the profits of the existing firms, which in turn would increase their stock market value and attract entry. The temporary increase in the number of firms would strengthen competition so as to reduce the mark up, enhance production and increase the real wages (while dampening the impact on the profits). The proportional allocation of output between consumption and savings, which are invested in business creation, contributes to spread gradually the effects of the shock over time.

More formally, we can derive the impulse response function of the number of firms by log-linearizing around the steady state the equation of motion (2.70). Taking the logs of both sides, differentiating with respect to the time-varying variables, and evaluating them at their steady state levels, we obtain:

$$\hat{N}_{t+1} = \left(1 - \frac{\tilde{N}\delta_N}{\tilde{N} - 1 + s}\right)\hat{N}_t + \delta_N \hat{A}_t \tag{2.74}$$

$$= \frac{s\left(1-\delta_N\right)^2 AL - (1-s)\delta_N^2\eta}{s\left(1-\delta_N\right)AL}\hat{N}_t + \delta_N \hat{A}_t \tag{2.75}$$

where $\hat{X}_t \equiv dX_t/\tilde{X}$ is the percentage distance from the steady state value of a variable X_t. Log-linearizing (2.69) around the steady state we obtain also:

$$\hat{Y}_t = \left\{1 - \frac{\eta\delta_N}{s\left[(1-\delta_N)AL - \eta\delta_N\right]}\right\}\frac{\eta\delta_N}{(1-\delta_N)AL}\hat{N}_t + \hat{A}_t$$

The response functions show that a one-shot increase in productivity increases the number of firms on impact. Afterward, even if productivity goes back to its initial value, the number of firms and output remain above their steady state values. They gradually decreases over time because of the increased competition and lower mark ups. Notice that the impact of the shock on the aggregate variables operates through the stock market, which reflects the value of the firms, the incentives to enter in the market and the impact on competition and on the mark ups. The dynamics of the stock market are due to the presence of imperfect competition between the firms, which generates large operative profits whose expected discounted value is affected by the shocks and affects the entry process. Under perfect competition (for $\eta \to 0$) any additional propagation mechanism would disappear.

In case of a temporary but persistent technology shock, the effects are much stronger. The impulse response of the number of firms becomes hump shaped when the autocorrelation of the shock is high enough, savings are high enough and the exit rate is low enough. In this case, the shock induces a gradual increase of the stock market value of the firms and of their number, associated with a gradual reduction of the mark ups: only after a few periods these variables start returning toward their initial levels. Nevertheless, the impact on output and consumption follows closely the behavior of the technology parameter - for this reason the performance of the model can be improved introducing endogenous consumption and labor choices.

Analogous effects would derive from temporary shocks to the size of the entry cost (which could be interpreted as product market reforms for liberalization or deregulation or as introduction of cost reducing general purpose technologies) or even to the savings rate (which could be interpreted as demand shocks). A positive shock to the exit rate could be interpreted as a crisis leading to a chain of bankruptcies, and would have the consequence of reducing the number of firms and the output level, which would return only gradually to their steady state levels.

In Chapter 3 we will augment this same model with endogenous savings decision and also with endogenous labor supply: the former will introduce a new propagation mechanism based on the positive effect of competition on demand (as we have already seen in our two period partial equilibrium model), the latter will strengthen the propagation of the shock through a mechanism of intertemporal substitution of labor supply due to the impact of shocks on real wages through a general equilibrium effect (already seen in the previous section). In this context, we will also analyze fiscal policy and, under nominal price rigidities, monetary policy: we will suggest that the optimal policies should be aimed at stabilizing the business creation process around the efficient allocation of resources, with countercyclical tax rates and interest rate rules aimed at equity price stabilization.

2.8 Trade

A second way to use the EMSs approach is to augment closed economy models as the one used until now with trade with other countries. When countries open up, the international EMSs are affected in an interesting way that leads to a new source of gains from trade. This depends on the reduction of the mark ups and of the prices, contrary to what happens in the neoclassical approach and in the new trade theory with monopolistic behavior (Krugman, 1980), in which prices are not affected by trade between identical countries.

We can use our simple model of business creation to verify the impact of opening up to trade with another country and evaluate the effect of increasing the size of the market. Traditional models of intra-industry trade based on monopolistic behavior of the firms usually emphasize the impact of openness on the number of varieties produced and traded across countries and determine the gains from trade on the basis of this variety effect. When strategic interactions play a role, however, openness has the additional effect of strengthening competition and reducing the mark ups. This phenomenon leads to a reduction of the international prices which creates a second form of gains from trade.

Imagine that the closed economy considered above opens up to trade with another identical economy in the absence of trade frictions. Since the size of the market doubles, the new steady state number of firms from both countries in the joint market becomes:

$$\tilde{N} + \tilde{N}^* = 1 + s\left[\frac{2\left(1 - \delta_N\right)AL - \eta\delta_N}{\eta\delta_N}\right] \tag{2.76}$$

The substantial increase in the number of firms strengthens competition and reduces the global mark up to the following steady state level:

$$\tilde{\mu} = \frac{2s\left(1 - \delta_N\right)AL + (1 - s)\eta\delta_N}{2s\left(1 - \delta_N\right)AL - s\eta\delta_N} \tag{2.77}$$

which corresponds to an increase in steady state output in both countries.

In conclusion, in this model the gains from trade do not derive from the variety effect, as in the model of Krugman (1980): this effect is absent here because goods are homogenous. The gains from trade derive uniquely from the reduction in the price level. Of course, if we introduce product differentiation the gains from trade would derive from both sources: lower prices and more varieties (a similar point is made by Devereux and Lee, 2001). Notice that these dynamic models can be used to examine the reaction to shocks in a dynamic open economy framework and to explain a number of stylized facts without explanation in the neoclassical approach. Ghironi and Melitz (2005) have provided a fundamental contribution in this direction which will be discussed in Chapter 4.

Another consequence of the reduction in the mark ups is that gross profits increase less than proportionally with the increase in the size of the integrated market, therefore the endogenous number of firms active in each country tends to decrease. This phenomenon is quite evident in a static environment as the one considered in the new international trade literature associated with Krugman (1980). In such a case, globalization generates business destruction due to the reduction in the global prices. This phenomenon depicts a well known fear associated with our times (which has induced widespread support to novel forms of protectionism against globalization).

In Chapter 4 we will examine open economy issues and we will also use the EMSs approach to analyze the role of trade policy for globalized markets. Such an analysis is crucial to understand a world where competition takes place at the global level and most firms are active in domestic and foreign markets. Most importantly, our analysis will emphasize a result in sharp contrast with the traditional results. In particular, contrary to a standard outcome of neoclassical trade policy for which export taxes are always optimal to improve the terms of trade, the EMSs approach shows that export subsidies are always the optimal unilateral policy because they are the only way to provide a strategic advantage to the domestic firms active in international markets where entry is endogenous. This happens not only in certain markets characterized by Cournot competition as noticed in the literature on strategic trade policy starting with Brander and Spencer (1985), but under any form of competition including Bertrand competition. The optimal unilateral policy always requires policies that turn domestic exporters into aggressive leaders conquering larger market shares abroad. The result has also implications for exchange rate policy and R&D policy.

2.9 Growth

Finally, we can switch our attention to the process of business creation as a source of growth. As we have seen, our simple model confirms that the growth rate should be declining toward its steady state level (because of the decreasing marginal incentive to enter), and that only an exogenous growth of total factor productivity could generate long run growth, exactly as in the neoclassical approach of Solow (1956). However, endogenous growth can emerge when the creation of new firms is associated with an increase in total factor productivity.

Long run growth can be seen as the result of externalities in the accumulation of knowledge, as in Romer (1986). For instance, imagine that the productivity parameter A_t increases with the number of firms active in the market because each one brings new knowledge and experience to the production process with spillovers on the whole sector (possibly thanks to the investment in sunk costs of production, which could be seen as an investment in R&D). In particular, assume that $A_t = BN_t$ with $B > 0$. Then, the equation of accumulation of the number of firms (2.70) becomes:

$$N_{t+1} = (1 - \delta_N) N_t + \frac{s(1 - \delta_N)BLN_t^2}{\eta N_t - \eta (1 - s)} \tag{2.78}$$

which implies a process of perpetual business creation in which the growth rate of N_t converges to a constant and stable steady state level. This process is associated with a dynamic of the growth rate of income that converges to:

$$\tilde{g} = \frac{s(1 - \delta_N)BL}{\eta} - \delta_N \tag{2.79}$$

which is positive as long as the savings rate is high enough or the rate of exit is low enough. Notice that, as in the Romer model with externalities in capital accumulation with growth rate (1.39), also here the long run growth rate is increasing in the size of the labor force: scale effects take place here (which implies that opening up to trade would lead to larger growth rates, rather than larger output levels). Moreover, the growth rate is increasing in the savings rate and decreasing in the rate of business destruction and in the size of the costs of entry.

However, notice that, contrary to the traditional result of the endogenous growth theory (Romer, 1986), the endogeneity of the market structure generates a gradual convergence of the growth rate to its long run level, which is empirically plausible. At the beginning of the growth process the incentives to create new firms are high and the rate of increase in the number of firms is high. While firms enter and competition becomes more intense, the rate of entry decreases and the growth rate of production decreases with it. In the long run, the growth rate remains constant because the increase in productivity associated with business creation maintains high the incentives to create new firms. A similar growth process characterized by EMSs in the competition in the market emerges in the model of Peretto (1996, 1999).

There is a deeper way in which the creation of new business augments total factor productivity. As suggested by the recent revival of the Schumpeterian tradition (Aghion and Howitt, 1992), this takes place when firms invest not just to create new products, as we assumed until now, but to create them at a lower cost. Of course, this allows us to increase total production through innovations, which is the essence of growth driven by endogenous technological progress. Such a mechanism requires a system of intellectual property rights which preserves the incentives to undertake R&D investments with an uncertain return, and its evolution relies on the market structure of the innovative activity (rather than the market structure of the productive activity, on which we focused until now). In Chapter 5 we will study a model of endogenous growth of this kind and we will analyze the EMS of the innovative sector with particular reference to the role of technological leaders and policy issues.

2.10 Conclusions

In this chapter we have introduced the concept of EMSs in a partial equilibrium context built on the basis of the industrial organization literature. We have adopted a demand structure that derives from the Dixit-Stiglitz utility function, a framework that we will keep using in the following chapters for its tractability and large generality, but many of our conclusions would apply to different frameworks. We have also extended the basic model to a simple two-periods framework to show a basic mechanism of propagation of the shocks in the presence of EMSs, and to a general equilibrium framework to show the mechanism of transmission of a shocks to the labor market. Finally, we have developed a fully dynamic model inspired by the Solow model but augmented with imperfect competition and gradual business creation. The reader should keep in mind that the models we presented in this chapter were largely explorative and they may serve mainly as prototypes. In the next chapters we will introduce more complex and realistic models concerning the main fields of the macroeconomy.

The aim of this chapter was to provide a simple introduction to these topics, and to support the idea that it is important to introduce in new fields the study of market structures characterized by strategic interactions between a limited number of firms and endogenous entry determining this limited number of firms. In this book we will focus on the growing literature introducing EMSs in the study of business cycle, trade and growth, but the EMSs approach can be useful also in other microeconomic and macroeconomic fields that are not discussed here. Etro (2007,a) has reviewed a large number of applications to industrial organization and industrial policy. A lot of work has taken place independently in the last years also in the theory of auctions with endogenous entry of strategic bidders,[29] in the theory of competition under asymmetric information with endogenous entry,[30] in the theory of tax incidence and optimal taxation for markets with endogenous

[29] Palfrey and Pevnitskaya (2008) have presented a model of endogenous entry in first-price auctions with heterogeneous risk averse bidders and have tested its predictions with an experimental study.

[30] The important work by Creane and Jeitschko (2009) shows that endogenous entry overturn the collapse of markets with adverse selection. They consider a market in which each firm can pay an observable fixed costs of entry that generates a product of a quality that becomes known only to the firm. Entry has the tendency to lower prices, which may lead to exit of high quality products. However, the implied price collapse endogenously limits the amount of entry, so that high mark ups are supported in the market equilibrium.

entry,[31] in the theory of regulation with endogenous entry,[32] in the theory of political competition[33] and in other fields as well. Hopefully, these further applications confirm the utility of the joint analysis of strategic interactions and endogenous entry, and the necessity, that cannot be postponed further, of introducing EMSs in a systematic way within mainstream economic theory.

[31] See Wu and Zhang (2000) and Tamai (2006) on taxation in dynamic models and McCracken and Stähler (2007) on international tax competition. Katsoulacos and Xepapadeas (1995) is an early application to environmental policy.

[32] Gautier, Dam and Mitra (2007) have introduced the first analysis of endogenous entry in a model of regulated competition in differentiated retail goods and services between an incumbent leader, who owns a network good (an essential input) and potential entrants, whose cost of production is private information. The regulator sets the retail prices and the access charge that the entrant pays to the incumbent, but entry is endogenous.

[33] See Mulligan and Tsui (2009).

3. Endogenous Market Structures and Business Cycles

> To me what's really amazing is that for every job there is in the world, there is someone willing to do it... I think that a lot of people that are unemployed are not really unable to find work, they are just easily disgusted: "Yes I'm starving, and my family has no clothing or shelter, but I'm not cleaning that up!" **Jerry Seinfeld**

The standard theory of the Real Business Cycle, which we have reviewed in Chapter 1, is based on perfect competition, constant returns to scale and the absence of fixed costs of production and entry. In this neoclassical environment goods are priced at the marginal cost, there is no room for profits and the structure of the markets is indeterminate - i.e.: the number of firms and their individual production are irrelevant. However, a wide theoretical and empirical literature has emphasized the importance of market power to explain the behavior of the economy along the business cycle; in particular, Hall (1986, 1990), Bils (1987), Rotemberg and Woodford (1992) and more recently Galì (2007,a) have provided evidence of countercyclical mark ups and procyclical profits.

As we have seen in Chapter 1, the New-Keynesian theory, starting with Blanchard and Kiyotaki (1987), has introduced product differentiation and imperfect competition in general equilibrium models. Most of this literature departed from the neoclassical framework assuming monopolistic behavior as in Dixit and Stiglitz (1977) for a continuum of firms producing differentiated goods. This approach rapidly became the standard framework for the analysis of macroeconomic policy, with a special focus on monetary policy in the presence of simple forms of price stickiness. Nevertheless, the monopolistic approach leads to an exogenous market structure: the number of firms is exogenous and prices are given by a constant mark up on the marginal cost, so that the individual production is completely determined by the demand side. As such it neglects the role of strategic interactions between firms of the same sectors, it neglects the endogeneity of the number of firms, and it neglects the impact of entry on the same strategic interactions. The result is that the structure of the sectors of the economy remains a sort of "black

F. Etro., *Endogenous Market Structures and the Macroeconomy*,
DOI: 10.1007/978-3-540-87427-0_3,

box" whose main components, the mark ups and the number of competitors, are substantially exogenous.

In this chapter, largely based on Etro and Colciago (2007), we open the "black box" of the market structure and link the endogenous behavior of firms at the sectorial level with the general equilibrium properties of the economy, and in particular with its business cycle properties. We consider distinct sectors, each one characterized by many firms supplying homogenous goods (as in the basic neoclassical framework) or differentiated goods (as in the New-Keynesian literature), taking strategic interactions into account and competing either in prices (Bertrand competition) or in quantities (Cournot competition).[1]

Building on the recent important work by Ghironi and Melitz (2005), we introduce fixed costs of entry to endogenize the number of firms in each sector. However, in our model the number of firms affects equilibrium mark ups, firms' individual production and (stock market) value in each sector. In particular, the mark ups are endogenous and, in the short run, they depend on the form of competition, on the degree of substitutability between goods and on the number of firms. For instance, in the presence of homogenous goods Cournot competition allows to preserve substantial mark ups as long as the EMSs are concentrated enough.

The rest of the economy operates as in a standard dynamic flexible price model, and converges to a long run equilibrium where the steady state EMSs are characterized by a number of firms in each sector, individual production levels and mark ups that depend on the form of competition, on the degree of substitutability between goods and also on structural parameters as the discount factor, the aggregate productivity of labor, and the parameters governing entry and exit rates. A crucial aspect of these steady state EMSs is that they are characterized by excessive business creation and possibly also by dynamic inefficiency: we derive the market conditions under which this phenomenon emerges and discuss the possible solutions.

In this context, a temporary supply shock induces a novel propagation mechanism: it initially increases profits, so as to attract entry of firms, which in turn strengthens competition, reduces the mark ups and increases the real wages. The associated temporary reduction of the mark ups induces a stronger intertemporal substitution effect in favor of current consumption (and the increase in the real wages induces a stronger intertemporal substitution effect in favor of current work as well), which magnifies the effect of the shock compared to a perfectly competitive model. Finally, the temporary increase in demand has a positive feedback on profits which keeps the propagation mechanism alive. Notice that the stock market affects the real economy not only because it reflects the productivity level in the economy,

---

[1] We also consider a conjectural variation model that generalizes competition in quantities to forms of imperfect collusion and Stackelberg competition with endogenous entry in the long run (where a leader chooses its production before the other firms in each period).

but also because it reflects the strategic interactions between firms and the entry/exit process generated by the shock, an aspect that is totally absent in the neoclassical approach. Compared to a standard RBC model, we argue that the EMSs framework can perform better at matching U.S. data in terms of output, labor and consumption volatility, and it allows one to generate procyclical movements of aggregate profits and entry, and countercylical movements of the mark ups.

Notice, finally, that the endogeneity of the market structures allows one to reproduce realistic impulse response functions for demand shocks as well. A temporary increase in government spending creates a boom as in the RBC framework, but, contrary to the latter, it increases profits on impact, attracts entry, reduces the mark ups, increases the real wages, and allows to obtain a delayed consumption boom as well.

An important early work on endogenous mark ups in a dynamic stochastic general equilibrium (DSGE) model is by Rotemberg and Woodford (1992), who rely on a perfectly collusive framework and generate countercylical mark ups; nevertheless, their model does not endogenizes the entry process.[2]

The first attempts to endogenize entry in models of the business cycle are due to Chattejee and Cooper (1993), Devereux, Head and Lapham (1996) and Bénassy (1996,a,b).[3] These important works focused their analysis on fixed costs of production in each period and therefore on endogenous entry in each single period, which has the disadvantage of eliminating any dynamic process for the profits of the firms. The recent work by Ghironi and Melitz (2005), extended by Bilbiie, Ghironi and Melitz (2007, 2008,a,b), has provided an important contribution to model endogenous entry in a DSGE model with sunk costs of entry. However, this line of research does not take in consideration the strategic interactions between firms and the impact of entry on them, but focuses on the traditional case of constant mark ups due to monopolistic behavior *à la* Dixit-Stiglitz.[4]

---

[2] An alternative explanation of countercyclical mark ups is due to Ravn, Schmitt-Grohe and Uribe (2006) who have introduced habits over individual varieties of goods (as opposed to over a composite consumption good). This implies that the demand function faced by individual producers depends on past sales and generates monopolistic pricing with countercyclical mark-ups. Under these deep habits, consumption and wages respond procyclically to government-spending shocks. Of course, this rationale for countercylical mark ups is unrelated to the endoegenity of entry, but can be seen as complementary to it.

[3] Cooper (1999) surveys this early literature. Chatterjee, Cooper and Ravikumar (1993) endogenize entry as well, but their focus is on sunspots equilibria in an OLG model. See Kim (2004) for an interesting discussion.

[4] Bilbiie, Ghironi and Melitz (2007) have used translog preferences (introduced by Feenstra, 2003) to derive an elasticity of substitution between products that depends on the number of firms. As long as entry increases this elasticity, it generates mark ups decreasing in the number of goods. However, this *ad hoc* explanation is unrelated to endogenous motivations on the supply side.

More recently, Cook (2001), Wu and Zhang (2000, 2001), Devereux and Lee (2001) and Jaimovich (2007) have augmented the model of Devereux, Head and Lapham (1996) with mark ups depending on the number of firms.[5] However, these models endogenize entry again through fixed costs of production in each period, so that profits remain zero at all times and not procyclical as in our framework, and they focus on different issues. Nevertheless, these works belong to the EMSs approach suggesting a crucial role for market structures in explaining the business cycle.

The main objective of this chapter is the presentation of a realistic and tractable way of modeling EMSs in a DSGE framework. Of course, the success of this attempt depends on the applications that can be derived from our framework. For this reason, we dedicate a large space to the discussion of fiscal and monetary policy in the presence of EMSs, and to a preliminary investigation of the extensions to imperfections in the labor and credit markets. Hopefully, this analysis will stimulate future research.

The chapter is organized as follows. Section 3.1 provides empirical evidence on the cyclical behavior of mark ups, entry and profits. Section 3.2 describes the model and its dynamic properties under competition in quantities and in prices. Section 3.3 analyzes the steady state EMSs and derives the conditions for dynamic inefficiency. Section 3.4 calibrates and simulates the model. Section 3.5 extends the model to the accumulation of physical capital. Sections 3.6 and 3.7 apply the model to the analysis of fiscal and monetary policy, while Section 3.8 discusses extensions to imperfections in the labor and credit markets. Section 3.9 concludes.

3.1 Empirical Evidence on EMSs over the Business Cycle

The EMSs approach to the business cycle puts a lot of emphasis on the countercyclicality of mark ups and on the procyclicality of profits, two real world phenomena still in search for a convincing joint explanation in macroeconomic analysis. Our explanation is based on the endogenous entry/exit process which affects the degree of competition between firms over the business cycle, and allows us to match another stylized fact concerning the strong procyclicality of firms entry and its important role in determining output and employment fluctuations.[6]

In this section, we review the basic empirical evidence on the cyclical behavior of the main aggregate variables, of the corporate profits and of the

[5] Brito, Costa and Dixon (2008) extend the same model introducing competition in quantities to study complex dynamics.

[6] Davis, Haltiwanger and Schuh (1996) estimate from U.S. manufacturing data (between 1972 and 1986) that 25 % of annual gross job destruction can be attributed to establishment deaths and 20 % of annual gross job creation to establishment births.

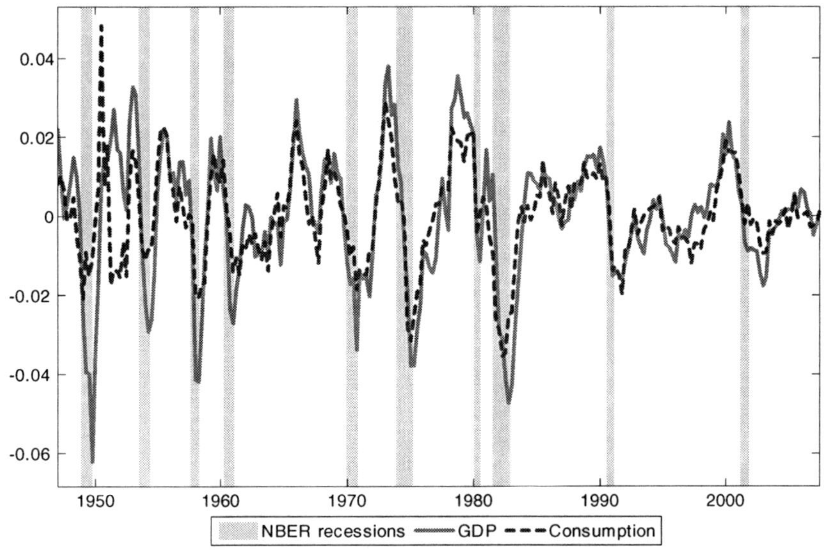

Fig. 3.1. U.S. Consumption/GDP (1948-2008)

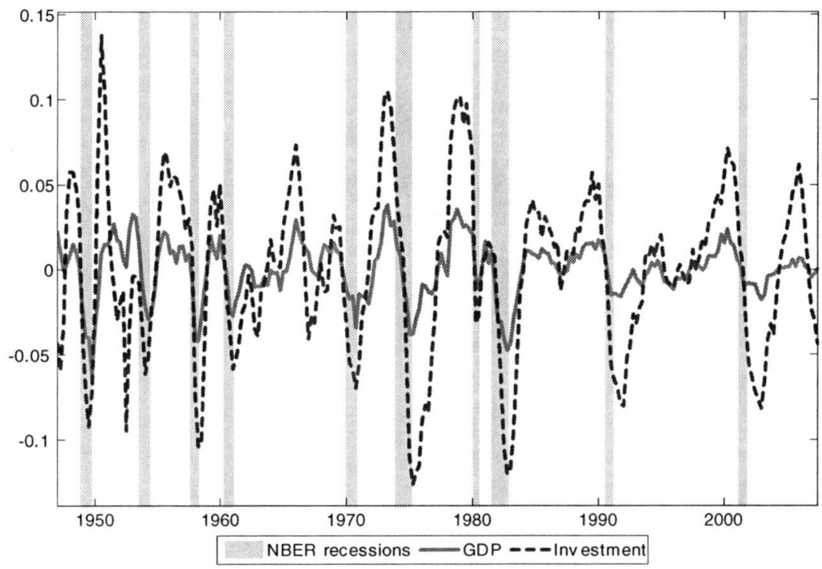

Fig. 3.2. U.S. Investment/GDP (1948-2008)

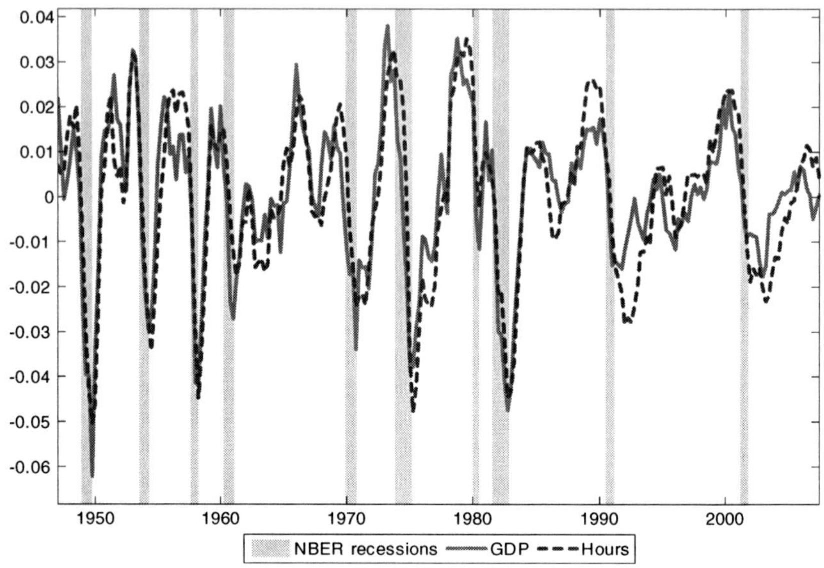

Fig. 3.3. U.S. Hours of work/GDP (1948-2008)

mark ups for the U.S. economy during the last sixty years, first with a simple analysis of the detrended behavior of the U.S. aggregate data and then with a VAR analysis of the reaction of the U.S. economy to a productivity shock.[7]

3.1.1 Business cycle fluctuations in the U.S.

Figures 3.1-3.5 show the detrended behavior of consumption, investment, hours of work, profits and price mark ups and compare each one with the detrended behavior of GDP over the period 1948-2008. The data are derived from FRED, the Federal Reserve Economic Database of the Federal Reserve Bank of St. Louis.[8] All the variables have been logged and detrended using the Hodrick-Prescott filter with a smoothing parameter equal to 1600

[7] I am extremely thankful to Andrea Colciago for preparation of these data.

[8] Below, we report in brackets the mnemonics of each series used here and in other occasions in the book:

Gross Domestic Product (GDP): Billions of Dollars, Quarterly.

Compensation of Employees (COE): Billions of Dollars, Quarterly, Seasonally Adjusted Annual Rate.

Proprietors' Income with inventory valuation adjustment (IVA) and capital consumption adjustment (PROPINC): Billions of Dollars, Quarterly.

Personal Consumption Expenditure (PCEC): Billions of Dollars.

Corporate Profits with inventory Valuation Adjustment (IVA) and Capital Consumption Adjustment (CPROFIT): Billions of Dollars, Quarterly.

(given quarterly frequency). The figures report the periods of recession for the American economy according to the definition of the *National Bureau of Economic Research*: one can easily see the shaded areas corresponding to the recessions associated with the first and second oil shocks, and the others taking place at the beginning of each one of the last four decades.

Figures 3.1-3.3 show the strong procyclicality of consumption, investment and labor over the last sixty years. As well known, consumption is less volatile than GDP, while investment is much more volatile, and labor supply is approximately as volatile as GDP. As we noticed in Table 1.1. (where we reported second moments derived from the same data and from a standard RBC model), the basic neoclassical approach to the business cycle fails to explain part of the cyclical behavior of GDP, and especially of consumption and labor supply, while it captures the large variability of investment, which is about three times as volatile as GDP. A part from these limits, which can be partially solved with extensions of the model in a number of realistic directions, the neoclassical approach has nothing to say about the cyclical behavior of the extra-profits of the firms and of the mark ups.

Figure 3.4 plots HP-filtered GDP together with the HP-filtered series of real corporate profits for 1948-2008.[9] The contemporaneous correlation is positive and equal to 0.67. Moreover, the picture documents that profits are extremely volatile at the business cycle frequencies, almost five times more than GDP.

Figure 3.5 plots the HP-filtered series of GDP and price mark up. Of course, mark ups are not observable and we had to measure them indirectly. We constructed a labor-share based measure of the mark ups along the lines suggested by Rotemberg and Woodford (1999). To derive the empirical measure of the mark up we used the pricing equation $p_t = \mu_t(W_t/A_t)$ under the assumption of homogenous goods to obtain $\mu_t = A_t/w_t = Y_t/w_tL_t$. Following Rotemberg and Woodford (1999) and assuming the existence of overhead labor, so that effective labor for the production of consumption goods is $L_t^c = L_t - L_t^o$, one can obtain:

$$\mu_t = \frac{Y_t}{w_t L_t} \frac{L_t}{L_t - L_t^o}$$

whose log-linearization is:

$$\widehat{\mu}_t = -\frac{l^o}{1-l^o}\hat{L}_t - \widehat{1-\alpha_t}$$

Gross Domestic Product Implicit Price Deflator (PCEC): Index 2000=100, Quarterly, Seasonally Adjusted.

Hours of all Persons, nonfarm business sector (HOANBS): Index 1992=100, Quarterly, Seasonally Adjusted.

Fixed Private Investment (FPI): Billions of Dollars, Quarterly.

[9] It would be interesting to separate the normal remuneration of capital from the extra profits, but this not likely to change the qualitative picture obtained from the full corporate profits.

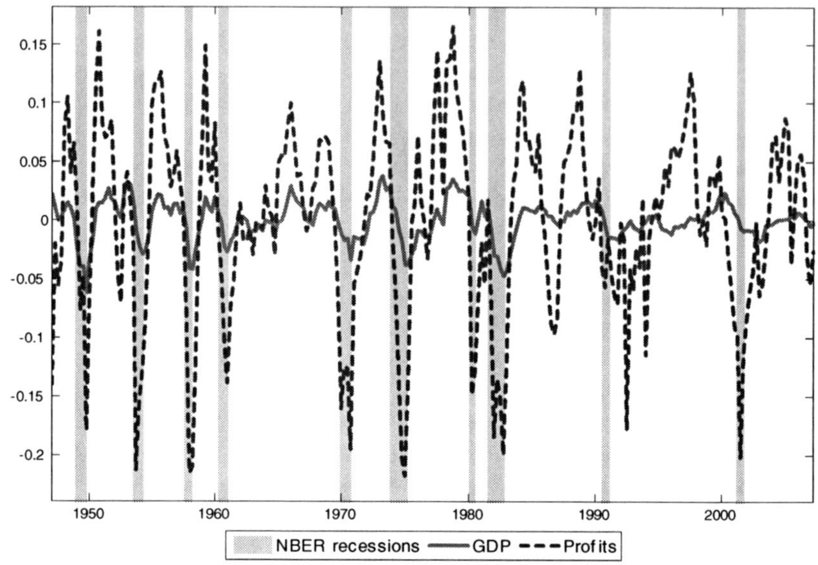

Fig. 3.4. U.S. Profits/GDP (1948-2008)

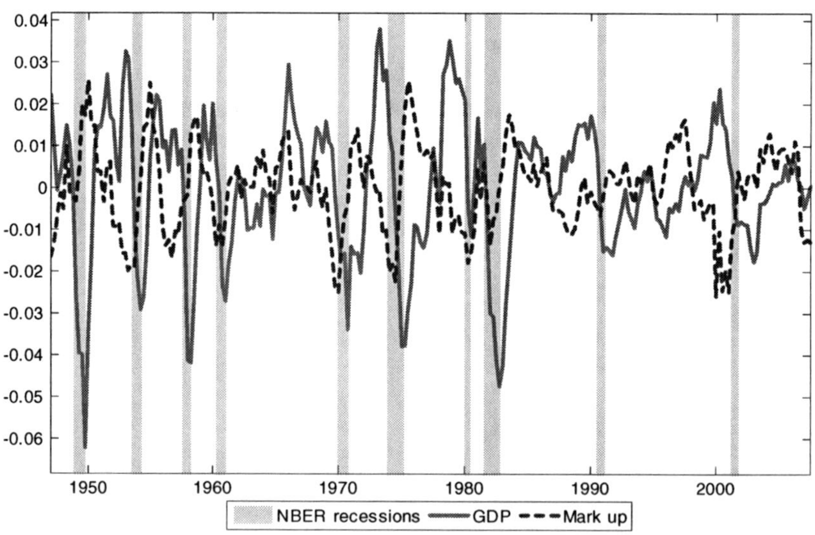

Fig. 3.5. U.S. Mark ups/GDP (1948-2008)

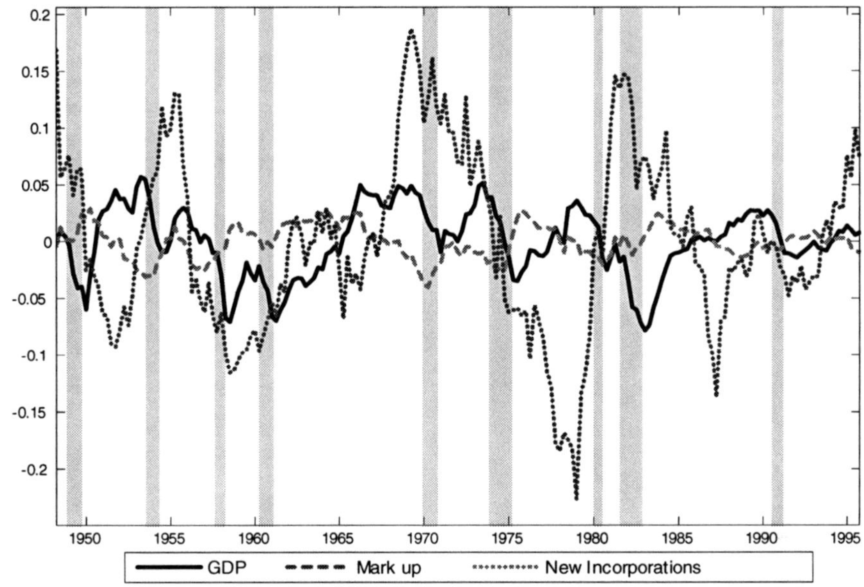

Fig. 3.6. Cyclical component of GDP, New Incorporations and Price mark up. Shaded areas represent NBER recessions.

where $l^o \equiv L_t^o/L_t$ represents the average share of overhead labor over total labor input (assumed to be equal to 0.2), $1 - \alpha_t \equiv w_t L_t / A_t (L_t - L_t^o)$ is the labor share of income with $w_t = W_t/p_t$ real wage, and hatted variables indicate percentage deviations from the HP trend. Second moments derived from this measure of the mark ups were already reported in Table 1.1. The contemporaneous correlation between detrended GDP and mark ups is -0.28, which confirms the countercyclicality of the mark ups.[10]

3.1.2 VAR analysis

To provide further support to these empirical findings we performed an additional VAR analysis on U.S. We used data on the number of new incorporations supplied by the Brad & Broadstreet corporation. The data run from 1948 to 1995, for this reason we restrict our empirical analysis to this period.[11] Figure 3.6 plots the series of detrended GDP, the number of new incorporations (in logs) and the price mark up. Since we use these variables in subsequent time series analysis we opt for a polynomial of time to detrend

[10] Wilson and Reynolds (2005) have provided additional time series evidence on the countercyclical movement of mark ups at the industry level.

[11] We thank Vivien Lewis for providing the series on firms' data.

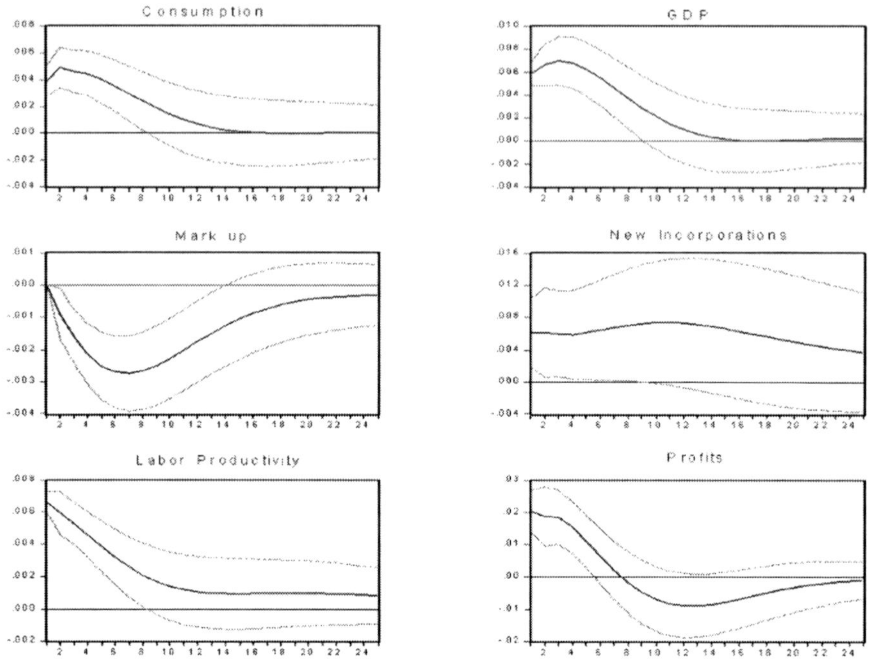

Fig. 3.7. VAR analysis. Dynamic response to a one standard deviation technology shock (with 95 % confidence intervals)

variables instead of the HP filter, as suggested by Galì *et al.* (2007,b). However the signs of contemporaneous correlations between our measures of entry and the price mark up with output are unchanged when variables are detrended using the HP filter. In line with the evidence reviewed above net business formation is procyclical and negatively correlated with the price mark up.[12] The contemporaneous correlation between GDP and new incorporations is positive and equals 0.11,[13] while that between GDP and the price mark up is -0.42.

We consider a VAR(2) model including our mark up measure, labor productivity (which we take as a measure of technology), consumption, GDP, profits, and the number of new incorporations. We identify the technology shock by means of a Cholesky decomposition. In our baseline specification

[12] For additional evidence on the crucial role of entry along the business cycle see the microeconomic works by Bernard, Redding and Schott (2008) and Broda and Weinstein (2009) and the discussion in Chapter 6.

[13] The contemporaneous correlation between GDP and the index of net business formation (from the same source) is larger and equal to 0.45.

variables are ordered as listed above,[14] and the ordering is naturally suggested by the EMSs approach. The technology shock affects output and consumption, which leads, for given mark up, to a change in profits affecting the entry decision of firm: the resulting number of competitors finally leads to a mark up change.

Figure 3.7 shows the estimated responses to an expansionary shock to labor productivity. The responses of output and consumption are similar to those found in the literature.[15] Most relevantly for our purposes, both profits and firm entry respond positively to the technology shock, while the price mark up declines significantly.

Monacelli and Perotti (2008) have added further VAR evidence on the impact of government spending shocks: these tend to increase output and, to a lower extent, consumption, and to increase the real wage while reducing the mark up, a pattern that is in contrast with the outcome of the neoclassical model (at least for the last three variables).

Lewis (2007,a) and Bergin and Corsetti (2008) have added VAR evidence on the impact of monetary shocks: these tend to generate a delayed but positive impact on new incorporations, suggesting the validity of a business creation channel for monetary policy.

Modeling countercyclical mark ups and procyclical profits together with procyclical entry of firms is an important task for the business cycle literature. In the rest of the chapter we present a model that accomplishes this purpose through an accurate formalization of the EMSs in the production process.

3.2 A Model of the Business Cycle with EMSs

In this section we introduce a DSGE model where the structure of the goods markets is neither perfectly competitive or monopolistic, but is characterized by imperfect competition and endogenous entry. The model, based on Ghironi and Melitz (2005), Etro and Colciago (2007) and Bilbiie, Ghironi and Melitz (2007), extends the rudimentary dynamic model of business creation of Section 2.6 introducing endogenous savings, labor supply and product differentiation.

The structure of the economy is extremely simple. Consider a representative agent with utility:

---

[14] The lag lenght has been selected according to a sequential Loglikelihood Ratio test starting with a maximum number of lags equal to 8. However, results do not change relevantly if we consider 3 or 4 lags. For robustness we experimented with alternative orderings of the variables ordered after the technology shock. We have also estimated the dynamic behavior of the model including the index of net business formation described above instead of the number of new incorporations. In all these cases just minor changes apply to our baseline results.

[15] See Dedola and Neri (2007).

$$U = E_t \sum_{t=0}^{\infty} \beta^t \left\{ \int_0^1 \log C_{jt} dj - \frac{\upsilon L_t^{1+1/\varphi}}{1+1/\varphi} \right\} \quad \upsilon, \varphi \geq 0 \tag{3.1}$$

where $\beta \in (0, 1)$ is the discount factor, L_t is labor supply and C_{jt} is a consumption index for a continuum of goods produced in sectors $j \in [0, 1]$. The agent maximizes (3.1) choosing how much to work and how much to invest in bonds and risky stocks. Without loss of generality, bonds and stocks are denominated in terms of good 1. Therefore, the budget constraint expressed in nominal terms is:

$$P_{1t}B_{t+1} + \int_0^1 P_{1t}V_{kt}(N_{kt} + N_{kt}^e)s_{kt+1}dk + \int_0^1 P_{kt}C_{kt}dk =$$

$$= W_tL_t + (1+r_t)P_{1t}B_t + \int_0^1 P_{1t}\left[\pi_{kt}(\theta, N_t) + V_{kt}\right] N_{kt}s_{kt}dk - P_{1t}T_t \tag{3.2}$$

where B_t is net bond holdings with interest rate r_t, V_{kt} is the value of a firm from sector k, N_{kt} and N_{kt}^e are the active firms in sector k and the new firms founded in this sector at the end of the period, s_{kt} is the share of the stock market value of the firms of sector k that are owned by the agent, W_t is the nominal wage and T_t are lump sum taxes. The intratemporal optimality conditions for the optimal choices of C_{jt} and L_t require:

$$P_{jt}C_{jt} = E_t \quad \text{for any } j \tag{3.3}$$

$$L_t = \left(\frac{W_t}{\upsilon P_{jt}C_{jt}} \right)^{\varphi} \tag{3.4}$$

Here, E_t is total expenditure allocated to the goods produced in each sector in period t and P_{jt} is the price index for consumption in sector j: due to the unitary elasticity of substitution, total expenditure is equally shared between the sectors. The choice of labor supply follows the same optimality condition as in the two period model of Section 1.1, that is (1.6). The intertemporal optimality conditions will be discussed later on.

Each sector j is characterized by different firms $i = 1, 2, ..., N_{jt}$ producing the same good in different varieties in each period, and the consumption index C_{jt} of sector j at time t is:

$$C_{jt} = \left[\sum_{i=1}^{N_{jt}} C_{jt}(i)^{\frac{\theta-1}{\theta}} \right]^{\frac{\theta}{\theta-1}} \tag{3.5}$$

where $C_{jt}(i)$ is the production of firm i, and $\theta > 1$ is the elasticity of substitution between the goods produced in each sector. The distinction between different sectors and different goods within a sector allows us to realistically separate limited substitutability at the aggregate level, and high substitutability at the disaggregated level. Contrary to many macroeconomic models with imperfect competition, our focus will be on the market structure of disaggregated sectors: *intrasectoral substitutability* (between goods produced by firms of a same sector) is high (or perfect when $\theta \to \infty$), while *intersectoral substitutability* is low.[16]

Each firm i in sector j produces a good with a linear production function. Labor is the only input, and output of firm i in sector j is:

$$x_{jt}(i) = A_t l_{ijt} \tag{3.6}$$

where A_t is total factor productivity at time t, and l_{ijt} is total labor employed by firm i in sector j. This implies that the production of one good requires $1/A_t$ units of labor, and the marginal cost of production is W_t/A_t. Since each sector can be characterized in the same way, in the rest of the chapter we will drop the sector index j and refer to the representative sector.

In each period, the same expenditure for each sector E_t is allocated across the available goods according to the standard direct demand function derived from the maximization of the consumption index (3.5), as shown in detail in Chapter 2:

$$C_t(i) = \frac{p_t(i)^{-\theta} E_t}{P_t^{1-\theta}} \qquad i = 1, 2, ..., N_t \tag{3.7}$$

where P_t is the standard price index:

$$P_t = \left[\sum_{j=1}^{N_t} p_t(j)^{-(\theta-1)} \right]^{\frac{-1}{\theta-1}} \tag{3.8}$$

such that total expenditure satisfies $E_t = \sum_{j=1}^{N_t} p_t(j) C_t(j) = C_t P_t$. Inverting the direct demand functions, we can derive the system of inverse demand functions:

$$p_t(i) = \frac{x_t(i)^{-\frac{1}{\theta}} E_t}{\sum_{j=1}^{N_t} x_t(j)^{\frac{\theta-1}{\theta}}} \qquad i = 1, 2, ..., N_t \tag{3.9}$$

where $x_t(i)$ is the quantity of good i.

As usual and realistic in dynamic macroeconomic models, we assume that firms cannot credibly commit to a sequence of strategies, therefore their behavior is equivalent to maximize current profits in each period taking as given

[16] For an empirical support of our modeling strategy see Broda and Weinstein (2009). The same assumption has been independently proposed by Atkeson and Burnstein (2009) in an unrelated context. Also their numerical results are obtained under the assumption that the intersectoral elasticity of substitution is unitary.

the strategies of the other firms. A main interest of this model is in the comparison of equilibria where in each period firms compete in quantities and in prices, taking as given their marginal cost of production and the aggregate expenditure of the representative consumer.[17] Under different forms of symmetric competition we obtain equilibrium prices satisfying:

$$p_t = \frac{\mu(\theta, N_t)W_t}{A_t} \tag{3.10}$$

where $\mu(\theta, N_t) > 1$ is the mark up depending on the degree of substitutability between goods θ and on the number of firms N_t. In the next subsections we characterize this mark up under competition in quantities and in prices (as in the static model of Chapter 2), and we sketch how to consider other forms of competition including conjectural variations and Stackelberg models.

3.2.1 Cournot competition

First, let us consider competition in quantities. Using the inverse demand function (3.9), we can express the profits of a firm i as a function of its output $x_t(i)$ and the output of all the other firms:

$$\Pi_t\left[x_t(i)\right] = \frac{x_t(i)^{\frac{\theta-1}{\theta}} E_t}{\sum_{j=1}^{N_t} x_t(j)^{\frac{\theta-1}{\theta}}} - \frac{W_t x_t(i)}{A_t} \tag{3.11}$$

Assuming that each firm chooses its production $x_t(i)$ to maximize profits taking as given the other strategies, we can characterize the Cournot equilibrium as in Chapter 2. In particular, the equilibrium mark up at time t is:

$$\mu^Q(\theta, N_t) = \frac{\theta N_t}{(\theta-1)(N_t-1)} \tag{3.12}$$

where the index Q stands for competition in quantities. Notice that the mark up is decreasing in the degree of substitutability between products θ and in the number of firms active in the same period N_t. Given the nominal profits $\Pi_t^Q = (p_t - W_t/A_t)x_t$, the individual profits in real terms at time t can be expressed as:

$$\pi_t^Q(\theta, N_t) = \frac{(N_t+\theta-1)C_t}{\theta N_t^2} \tag{3.13}$$

which is decreasing in the number of active firms at time t and in the degree of substitutability, and increasing in the aggregate demand of consumption.

---

[17] Of course, both of them are endogenous in general equilibrium, but it is reasonable (and common in the macroeconomic literature) to assume that firms do not perceive marginal cost and aggregate expenditure in the sector as affected by their choices.

3.2.2 Bertrand competition

Let us now consider competition in prices. In each period, the gross profits of firm i can be expressed as:

$$\Pi_t\left[p_t(i)\right] = \frac{\left[p_t(i) - W_t/A_t\right] p_t(i)^{-\theta} E_t}{\left[\sum_{j=1}^{N_t} p_t(j)^{-(\theta-1)}\right]} \tag{3.14}$$

Firms compete by choosing their prices. Contrary to the traditional Dixit-Stiglitz approach which neglects strategic interactions between firms, we take these into consideration and derive the exact Bertrand equilibrium. Each firm i chooses the price $p_t(i)$ to maximize profits taking as given the price of the other firms.[18] As we have seen in the static version of this model studied in Chapter 2, in the symmetric Bertrand equilibrium we obtain the following mark up at time t:

$$\mu^P(\theta, N_t) = \frac{1 + \theta(N_t - 1)}{(\theta - 1)(N_t - 1)} \tag{3.15}$$

This mark up is decreasing in the degree of substitutability between products θ and in the number of firms N_t.[19] In conclusion, with competition in prices the individual profits in period t can be expressed in real terms as:

$$\pi_t^P(\theta, N_t) = \frac{C_t}{1 + \theta(N_t - 1)} \tag{3.16}$$

which is again a decreasing function of the number of active firms and of the substitutability between goods, and increasing in aggregate demand.

[18] Since total expenditure E_t is equalized between sectors by the consumers, we assume that it is also perceived as given by the firms. Under the alternative hypothesis that consumption C_t is perceived as given, we would obtain the higher mark up:

$$\tilde{\mu}^P(\theta, N_t) = \frac{\theta(N_t - 1)}{(\theta - 1)(N_t - 1) - 1}$$

which leads to similar qualitative results. This case would correspond to the equilibrium mark up proposed by Yang and Heijdra (1993). A similar version is also adopted by Jaimovich and Floetotto (2008), whose mark up, however, depends on the degree of substitutability between (intermediate) goods produced in different sectors as well.

[19] As noticed in the previous chapter, entry decreases mark ups faster under competition in quantities compared to competition in prices. These results will play a crucial role in our subsequent analysis of the propagation mechanism of the business cycle under different forms of competition.

3.2.3 Stackelberg competition and imperfect collusion

Our framework can be used to study other forms of competition. To give the flavor of these possibilities we briefly report two simple extensions for the case of homogenous goods ($\theta \to \infty$).

Our first extension of the Cournot model introduces asymmetries between firms building on the theory of Stackelberg competition. Let us assume that a single leader is always active and N_t followers are active in each period. In Stackelberg equilibrium the mark up is:

$$\mu^S(N_t) = \frac{N_t}{N_t - 1/2} \tag{3.17}$$

which is lower compared to the mark up under pure Cournot competition. The profits of the leader and the representative follower are respectively larger and smaller than the profits under Cournot competition.

The second extension belongs to the traditional *conjectural variations approach* and allows us to reproduce forms of imperfect collusion between the active firms. Assuming that each firm takes as given the differential impact of its output choice on the output choice of the other firms $\lambda \equiv \partial x_t(j)/\partial x_t(i)$, the equilibrium mark up can be derived as:

$$\mu^{CV}(N_t) = \frac{N_t}{(N_t - 1)(1 - \lambda)} \tag{3.18}$$

which nests the case of Cournot competition in quantities for $\lambda = 0$ and tends to the (indeterminate) case of perfect collusion for $\lambda \to 1$. Intermediate values $\lambda \in (0, 1)$ describe cases of imperfect collusion between the firms which achieve mark ups and profits above the Cournot level but below the perfect collusion level.

3.2.4 Short run EMSs

Households choose how much to save in riskless bonds and in the creation of new firms through the stock market. Their choice follows a standard Euler equation - as (1.3) - for the choice of the safe asset, and a standard asset pricing equation - as (1.7) - for the choice of the uncertain one:

$$\frac{1}{C_t} = \beta(1 + r_{t+1})E\left(\frac{1}{C_{t+1}}\right) \tag{3.19}$$

$$\frac{1}{C_t} = \beta E\left[\left(\frac{[\pi_{t+1}(\theta, N_{t+1}) + V_{t+1}]\, N_{t+1}}{V_t\,(N_t + N_t^e)}\right)\left(\frac{1}{C_{t+1}}\right)\right] \tag{3.20}$$

The (average) number of firms per sector follows the equation of motion:

$$N_{t+1} = (1 - \delta_N)(N_t + N_t^e) \tag{3.21}$$

where N_t^e is the (average) number of new firms and $\delta_N \in (0,1)$ is the exogenous rate of exit.[20] The real value of a firm V_t is the present discounted value of its future expected profits. Using the two intertemporal conditions and the equation of motion for the number of firms, this value can be written in recursive form as follows:

$$V_t = (1-\delta_N)E\left[\frac{V_{t+1}+\pi_{t+1}(\theta,N_{t+1})}{1+r_{t+1}}\right] \tag{3.22}$$

Solving (3.22) one obtains:

$$V_t = E\left[\sum_{s=1}^{\infty}\frac{(1-\delta_N)^s\,\pi_{t+s}(\theta,N_{t+s})}{\prod_{q=1}^{s}(1+r_{t+q})}\right] + \lim_{s\to\infty} E\frac{(1-\delta_N)^s\,V_{t+s}}{\prod_{q=1}^{s}(1+r_{t+q})} \tag{3.23}$$

where the second term equals zero if we impose the transversality condition that avoids stock market bubbles (requiring that the average rate of capital gain does not exceed the average discount rate).[21]

Notice that the equilibrium interest rate is not governed by the marginal productivity of capital as in the neoclassical approach, but by the dynamics of the stock market value, in particular by the stock market return in terms of capital gains and dividends which depend on the entry and competition process. Any shock that increases (decreases) the return on the investment in the stock market through a positive (negative) impact on short run profits or long run capitalization induces a positive (negative) effect on consumption growth through the impact on the interest rate. Notice that the stock market affects the real economy not only because it reflects the productivity level in the economy, but also because it reflects the strategic interactions between firms and the entry/exit process due to various shocks.

The labor market is frictionless. The real wage can be derived from the equilibrium pricing relation (3.10) as:

$$w_t = \frac{A_t p_t}{\mu(\theta,N_t)P_t} = \frac{A_t N_t^{\frac{1}{\theta-1}}}{\mu(\theta,N_t)} \tag{3.24}$$

where we used the fact that $P_t = p_t N_t^{1/(1-\theta)}$ in the symmetric equilibrium. The representative agent supplies labor according to the equilibrium optimality condition (3.4), which, using the equilibrium real wage (3.24), becomes:

$$L_t = \left[\frac{A_t N_t^{\frac{1}{\theta-1}}}{\upsilon\mu(\theta,N_t)C_t}\right]^{\varphi} \tag{3.25}$$

---

[20] It would be interesting to endogenize the exit rate as a countercyclical factor: this would strengthen our propagation mechanism, since it would enhance the countercyclicality of mark ups.

[21] It follows that (the stock market) value of the representative firm equals the present value of its discounted future dividends.

Labor demand depends on the production and business creation activities.

The credit market is frictionless and perfectly competitive, which guarantees that the Modigliani-Miller theorem holds (Miller and Modigliani, 1961), so that we can assume without loss of generality that profits are distributed in forms of dividends in each period, and that the Ricardian equivalence holds (Barro, 1974) in the presence of lump sum taxes, so that fiscal policy (for a given path of public spending) does not affect the real economy. As a consequence, neither the dividend policy or the public debt policy affect the value of the firms and the process of business creation.

The creation of a new firm requires an initial fixed cost F_t at time t, and endogenous entry occurs until the gross value of a firm V_t given by (3.22) is equal to this fixed cost:

$$V_t = (1-\delta_N)E\left[\frac{V_{t+1}+\pi_{t+1}(\theta, N_{t+1})}{1+r_{t+1}}\right] = F_t \tag{3.26}$$

There are two ways of modeling the entry cost, according to whether the development of a different product requires an additional labor activity or whether the creation of an entirely new product requires a genuine monetary investment in R&D. In the first case we should adopt a fixed cost in terms of labor requirement, which diverts part of the labor force from the production activity toward the business creation activity. In the second case we should assume a monetary cost of business creation, which diverts some of income for the creation of new products.[22] In this chapter dedicated to the analysis of the business cycle (and in the next one dedicated to trade) we adopt the first approach, while in Chapter 5 dedicated to growth through innovation we will adopt the second approach.

---

[22] In case of a monetary fixed investment, the aggregate budget constraint reads as $V_t N_t^e + C_t = Y_t$, where $V_t = F_t$, which implies $N_t^e = (Y_t - C_t)/F_t$. The equilibrium system is then given by the asset pricing equation:

$$F_t = \beta(1-\delta_N)E_t\left\{\left(\frac{C_{t+1}}{C_t}\right)^{-1}\left[F_{t+1}+\left(1-\frac{1}{\mu(\theta, N_{t+1})}\right)\frac{A_{t+1}L_{t+1}}{N_{t+1}}\right]\right\}$$

and by the equation of motion for the number of firms:

$$N_{t+1} = (1-\delta_N)\left(N_t + \frac{A_t L_t - C_t}{F_t}\right)$$

where the labor supply is given by (3.25). In the stationary case with Cournot competition, homogenous goods and unitary labor supply, the steady state number of firms is $\tilde{N}^Q = \sqrt{\beta(1-\delta_N)A/[1-\beta(1-\delta_N)]F}$, and the mark up is (Etro, 2009,a):

$$\tilde{\mu}^Q = \frac{\sqrt{\beta(1-\delta_N)A}}{\sqrt{\beta(1-\delta_N)A} - \sqrt{[1-\beta(1-\delta_N)]F}}$$

Let us assume that entry requires a fixed cost of production equal to η/A_t units of labor, with $\eta > 0$. The real cost of a unit of labor is given by (3.24), therefore the endogenous value of a single firm must be equal to:

$$V_t = \frac{\eta N_t^{\frac{1}{\theta-1}}}{\mu(\theta, N_t)} \tag{3.27}$$

Notice that both the real wage and the equilibrium value of the firms are increasing with the number of firms.

Public consumption G_t is spent exactly as private consumption, and without loss of generality we assume budget balance in every period, $T_t = G_t$. This and goods market clearing implies the equilibrium aggregate resource constraint:

$$C_t + G_t + N_t^e V_t = N_t \pi_t(\theta, N_t) + w_t L_t$$

As it can be verified, market clearing in the labor market holds by Walras' law. Substituting all the equilibrium relations in the equation of motion of the number of firms (3.21) and in the endogenous entry condition (3.27), one can summarize the short run EMSs with a system of two equations for the evolution of N_t and C_t (eventually depending on the evolution of total factor productivity A_t and government spending G_t):

$$N_{t+1} = (1 - \delta_N) \left[N_t + \frac{A_t^{1+\varphi}}{\eta} \left(\frac{N_t^{\frac{1}{\theta-1}}}{\upsilon \mu(\theta, N_t) C_t} \right)^{\varphi} - \frac{C_t + G_t}{\eta N_t^{1/(\theta-1)}} \right] \tag{3.28}$$

$$\frac{\eta N_t^{\frac{1}{\theta-1}} \mu(\theta, N_{t+1})}{\mu(\theta, N_t)} = \beta(1-\delta_N) E \left[\frac{C_t}{C_{t+1}} \left(\eta N_{t+1}^{\frac{1}{\theta-1}} + \frac{C_{t+1} + G_{t+1}}{\frac{N_{t+1}}{\mu(\theta, N_{t+1})-1}} \right) \right] \tag{3.29}$$

One can show the (saddle path) stability of the deterministic version of this bidimensional system,[23] with monotonic convergence to steady state values of consumption and of the number of firms. Consequently, also the other elements of the EMS, the mark up and the production per firm, follow a monotonic path toward their steady state levels. We can summarize this in:

Proposition 3.1. **The deterministic general equilibrium model is saddlepath stable and convergent to a steady state endogenous market structure.**

As we will see graphically in Figures 3.8 and 3.9, starting from a situation with a low number of firms, the equilibrium implies monotonic convergence to the steady state through an increase of both consumption and the number

[23] The Jacobian has two eigenvalues, one larger and one smaller than one (see Colciago and Etro, 2008).

of firms.[24] The dynamic system can be used to analyze the business cycle properties of our model. However, before doing that, we derive analytically some properties of the long run equilibrium.

3.3 Steady State EMSs

In this section we analyze the dynamic properties of the model in the long run. For simplicity, we assume inelastic labor supply ($\varphi = 0$) and focus on the deterministic model with $G_t = 0$ and $A_t = A$ for any t. Under these assumptions, the equation of motion for the number of firms (3.28) does not depend on the form of competition and boils down to:[25]

$$N_{t+1} = (1 - \delta_N) \left(N_t + \frac{A}{\eta} - \frac{C_t}{\eta N_t^{1/(\theta-1)}} \right) \tag{3.30}$$

which in steady state provides a relation between the number of firms $\tilde{N}$ and the consumption index $\tilde{C}$. Solving the steady state relation for consumption, we obtain:

$$\tilde{C} = A\tilde{N}^{\frac{1}{\theta-1}} - \frac{\delta_N \eta \tilde{N}^{\frac{\theta}{\theta-1}}}{1 - \delta_N} \tag{3.31}$$

At least for low levels of substitutability (low θ), this expression for $\tilde{C}$ is an inverse-U relation in $\tilde{N}$ (see Figure 3.8): with few firms in steady state, the consumption index increases with the number of firms because of the love of variety effect, but with a large number of firms the index is negatively affected by a further increase in the number of firms due to the high savings necessary to create new firms. When substitutability between goods is high, the second effect always prevails and the above relation is monotonically decreasing; for instance, when goods are homogenous ($\theta \to \infty$), we have $\tilde{C} = A - \delta_N \eta \tilde{N}/(1 - \delta_N)$, which is linearly decreasing in the number of firms.

In the general case, the steady state number of firms that maximizes steady state consumption (3.31), and therefore maximizes steady state utility, can be derived as (the maximum between 1) and:

$$\tilde{N}^{GR} = \frac{(1 - \delta_N)A}{\delta_N \eta \theta} \tag{3.32}$$

where GR refers to the *golden rule* number of firms/goods. This implies the following comparative statics result:

---

[24] The cases of imperfect collusion and Stackelberg competition can be analyzed similarly.

[25] Further differences between the two forms of competition emerge in case of endogenous labor supply.

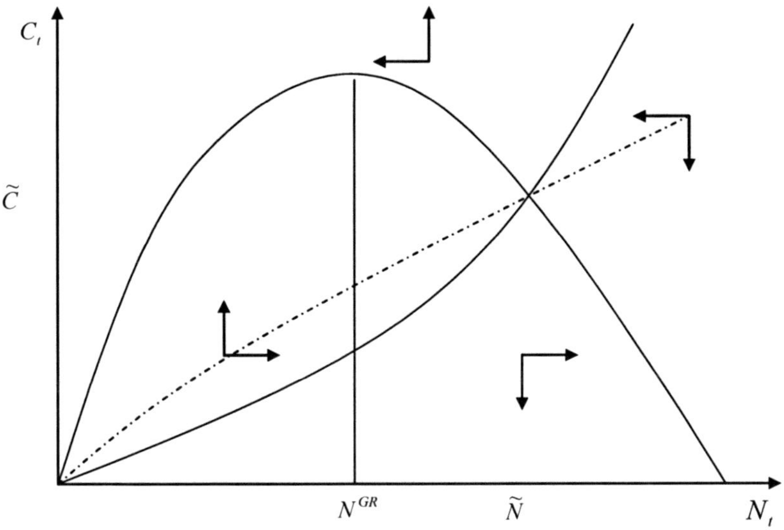

Fig. 3.8. Phase Diagram in (N_t, C_t): competition in quantities.

PROPOSITION 3.2. **The endogenous market structure that maximizes steady state utility implies a number of firms increasing in the productivity level A and decreasing in the degree of substitutability between goods θ, in the rate of exit of the firms δ_N and in the parameter of the entry cost η.**

Any steady state with a number of goods larger than $\tilde{N}^{GR}$ would be dynamically inefficient, in the sense that higher levels of long run utility could be permanently reached by reducing entry of firms and therefore reducing savings and increasing consumption at any time. Notice however, that the golden rule number of firms is not necessarily optimal, because the utility maximizing number of firms may be lower for impatient consumers. We derive the optimal allocation of resources in Section 3.4.

We now complete the characterization of the equilibrium in the two cases of competition in quantities and prices.

3.3.1 Dynamics and steady state under Cournot competition

Under competition in quantities the deterministic equilibrium system is given by (3.30) and:

$$C_{t+1} = C_t \left(\frac{\beta(1-\delta_N) N_t^{(\theta-2)/(\theta-1)}}{\eta(\theta-1)(N_t-1)} \right) \cdot$$
$$\cdot \left(\frac{\eta(\theta-1)(N_{t+1}-1)}{N_{t+1}^{(\theta-2)/(\theta-1)}} + \frac{(N_{t+1}+\theta-1)C_{t+1}}{N_{t+1}^2} \right)$$

which, in steady state, becomes:

$$\tilde{C} = \frac{\eta[1-\beta(1-\delta_N)](\theta-1)\tilde{N}^{\frac{\theta}{\theta-1}}}{\beta(1-\delta_N)\left[1+\theta/(\tilde{N}-1)\right]} \tag{3.33}$$

This expression represents a positive and convex relation between the number of firms and the consumption index, as the one shown in Figure 3.8. Two effects operate in the same direction: on one side more firms produce more in total and create larger gains from variety, on the other side they are more competitive and each one of them produces more of its variety. Notice that when the discount factor β increases, this relation is shifted downward: more patient agents save more and are able to consume less in steady state.

Conditions (3.31) and (3.33) jointly define the consumption index and the number of firms in steady state, and therefore the endogenous long-run mark up. It is immediate to verify that the steady state can be characterized by dynamic inefficiency when the discount factor is high enough.[26] We have represented this case in Figure 3.8, where we show the phase diagram of the model with the two steady state relations in solid lines (crossing beyond the golden rule cut-off) and the saddle-path in dashed line. Of course, if the discount factor is low enough, the propensity to save is limited and the steady state exhibits dynamic efficiency. One can easily verify from the two steady state conditions that the endogenous mark up must be decreasing in labor productivity A, in the discount factor β, and in the degree of substitutability θ, and increasing in the rate of exit of the firms δ_N and in the parameter of the entry cost η. The associated comparative statics for the number of firms is summarized as follows:

PROPOSITION 3.3. **The steady state endogenous market structure with competition in quantities is characterized by a number of firms increasing in the discount factor and in total factor productivity, and decreasing in the exit rate, in the entry cost parameter and in the degree of substitutability.**

The intuitions for the impact of the five structural parameters determining the steady state EMSs are the following. Higher productivity leads to more business creation, which increases the steady state number of firms and enhances competition while reducing the mark ups. The second determinant of the steady state EMSs is the size of the costs of entry: when these are

[26] We will find a similar result in Chapter 5, in a dynamic general equilibrium model of growth with EMSs in the competition for the market.

high, profitability is low and the long run equilibrium is characterized by high concentration and high mark ups. To the extent that the costs of entry are artificial, in the sense that there are barriers to entry due to product market regulations, reforms leading to deregulation reduce concentration and mark ups in the long run. The third factor is the way people discount future utility: when agents are more patient, the interest rate is lower and the discounted sum of future profits is higher, which attracts more entry, strengthens competition and ultimately reduces the mark ups; therefore more patient agents lead to a higher number of firms in steady state. The fourth element is the rate of business destruction due to exogenous reasons: when the risk of bankruptcy is high, there are only few firms in the long run (but with a high rate of turnover), and they apply a high mark up to their goods. The last determinant of the long run equilibrium emphasized in our framework is the degree of substitutability between goods: higher substitutability (homogeneity) between goods induces stronger competition and lower mark ups, with a negative impact on the number of firms.

Finally, notice that in case of high levels of substitutability between goods (high θ) we obtain a simpler situation. As we have seen, long run utility is always decreasing in the number of firms, because business creation employs savings without generating substantial gains from variety. However, larger consumption levels require the creation of many firms, which leads to inefficient entry levels. For instance, in the case of homogenous goods ($\theta \to \infty$) we have $\tilde{C} = \eta[1/\beta(1-\delta_N) - 1]\tilde{N}(\tilde{N}-1)$. In such a case the inefficient entry process generates an equilibrium with an excessive number of firms:

$$\tilde{N}^Q = \frac{\eta(1-\beta) + \Xi(\beta, \eta, A, \delta_N)}{2\eta\left[1-\beta(1-\delta_N)\right]} \tag{3.34}$$

where $\Xi \equiv \sqrt{\eta^2(1-\beta)^2 + 4\beta\eta(1-\delta_N)\left[1-\beta(1-\delta_N)\right]A}$. This number of firms is associated with the equilibrium mark up:

$$\tilde{\mu}^Q = \frac{\Xi(\beta, \eta, A, \delta_N) + \eta(1-\beta)}{\Xi(\beta, \eta, A, \delta_N) - \eta(1-\beta+2\beta\delta_N)} \tag{3.35}$$

As in the static model of EMSs of Chapter 2, the number of firms is increasing less than proportionally with the size parameter A/η, and the mark up is correspondingly decreasing, while the production of each firm $x = C/N$ is increasing in the size of the economy.

Summing up the welfare properties of the model with competition in quantities we have:

PROPOSITION 3.4. **The steady state endogenous market structure with competition in quantities is characterized by dynamic inefficiency if the discount factor is high enough or if the degree of substitutability between goods is high enough.**

Reintroducing the endogenous labor supply, one can study the impact of another structural parameter on the steady state EMSs, the elasticity of labor supply φ. When this is larger, agents tend to work more, which tends to allow entry of a larger number of firms and to strengthen competition. At least for low values of φ, its increase is associated with an increase in the steady state consumption level, and with a reduction of the mark ups.

3.3.2 Dynamics and steady state under Bertrand competition

Under competition in prices the deterministic equilibrium system is given by (3.30) and:

$$C_{t+1} = \frac{\beta(1-\delta_N)\left[1+\theta(N_t-1)\right]C_t}{\eta(\theta-1)(N_t-1)N_t^{1/(\theta-1)}} \cdot \\ \cdot \left(\frac{\eta(\theta-1)(N_{t+1}-1)N_{t+1}^{1/(\theta-1)}}{1+\theta(N_{t+1}-1)} + \frac{C_{t+1}}{1+\theta(N_{t+1}-1)} \right)$$

which, in steady state, becomes:

$$\tilde{C} = \frac{(\theta-1)\eta[1-\beta(1-\delta_N)]\tilde{N}^{\frac{1}{\theta-1}}(\tilde{N}-1)}{\beta(1-\delta_N)} \tag{3.36}$$

The two steady state expressions (3.31) and (3.36) can be easily represented in a phase diagram (N_t, C_t), as we do in Figure 3.9. The first one is the same hump shaped relation that we used before, and the considerations made earlier apply here as well. The second expression is again a positive and convex relation due to the role of firms in producing consumption goods. Solving for the steady state EMSs we obtain the number of firms:

$$\tilde{N}^P = \frac{\beta(1-\delta_N)A + \eta\,(\theta-1)\left[1-\beta(1-\delta_N)\right]}{\eta\left\{(\theta-1)\left[1-\beta(1-\delta_N)\right]+\beta\delta_N\right\}} \tag{3.37}$$

which is associated with the endogenous mark up:

$$\tilde{\mu}^P = \frac{\left(\frac{\theta}{\theta-1}\right)\beta(1-\delta_N)A + \eta\,(1-\beta)}{\beta(1-\delta_N)A - \eta\beta\delta_N} \tag{3.38}$$

that exhibits the same comparative statics properties as under Cournot competition. These results can be summarized as follows:

PROPOSITION 3.5. **The steady state endogenous market structure with competition in prices is characterized by a number of firms increasing in the discount factor and in total factor productivity, and decreasing in the exit rate, in the entry cost parameter and in the degree of substitutability.**

The intuitions for the impact of the structural parameters is analogous to the case of competition in quantities. As in the static model of EMSs

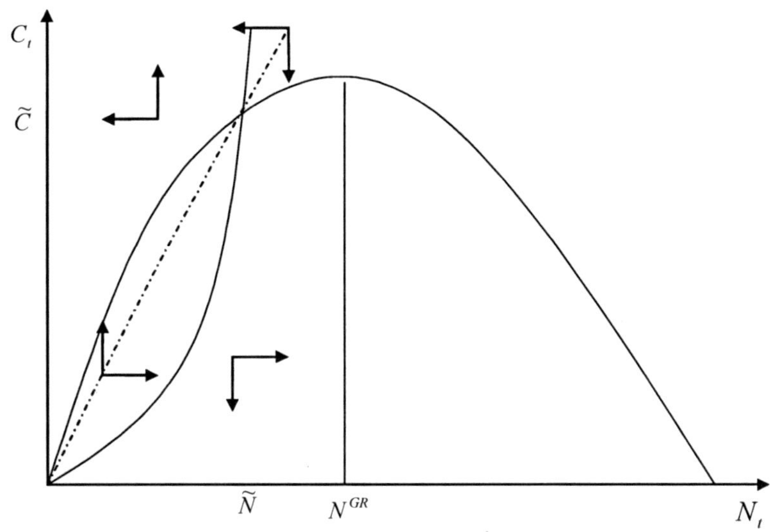

Fig. 3.9. Phase Diagram in (N_t, C_t): competition in prices.

of Chapter 2 with Bertrand competition, the number of firms is linearly increasing in the size parameter A/η, while the mark up is decreasing in it.

More importantly, it can be verified that the steady state number of firms is always increasing in β. When the discount factor is small $\tilde{N}^P \leq \tilde{N}^{GR}$, and when $\beta \to 1$ the steady state number of firms converges to $\tilde{N}^{GR} + 1 - 1/\theta$. Therefore, taking the integer constraint in consideration, it turns out that the steady state number of firms can be larger than the golden rule number by at most one firm, and only in case of high substitutability. Summing up:

PROPOSITION 3.6. **The steady state endogenous market structure with competition in prices is characterized by dynamic efficiency (or at most one firm more than the golden rule).**

Using the results of Chapter 2, one can also conclude that, given the same structural parameters, the steady state EMSs under competition in prices are characterized by smaller mark ups and less firms: $\tilde{\mu}^Q > \tilde{\mu}^P$ and $\tilde{N}^Q > \tilde{N}^P$. Even if, *ceteris paribus*, Cournot competition generates higher mark ups, endogenous entry attracts more firms and strengthens competition. This result allows us to compare the two forms of competition from a welfare point of view. Consider the case in which competition in quantities generates dynamic efficiency, which requires imperfect substitutability and low discounting. In such a case, it is immediate to derive that steady state utility is higher under competition in quantities: this generates more firms and a larger consumption index. Notice that this does not mean that competition in quantities creates always higher welfare because it may require excessive savings along the transitional path. Moreover, when the equilibrium with competition in quantities is dynamically inefficient, unambiguous comparisons are not possible. A full fledged welfare analysis is presented in Section 3.6.

In conclusion, notice that the steady state structure of the markets determines the steady state consumption index as well. The impact of changes of the structural parameters on the long run value of the consumption index under competition in prices is the following: larger labor productivity A, larger substitutability between the goods θ and a higher discount factor β lead to a larger consumption index,[27] while higher costs of entry η and a higher exit rate δ_N lead to a lower one in the long run.

3.4 Impulse Response Functions and Second Moments

This section has multiple purposes. First of all, we wish to evaluate the relative success of the models considered above at replicating the empirical facts described in Section 3.1, namely countercyclical markups together with procyclical profits and procyclical firms' entry. Second, we want to identify the extent to which the EMSs influence the propagation of technology and government spending shocks throughout the economy. Finally, we try to compare the performance of our model and that of a standard RBC model.

Calibration of structural parameters is standard and follows King and Rebelo (2000). The time unit is meant to be a quarter. The discount factor, β, is set to 0.99, while the rate of business destruction, δ_N, equals 0.025 implying an annual rate of 10 %. The value of υ is such that steady state labor supply is constant and equal to one. The Frish elasticity of labor supply is φ, and we fix it at four as in King and Rebelo (2000). We set steady state productivity to $A = 1$. The baseline value for the entry cost is set to $\eta = 1$. Notice that the combination of A and η affects the endogenous level of market power because a low entry cost compared to the size of the market leads to a larger number of competitors and thus to lower markups, and *viceversa*. However, the impulse response functions below are not qualitatively affected by values of η within a reasonable range.[28] Finally, we set the share of government spending over aggregate output to 20 % as in many other studies of the business cycle.

In what follows we first study the impulse response functions to temporary supply shocks and demand shocks - the reaction to permanent shocks would emphasize a gradual convergence toward the new steady state without

---

[27] However, notice that more patient agents will consume less of each variety. Comparative statics is not immediate under competition in quantities because this could induce dynamic inefficiency.

[28] When the steady state number of firms increases, the sensitivity of the mark ups to entry diminishes but it applies to a larger number of goods. Therefore the fundamental moments exibit minor changes.

intertemporal substitution effects.[29] Finally we evaluate the second order moments. Our model allows for a large variety of combinations of substitutability between goods (θ) and mark up (μ), which in turn depends on the mode of competition, but we limit the discussion to a few key cases.

3.4.1 Supply shocks

The EMSs approach allows us to study different forms of supply shocks: not only the traditional productivity shocks which affect the marginal cost of production, but also shocks to the fixed cost of entry, which may reflect deregulation through policy or the introduction of general purpose innovations which reduce up-front investments. For the sake of comparison with the traditional literature, in this section we focus on a standard shock to the technology parameter following the first order autoregressive process:

$$\hat{A}_{t+1} = \rho_A \hat{A}_t + \varepsilon_{At}$$

where $\rho_A \in (0,1)$ is the autocorrelation coefficient and ε_{At} is a white noise disturbance, with zero expected value and standard deviation σ_A. Figures 3.10-12 depict percentage deviations from the steady state of key variables in response to a 1 % technology shock with persistency $\rho_A = 0.9$ in case of alternative market structures; time on the horizontal axis is in quarters.

In Figures 3.10 and 3.11 we report the impulse response functions for different values of θ under respectively competition in quantities and in prices. To evaluate the results, let us consider the standard case of low substitutability between goods with $\theta = 6$, which is in line with the typical calibration for monopolistic competition.[30] Under competition in prices and in quantities the market structure is generated endogenously and the steady state mark ups are respectively 22 % and 35 %, both belonging to the empirically reasonable range. As well known, when firms compete in prices the equilibrium mark ups are lower, which in turn allows for a lower number of firms to be active in the market: this implies that the model is characterized by a lower number of goods compared to the model with competition in quantities. Since this requires a smaller number of new firms to be created in steady state, lower mark ups are associated with a lower saving rate as well. In spite of these substantial differences in the steady state features of the economy, Figures 3.10

[29] See Colciago and Etro (2008). The model has been solved using DYNARE a software devoloped by Michele Juillard. In the case of permanent shocks the impulse response functions are actually deterministic simulations. For a given path of the exogenous variable DYNARE provides the response of the whole system assuming that variables are initially at the steady state. This is done by stucking all the equations of the system for all periods and solving the resulting system using the Newton-Raphson algorithm

[30] The qualitative behavior of the impulse response functions is similar in case of $\theta = 3$, which delivers larger steady state mark ups (63% under competition in quantities and 40 % under competition in prices).

and 3.11 show that the quantitative reactions of the main aggregate variables to the shock are similar in these two models with low substitutability.

Under both frameworks, the temporary shock increases individual output and profits on impact, which creates large profit opportunities. This attracts entry of new firms, which in turn strengthens competition and reduces the equilibrium mark ups. Therefore, our model manages to generate individual and aggregate profits that are procyclical despite mark ups being countercyclical, in line with the empirical evidence on business cycles. Notice that the negative effect of competition on the mark ups is stronger under competition in quantities, where entry erodes profits margins faster than under competition in prices.[31] The number of firms remains above its steady state level along all the transition path. The stock market value (both at the firm and aggregate level) has a hump shaped reaction, and its temporary increase contributes to raise the real interest rate and induces the agents to temporarily increase consumption: as one can see, the reactions of consumption and of the stock market value are strictly related. While the shock vanishes and entry strengthens competition, output and profits of the firms drop and the incentives to enter disappear. At some point net exit from the market occurs and the mark ups start increasing toward the initial level.

The impact of these reactions on the real variables resembles that of a basic RBC model, even if it derives from largely different mechanisms. Aggregate output jumps up and gradually reverts to the steady state level, being initially fueled by the reduction in the mark ups associated with entry and by the increase in labor supply associated with higher wages. Part of the increase in income (from higher wages and profits) is saved for consumption smoothing purposes, which feeds the process of business creation. However, contrary to standard models, here the impact of the shock on consumption is strengthened by a new competition effect. Entry of new firms strengthens competition and temporarily reduces the mark ups, which in turn boosts consumption through an intertemporal substitution effect. In other words, the productivity shock reduces not only the marginal cost (as already happens in the RBC model), but also the equilibrium mark up (which is zero in the RBC model and constant in the models with monopolistic behavior), therefore the intertemporal substitution toward current consumption is stronger when the market structure is endogenous. Finally, it is important to notice that there is a second intertemporal substitution effect associated with the increase in competition: the temporary reduction of the mark ups is associated with a temporary increase in the real wages, which raises the labor supply and drives the strong impact on output. In conclusion, the impact of a temporary shock on consumption is magnified in the presence of EMSs.[32]

---

[31] The mark up elasticity to the number of firms is larger under competition in quantities.

[32] The intertemporal substitution effect is limited by the logarithmic preferences in consumption, which imply a unitary elasticity of intertemporal substitution.

This analysis makes clear that EMSs with low substitutability between products can provide reasonable qualitative responses to technology shocks and can also reproduce the dynamic behavior of profits and mark ups which is substantially ignored within the standard RBC framework. As noticed earlier, our general model should be interpreted as a model of a representative sector with a potentially high degree of substitutability between goods. When we increase the degree of substitutability (θ) the same qualitative results hold, at least under competition in quantities, and the impact of the shock on competition and mark ups is stronger. Consider the extreme case of homogenous goods ($\theta \to \infty$), that corresponds to the typical assumption of the RBC literature: in such a case, our model with competition in quantities is compatible with positive mark ups and, as we can see in Figure 3.10, it is able to reproduce a similar propagation mechanism to the one we have just seen, with a larger reduction in mark ups and a smaller impact on the stock market value. On the contrary, under competition in prices and homogenous (or highly substitutable) goods, the model collapses to one where mark ups vanish and entry does not take place because of the positive fixed costs of production (therefore we did not display this case in Figure 3.11). For this reason, and contrary to a long standing literature, we consider the model with competition in quantities as a better tool for macroeconomic analysis of the business cycles in the presence of realistic EMSs.

The above comparison between two models featuring the same structural parameters but different modes of competition can be interesting in its own, but its interpretation is limited by the fact that in different markets different forms of competition take place, and most of the times we are not even able to screen between them. An alternative comparison which can be useful to understand the implications of EMSs emerges when models with equal steady state mark ups are studied. In such a case all the aggregate ratios are the same as well, and different responses to a shock reveal fundamental differences of alternative modes of competition. To study a comparison of this second type, let us consider the model with competition in prices and $\theta = 6$ (Figure 3.11). This model is characterized by a steady state mark up of 22 %. Under our parametrization, the same mark up emerges endogenously in a model of competition in quantities when goods are homogenous, that is when $\theta \to \infty$ (Figure 3.10).

A comparison between the impulse response functions of these two cases with a mark up of 22% (and therefore with equal steady state values) shows that the effect of competition on the mark up is stronger in the case where firms compete in quantities and goods are homogenous. As a consequence, strategic interaction between firms selling homogeneous goods brings about a substantial competition effect on consumption and it may contribute to solve the low variability of consumption puzzle identified in standard RBC models.

With an isoelastic utility function, the competition effect would be stronger when the elasticity of substitution is larger than unity.

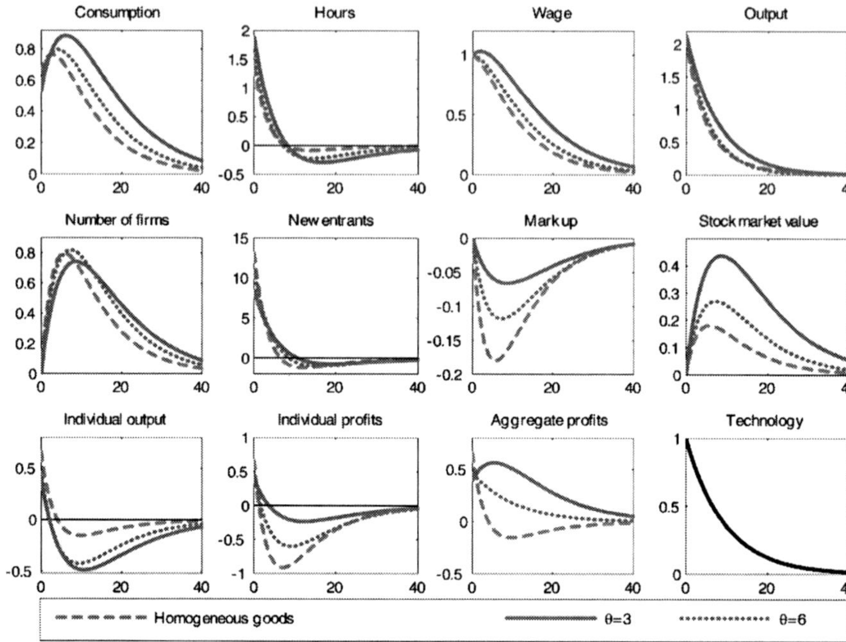

Fig. 3.10. Impulse response to a persistent technology shock. Competition in Quantities. Percentage deviations/quarters.

Finally, with the purpose of illustrating the potentiality of our approach to study the relation between market structures and the business cycle, in Figure 3.12 we present the impulse response functions for different models of competition in quantities with homogenous goods: the symmetric Cournot case (already present in Figure 3.10), a model with conjectural variations with $\lambda = 0.15$, that leads to imperfect collusion, and a model of Stackelberg competition. The endogenous mark ups are respectively 22 % for Cournot case, 35% for the conjectural variations and 15 % for the Stackelberg case - in which output, profits and stock market value of the leader are larger than those of the followers. The impulse response functions follow similar paths from a qualitative point of view, but there are substantial quantitative differences. For instance, compared to the Cournot case, consumption smoothing is more relevant when the markets are characterized by imperfect collusion and higher mark ups, and less relevant when they are characterized by a market leader whose overproduction reduces the mark ups.

The above results are largely in line with the VAR evidence presented before. A deeper evaluation of these models requires second moments analysis, which will be presented later. Before that, we need to evaluate qualitatively

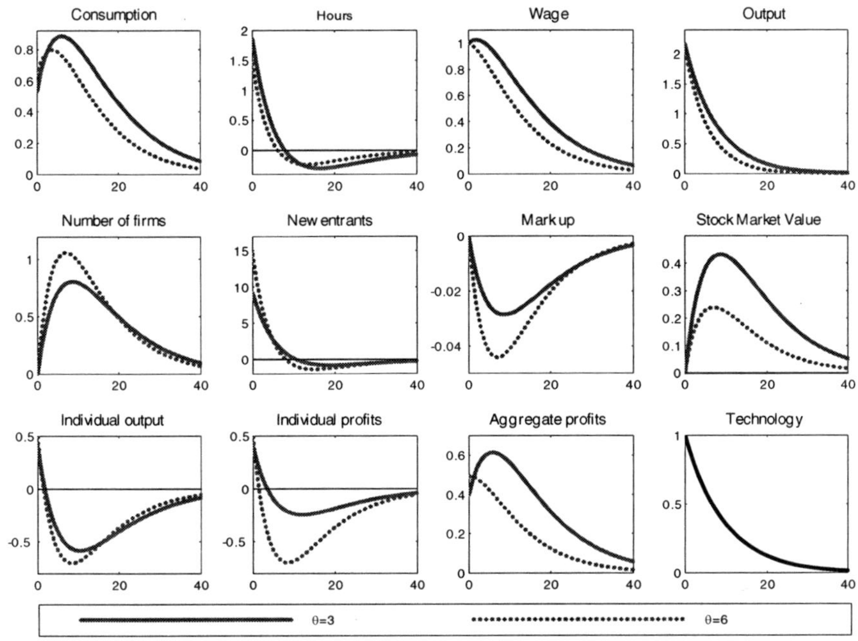

Fig. 3.11. Impulse response to a persistent technology shock. Competition in Prices. Percentage deviations/quarters.

the response of the models to a different kind of shock: an aggregate demand shock.

3.4.2 Demand shocks

We now consider the impact of a demand shock associated, as standard in the theory of business cycles, with an increase in government spending. We assume that government spending follows the first order autoregressive process:

$$\hat{G}_{t+1} = \rho_G \hat{G}_t + \varepsilon_{Gt}$$

where $\rho_G \in (0,1)$ is the autocorrelation coefficient and ε_{Gt} is a white noise disturbance, with zero expected value and standard deviation σ_G.

Figure 3.13 depicts the response of key variables to a one percent government spending shock with persistency $\rho_G = 0.9$. We report the case of competition in quantities under alternative parameterization for the elasticity of substitution between goods. Solid lines depict the case where $\theta = 3$ which delivers a steady state mark up equal to 63 %, dashed lines represent the case case with $\theta = 6$ and a mark up of 35 %, and finally dotted lines are relative to the case with homogeneous goods ($\theta \to \infty$) and a 22 % mark up.

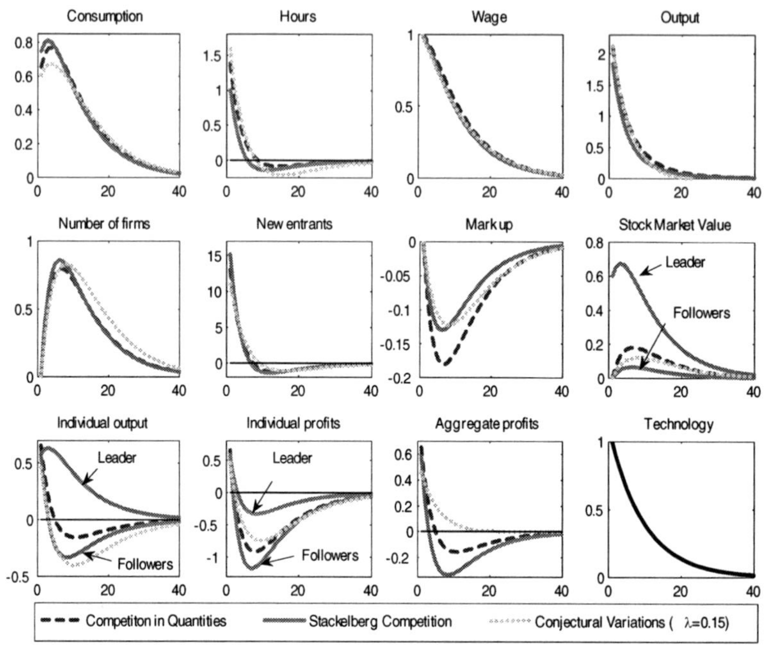

Fig. 3.12. Impulse response to a persistent technology shock. Homogenous goods. Percentage deviations/quarters.

As in the standard neoclassical model (Barro, 1981; Barro and King, 1984), the temporary shock to government spending creates a boom because the initial reduction in private consumption is more than compensated by the increase in public spending. As a consequence, labor demand for production increases. Consumers feel poorer, reduce their demand of leisure and increase their labor supply. In the RBC framework the net effect would be given by a reduction of the wage rate and by a reduction of consumption, with both remaining below the steady state levels along the entire transition path (meanwhile, the interest rate would jump up and gradually decrease toward its initial level). In our model, however, there are new mechanisms that substantially change the impact of the shock.

First of all, the demand shock increases individual output and profits on impact, and the stock market value in the subsequent quarters. This attracts entry of new firms, with two consequences. The first one is that the demand of labor for the creation of new firms goes up, which leads to a stronger increase in total labor demand and ultimately to an increase in the wage rate (the opposite compared to the RBC framework), which promotes consumption. The second (and possibly more important) consequence is that entry

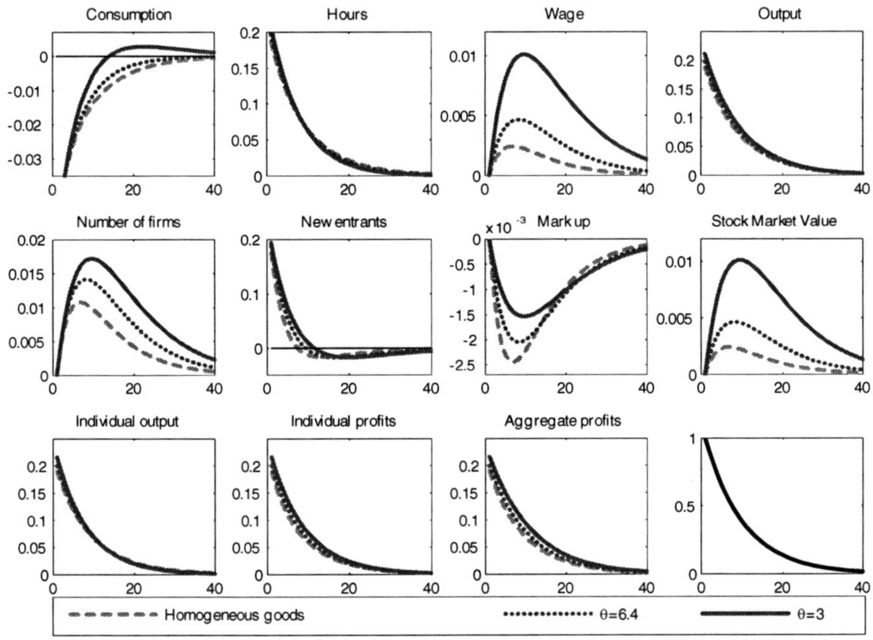

Fig. 3.13. Impulse response to a persistent government spending shock. Homogenous goods. Percentage deviations/quarters.

strengthens competition and endogenously reduces the equilibrium mark ups, which again makes current consumption more attractive for the consumers.[33]

The impact of these two mechanisms is to counterbalance the initial drop in consumption. When substitutability between goods is low, consumption goes above the steady state level after a few quarters and gradually returns toward its long run level from above, which is in sharp contrast with the dynamic response delivered by the standard RBC model (where the convergence of consumption to the steady state is monotonic), but is not too far from the available evidence on the reaction of private consumption to public spending shocks.[34] Overall, these dynamic paths are radically different from the standard RBC models and they are potentially more in line with the mixed empirical evidence on the impact of demand shocks - in particular with the procyclicality of profits and countercyclicality of markups.[35]

[33] Therefore the model is consistent with the suggested requirement of Rotemberg and Woodford (1992) of a real wage increasing after a positive demand shock. Nevertheless the reaction of the real wage is limited (a factor that could help to explain the substantial acyclicality of wages in the presence of multiple shocks).

[34] See also Galì *et al.* (2007,b) for a recent reference.

[35] See Monacelli and Perotti (2008).

Colciago and Etro (2008) have analyzed a preference shock, a different form of demand shock associated with a disturbance Θ_t of the subutility from consumption $\log(C_t - \Theta_t)$. A positive shock to Θ_t increases the marginal utility of consumption so as to increase consumption and to attract entry, to increase labor demand and real wages and to reduce the mark ups. As emphasized by Baxter and King (1992) and Wen (2006), taste shocks in standard neoclassical models generate countercyclical investment dynamics due to the crowding out effect of consumption on investment. This is not the case in our framework, as long as we allow for elastic labor supply. In such a case the model replicates the positive comovement between output, consumption and investment which characterizes a typical business cycle; moreover, aggregate profits remain procyclical and the mark up remains countercyclical.[36]

3.4.3 Second moments

To further assess the implications of EMSs for the business cycle, we compute second moments of the key macroeconomic variables. In this exercise we follow the RBC literature and assume that the only source of random fluctuations are technology shocks. We calibrate the productivity process as in King and Rebelo (2000), with persistence $\rho_A = 0.979$ and standard deviation $\sigma_A = 0.0072$. We reported in Table 1.1 the performance of the standard RBC model with respect to the statistics on U.S. data (1948-2008) for output Y, consumption C, investment I, labor force L, aggregate profits Π and the mark up μ.[37] As noticed, the main problems of the RBC model are the limited variability of output and especially consumption and labor force, and the lack of explanations for the cyclical movement of profits and mark ups. As we will see, our model allows us to improve the performance of the standard RBC model in all these dimensions.

Table 3.1 reports second moments of Y, C, $I \equiv N^e V$, L, Π, and μ for our model with competition in quantities and with competition in prices under the common parameterization with low substitutability between goods

[36] The intuition for this result goes as follows. When labor supply is rigid, an increase in investment in the aftermath of the shock would require a shift of the fixed quantity of labor from the production of the final good to the business creation. In this case, however, the household would not be able to satisfy the desire to consume. Thus, the increase in demand can be satisfied uniquely through disinvestment. This generates a perverse effect which by reducing the number of new entrants and the overall number of firms ultimately leads to a higher mark up. On the contrary, when labor supply is endogenous and enough elastic, agents have an additional channel through which they can react to the shock. By increasing total hours worked they can set up new firms without decreasing the production of the final good. The increase in hours allows a contemporaneous positive impact variation of investment, consumption and entry (and generates countercyclical mark ups once again).

[37] Profits include both the remuneration of capital and the extra-profits due to market power: while we could not distinguish between the two, future research may try to do it.

Variable	$\sigma(X)$	$\sigma(X)/\sigma(Y)$	$E(X_t, X_{t-1})$	$Corr(X,Y)$
Y	1.52, 1.51	1	0.68, 0.68	1
C	0.78, 0.78	0.51, 0.52	0.77, 0.76	0.94, 0.95
I	5.89, 7.56	3.87, 5.00	0.65, 0.64	0.97, 0.97
L	0.85, 0.77	0.56, 0.50	0.65, 0.64	0.96, 0.96
Π	0.70, 0.74	0.46, 0.49	0.71, 0.72	0.99, 0.98
μ	0.15, 0.13	0.10, 0.08	0.95, 0.94	−0.17, −0.17

Table 3.1. Second moments under low substitability. Left: Competition in quantities; Right: Competition in prices

($\theta = 6$), corresponding to the impulse response functions of Figures 3.10 and 3.11.[38] Both the competitive frameworks provide a similar performance at reproducing some key features of the U.S. business cycle. Imperfect competition, strategic interaction and endogenous entry allow the EMSs framework to outperform the standard RBC framework in a number of aspects.

First of all, endogenous mark up fluctuations together with endogenous entry deliver a substantially higher output volatility with respect to the RBC model (1.51/1.52 against 1.39), almost matching the one emerging from U.S. data. As emphasized above, we can also capture procyclical profits and entry together with countercyclical mark ups as in the data. This is obtained through the direct effect that entry has on the degree of competition rather that by resorting to an *ad hoc* functional form specification for preferences as in Bilbiie, Ghironi and Melitz (2007, 2008,a,b). Our model provides a good match for the correlation of profits and mark ups with output, but it underestimates their variability, emphasizing the need for further work on the microfoundation of the EMSs to better explain the high volatility of both profits and mark ups. Moreover, mark up countercyclicality allows us to strengthen the propagation of the shock on aggregate demand through the competition effect. Both models display an absolute and relative (with respect to output) variability of consumption larger than that delivered by the RBC model. Since low variability of consumption is a well known shortcoming of the RBC theory, the competition effect associated with strategic interactions and endogenous entry appears to be a relevant channel to overcome it. Finally, notice that, compared to the RBC framework, our model with EMSs slightly improves the performance in terms of variability of the labor force.

Even if we do not report a sensitivity analysis here, the variability of output increases further (but that of consumption goes down) when lower degrees of substitutability between goods are taken in consideration, while it decreases (and the variability of consumption goes up) for higher degrees of substitutability, under both forms of competition.

---

[38] As in Bilbiie, Ghironi and Melitz (2007) we report moments of data consistent variables, i.e. deflated using the average price index rather than the consumption based price index.

Table 3.2 reports second moments for the model with competition in quantities in the case of homogeneous goods (in the left columns), corresponding to the case presented in Figure 3.10. The relevance of this extreme model relies on the fact that it assumes perfect substitutability between goods exactly as in the standard RBC framework. The figures confirm that high elasticity of substitution coupled with market concentration enhances consumption volatility with respect to the case where goods are imperfectly substitutable. Also the contemporaneous correlation of the markup with output (-0.29) matches closely that assumed by the labor share-based measure of the markup reported in Chapter 1 (-0.28). This is however obtained at the cost of volatility of aggregate output and labor supply which are lower that in the cases considered above - but still in line with the results from the standard RBC model. In conclusion, the model with homogenous goods and competition in quantities (Cournot competition) is able to perform quite well in matching the cyclical properties of profits and mark ups, on which the neoclassical model is completely silent, and it provides a better approximation of the variability of consumption in front of real shocks.

$Variables$	$\sigma(X)$	$\sigma(X)/\sigma(Y)$	$E(X_t, X_{t-1})$	$Corr(X,Y)$
Y	$1.36, 1.69$	1	$0.67, 0.69$	1
C	$0.87, 0.59$	$0.64, 0.34$	$0.78, 0.77$	$0.94, 0.90$
I	$5.86, 4.58$	$4.31, 2.71$	$0.63, 0.68$	$0.92, 0.99$
L	$0.57, 1.30$	$0.42, 0.77$	$0.60, 0.66$	$0.93, 0.96$
Π	$0.70, 1.57$	$0.51, 0.93$	$0.63, 0.61$	$0.99, 0.95$
μ	$0.10, 0.11$	$0.07, 0.06$	$0.93, 0.93$	$-0.29, -0.37$

Table 3.2. Second moments. Cournot competition with homogeneous goods: baseline model (left) and model with physical capital (right)

Of course, the performance of the model can be improved further by adopting the usual extensions that are often added in the RBC literature (habits, indivisible labor, adjustment costs of investment,...), but in the next section we will augment our baseline model with capital accumulation to provide a more complete comparison with the baseline RBC framework.

3.5 EMSs and Accumulation of Physical Capital

One of the main contributions of Bilbiie, Ghironi and Melitz (2007) has been the introduction of physical capital in a model of business creation without strategic interactions but with endogenous entry. However, their model with identical production functions for the good market and the business creation market exhibits cycling convergence to the steady state and looses stability for reasonable values of the depreciation rate (because of increasing returns in the accumulated factor). We follow a different approach and assume that capital is used only in the production of final goods. This allows us to obtain a stable equilibrium and to nest both the standard RBC model and our model with homogenous goods and Cournot competition.

Let us assume that final goods are produced with a standard Cobb-Douglas production function of the form:

$$x_{jt}(i) = A_t k_{ijt}^{\alpha} l_{ijt}^{1-\alpha} \tag{3.39}$$

where k_{ijt} is the stock of capital of firm i in sector j at time t and $0 < \alpha < 1$. New firms are created with the same technology as before.

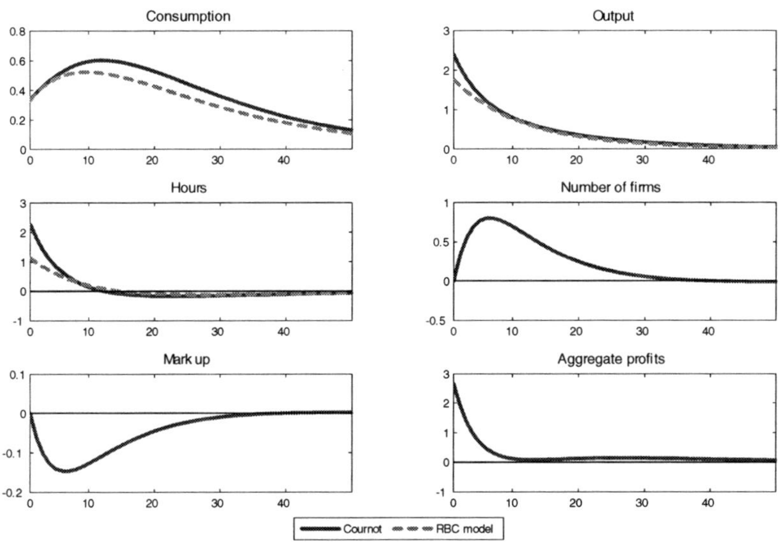

Fig. 3.14. Response to a temporary technology shock.

The representative household holds the aggregate stock of capital and rents it to the producers of the final good. In equilibrium each firm of each sector j employs the same stock of capital k_{jt}, and the number of firms is N_t in each sector. Therefore the aggregate stock of capital $K_t = \int_0^1 N_t k_{jt} dj$ evolves according to:

$$K_{t+1} = (1 - \delta_K) K_t + I_{Kt} \tag{3.40}$$

where I_{Kt} is time-t investment in physical capital and δ_K is its rate of depreciation. The household has a different intertemporal optimality condition with respect to the baseline model:

$$\frac{1}{C_t} = \beta(1 - \delta_K + r_{t+1}) E\left(\frac{1}{C_{t+1}}\right) \tag{3.41}$$

where r_t corresponds now to the rental rate of physical capital.

The symmetric Cournot equilibrium in the market for final homogenous goods leads always to the mark up $\mu_t = N_t/(N_t - 1)$. The demand of inputs in the good producing sector is obtained by cost minimization, which requires:

$$w_t = (1-\alpha) A_t \left(\frac{K_t}{L_t^c}\right)^{\alpha} \left(\frac{N_t - 1}{N_t}\right) \quad \text{and} \quad r_t = \alpha A_t \left(\frac{K_t}{L_t^c}\right)^{\alpha - 1} \left(\frac{N_t - 1}{N_t}\right)$$

where $L_t^c = C_t/A_t$ is labor employed in the production of final goods. These equations show that in this extended model both the wage and the interest rate are positively affected by an increase in the number of firms. The presence of capital accumulation implies that the new equation governing the dynamics of the number of firms reads as:

$$N_{t+1} = (1 - \delta_N)\left[N_t + \frac{A_t L_t}{\eta} - \frac{A_t}{\eta}\left(\frac{C_t + I_{Kt}}{A_t K_t^{\alpha}}\right)^{\frac{1}{1-\alpha}}\right] \tag{3.42}$$

where labor supply is given by (3.25), while the equation governing the dynamics of consumption remains unchanged, namely (3.29), with $\theta \to \infty$.

The system is stable and can be reduced to three equations that fully determine the dynamics of C_t, K_t, and N_t along a stable equilibrium path. The model nests our baseline framework for $\alpha \to 0$ and the standard RBC framework when $\eta \to 0$.

Figure 3.14 reports the impulse response functions to a temporary technology shock for the RBC model with perfect competition and our augmented model with Cournot competition - in both cases we assume homogenous goods and we calibrate the additional parameters as in King and Rebelo (2000), with $\alpha = 1/3$ and $\delta_K = 0.025$. The RBC model augmented with Cournot competition is clearly able to generate a larger impact of the shock on output, consumption and labor, together with the usual procyclical responses of entry and profits and the countercyclical response of mark ups.

Table 3.2 shows the second moments of the augmented model with Cournot competition, homogenous goods and physical capital (in the right columns). The performance of the model improves substantially for what concerns the second moments of both output and hours worked (output volatility goes up to 1.69, hours volatility to 1.30). However, since households can smooth consumption through the additional channel of investment in physical capital (beside investment in business creation), the relative variability of consumption decreases with respect to the model without capital accumulation. Both specifications deliver a contemporaneous negative correlation between output and markups similar to that found in the data. Profits' volatility increases to 1.57, but the low variability of both mark-ups and profits compared to the data remains an issue to be addressed in future research.[39]

3.6 Fiscal Policy

Our DSGE model can be used for the analysis of fiscal policy. First of all, notice that the assumption of a perfectly competitive credit market implies that all the conditions for Ricardian equivalence hold (Barro, 1974) even if there are imperfections in the goods market. Therefore, debt policy is irrelevant as long as lump sum taxes are used, as we will assume for now.

Even if lump sum taxes are used to finance public spending, fiscal policy has an important role in correcting the inefficiencies associated with the EMSs. In particular, the optimal fiscal policy has to neutralize the inefficiencies of the decentralized equilibrium analyzed above. As we have seen, this can be characterized by dynamic inefficiency associated with the entry process, which leads to the creation of too many firms in steady state. This problem can be addressed with a sale subsidy (redistributed with lump sum taxes) or a system of entry fees and profit taxes (financed with lump sum taxes again). However, the optimal fiscal policy should intervene also along the transition path to restore the efficient process of business creation and labor allocation outside of the steady state. This efficient process has been characterized by Bilbiie, Ghironi and Melitz (2008,a) in a framework that is general enough to apply to our model with strategic interactions. They notice that the equilibrium consumption index can be written as $C_t + G_t = y_t N_t p_t / P_t = A L_t^c N_t^{\frac{1}{\theta-1}}$, and the number of entrants must be $N_t^e = A_t \left(L_t - L_t^c \right) / \eta$, which, using the equation of motion for the number of firms, implies the following amount of labor used in the production of final goods:

$$L_t^c = L_t - \frac{\eta N_{t+1}}{(1-\delta_N)A_t} + \frac{\eta N_t}{A_t}$$

Consequently, introducing a logarithmic subutility in public consumption, the social planner problem can be expressed as follows:

$$\max_{\{N_{t+1}, L_t, G_t\}_0^\infty} E_t \sum_{t=0}^{\infty} \beta^t \left(\log C_t - \frac{\upsilon L_t^{1+1/\varphi}}{1+1/\varphi} + \alpha \log G_t \right)$$

$$s.v. \;\; C_t = A_t N_t^{\frac{1}{\theta-1}} \left[L_t - \frac{\eta N_{t+1}}{(1-\delta_N)A_t} + \frac{\eta N_t}{A_t} \right] - G_t \qquad (3.43)$$

39 Imperfect substitutability improves even more the performance of the Cournot model with physical capital. When $\theta = 6$, the variability of output and labor reach 1.74 and 1.54, and profits variability arrives to 1.70 without significant losses on consumption variability.

First of all, notice that the optimal choice of public spending follows a standard Samuelson rule which implies:

$$G_t: \quad G_t = \frac{\alpha A_t N_t^{\frac{1}{\theta-1}}}{1+\alpha}\left[L_t - \frac{\eta N_{t+1}}{(1-\delta_N)A_t} + \frac{\eta N_t}{A_t}\right] \tag{3.44}$$

where government spending is increasing in the current number of firms to remain a constant fraction of output.

The choice of the number of firms in each period must trade off the benefits in terms of increased variety with the costs in terms of resources spent to create new firms, while the choice of labor supply must trade off the benefits in terms of production possibilities with the costs in terms of disutility of labor. Given the separable utility, the two problems can be analyzed independently.

Consider first the case of exogenous labor supply. The optimality condition for the choice of N_{t+1} is:

$$N_{t+1}: \quad \frac{\eta N_t^{\frac{1}{\theta-1}}}{1-\delta_N} C_t^{-1} = \beta E\left[C_{t+1}^{-1}\left(\eta N_{t+1}^{\frac{1}{\theta-1}} + \frac{C_{t+1}+G_{t+1}}{N_{t+1}(\theta-1)}\right)\right] \tag{3.45}$$

whose left hand side can be interpreted as the social marginal cost of business creation due to the resources spent to increase the future number of firms, and whose right hand side can be interpreted as the social marginal benefit of business creation due to the future increase in the number of varieties and in total consumption. The optimality condition can be rewritten as:

$$\eta N_t^{\frac{1}{\theta-1}} = \beta(1-\delta_N)E\left[\left(\frac{C_{t+1}}{C_t}\right)^{-1}\left(\eta N_{t+1}^{\frac{1}{\theta-1}} + \frac{C_{t+1}+G_{t+1}}{N_{t+1}(\theta-1)}\right)\right] \tag{3.46}$$

which can be directly compared with its decentralized equilibrium counterpart (3.29). Assume that the initial number of firms is below the steady state level, so that it grows over time. This implies that the social marginal cost of business creation is higher than the decentralized marginal cost - the left hand side of (3.29), where $\mu(\theta, N_{t+1})/\mu(\theta, N_t) < 1$. Moreover, the social marginal benefit of business creation is smaller than the decentralized one - because $1/(\theta-1) < \mu(\theta, N_{t+1}) - 1$. Therefore, we can conclude that the optimal number of firms is always smaller than the equilibrium number. More precisely, we have:

PROPOSITION 3.7. **Starting from a number of firms below the steady state level, the equilibrium endogenous market structure implies too many firms along the transition path under both competition in quantities and prices.**

While this property holds outside of the steady state, it also holds in a stationary environment with constant total factor productivity, and allows us to derive the long run optimal EMS. In particular, the optimal number of firms in steady state is:

$$\tilde{N}^{MGR} = \frac{(1-\delta_N)A\beta}{\eta\left\{(\theta-1)\left[1-\beta(1-\delta_N)\right]+\beta\delta_N\right\}} \tag{3.47}$$

whose label reminds us that this is a modified golden rule. When the discount factor β approaches unity, $\tilde{N}^{MGR}$ converges to $\tilde{N}^{GR}$, otherwise it is always lower than that (and of the decentralized steady state as well).

The optimality can be restored with different instruments, as production subsidies or entry fees. Following Bilbiie, Ghironi and Melitz (2008,a) let us introduce an *ad valorem* sale subsidy at rate s_t^s financed with lump sum taxes on the firms. This changes the nominal profits into $\Pi_t = [(1+s_t^s)p_t - W_t/A_t]\,x_t - T_t$ where the lump sum taxes are $T_t = s_t^s x_t$. The subsidy induces higher production under both competition in quantities and prices, and the equilibrium mark up on the marginal cost becomes $\mu(\theta, N_t)/(1+s_t^s)$. This turns net profits into $\Pi_t = [1-(1+s_t^s)/\mu(\theta, N_t)]\,p_t x_t$, and the equilibrium asset pricing condition in:

$$\frac{(1+s_t^s)\eta N_t^{\frac{1}{\theta-1}}\mu(\theta, N_{t+1})}{\mu(\theta, N_t)} = \beta(1-\delta_N)\cdot$$

$$\cdot E\left[\frac{C_t}{C_{t+1}}\left((1+s_{t+1}^s)\eta N_{t+1}^{\frac{1}{\theta-1}} + \left[\mu(\theta, N_{t+1}) - 1 - s_{t+1}^s\right]\frac{C_{t+1}+G_{t+1}}{N_{t+1}}\right)\right]$$

Equating both sides of this equation to the corresponding parts of (3.46) one can find the optimal subsidy depending on the number of firms. In particular, we have:

PROPOSITION 3.8. **The optimal fiscal policy requires a countercyclical production subsidy at the rate $s_t^s = \frac{1}{N_t - 1}$ under competition in quantities, or $s_t^s = \frac{1}{\theta(N_t-1)}$ under competition in prices.**

Notice that these optimal instruments lead to a steady state characterized by the modified golden rule number of firms derived above. Moreover, the fiscal instruments are countercyclical, in the sense that in a boom the number of firms increases and the size of the optimal subsidies diminishes. Finally, the same result could be obtained with other tools constraining entry as a system of entry fees and profit taxes.

Consider the case of endogenous labor supply now. The optimality condition for the choice of N_{t+1} is the same as before, while the social planner first order condition for L_t is:

$$L_t: \quad L_t = \left(\frac{A_t N_t^{\frac{1}{\theta-1}}}{\upsilon C_t}\right)^{\varphi} \tag{3.48}$$

which can be compared with (3.25) to verify that labor is undersupplied in the decentralized equilibrium. Efficiency can be restored through a wage subsidy which turns the nominal wage into $(1 + s_t^l)W_t$, and which is financed with lump sum taxes on the representative agent. It is immediate to derive:

PROPOSITION 3.9. **The optimal fiscal policy requires a countercyclical wage subsidy at the rate $s_t^l = (N_t + \theta - 1)/(N_t - 1)(\theta - 1)$ under competition in quantities, or $s_t^l = N_t/(N_t - 1)(\theta - 1)$ under competition in prices.**

Lewis (2007,b) confirms these results in a related model with endogenous entry of firms active for a single period and with monopsonistic competition in the labor market. The presence of monopoly power in the supply of labor requires an even higher wage subsidy to restore optimality.

The result on the countercyclicality of fiscal policy is in line with a wide consensus in both the neoclassical and Keynesian approaches. However, here it derives from different reasons. In the neoclassical approach, a countercyclical fiscal policy is the result of the principle of tax smoothing (Barro, 1979) to minimize the tax distortions on the supply side. In the Keynesian approach it is the stabilizing role of government intervention on the demand side that leads to a countercyclical fiscal policy. In the EMSs approach the distortions are endogenous, they increase in recession, when competition is weaker, and they are reduced in boom, when competition is stronger. Therefore the optimal fiscal policy minimizes these market distortions with an expansive role in recessions and a (relatively) contractionary role in booms.

Until now, we have assumed the possibility of financing public spending through lump sum taxation. In this way we have abstracted from the important problem of choosing the optimal distortionary taxes needed to fund public spending and the optimal debt policy.[40] The solution of such a problem is inspired by the general principles of optimal taxation developed in Chapter 1. Of course, the necessity of a correction for the decentralized process of business creation avoids the optimality of a simple tax smoothing rule. In the presence of tax rates on sales τ_t^s and on labor income τ_t^l needed to finance public spending, the optimal fiscal policy would require tax rules $\tau_t^j = \tau^j(N_t)$ with $\partial\tau^j(N_t)/\partial N_t < 0$ for $j = s, l$. Therefore, not only debt policy should be countercyclical as in the neoclassical and Keynesian approaches, but also indirect taxes on sales and direct taxes on labor income should be countercyclical to stabilize the economy around the optimal steady state.[41]

40 As in the neoclassical model, the evolution of the debt follows (1.35) and fiscal policy must respect the IRC (1.36).

41 An early work on the role of distorsive taxation in a model with EMSs is by Wu and Zhang (2000).

3.7 Monetary Policy

One of the first contributions on the relation between inflation and the market structure is due to Wu and Zhang (2001), who introduced monetary policy in a model *à la* Devereux, Head and Lapham (1996) where entry drives profits to zero at all times and determines the mark ups according to an exogenous relation. Their simple model shows that inflation reduces business creation and increases the mark ups, which leads to larger welfare costs of inflation compared to a perfectly competitive environment.

A more recent literature has analyzed the role of monetary policy in New-Keynesian models with endogenous entry of forward looking firms under different theoretical approaches: Elkhoury and Mancini Griffoli (2007) adopted the money in the utility function approach[42] based on the same specification as in the standard RBC model, that is (1.64), Lewis (2007,b) adopted the cash-in-advance constraint approach, and Bilbiie, Ghironi and Melitz (2008,b) and Lewis (2007,a) abstracted from the particular source of money demand adopting the cashless approach of Woodford (2003).

Under all these approaches, the optimal monetary policy in the absence of price frictions tends to be characterized by the standard Friedman rule. However, price frictions introduce a new role for monetary policy and a slightly different characterization of the optimal monetary policy. Elkhoury and Mancini Griffoli (2007) have introduced sticky entry costs by assuming that there is a market for the creation of new firms (for instance through lawyers) whose prices (i.e. the legal fees of the lawyers) adjust in a sluggish way (as in Calvo, 1983), while the prices of the final goods are perfectly flexible. Under this assumption, a positive monetary shock reduces the real cost of entry and immediately attracts investments in business creation: this generates persistent and hump-shaped responses of consumption, investment, output and the number of firms. In the presence of strategic interactions, these patterns are associated with a strengthening of competition that reduces the mark ups and propagates the shock even further.

The Bilbiie-Ghironi-Melitz model. Bilbiie, Ghironi and Melitz (2008,b) have introduced price adjustment costs (as in Rotemberg, 1982,a,b) that are quadratic in the producer price inflation rate. In particular, assume for simplicity that monopolistic pricing takes place for all firms, and assume that the adjustment costs of firm i from a price $p_{t-1}(i)$ to $p_t(i)$ is a fraction $k\boldsymbol{\pi}_t(i)^2$ of the revenues, with $\boldsymbol{\pi}_t(i) \equiv [p_t(i)/p_{t-1}(i) - 1]$. Then, the forward looking problem of maximization of the discounted sum of profits leads firm i to adopt a time-dependent mark up affected by the size of nominal rigidities (k). Defining $\boldsymbol{\pi}_t$ as the inflation rate in producer prices, the symmetric equilibrium generates the mark up:

---

[42] The same approach is adopted by Etro (2007,c) and Bergin and Corsetti (2008).

$$\mu_t^M = \frac{\theta}{(\theta-1)(1-k\boldsymbol{\pi}_t^2)+2k\left\{\begin{array}{c}(1+\boldsymbol{\pi}_t)\boldsymbol{\pi}_t-\beta(1-\delta)\cdot\\ \cdot E_t\left[\frac{(1-k\boldsymbol{\pi}_t^2)}{(1-k\boldsymbol{\pi}_{t+1}^2)}\frac{N_t}{N_{t+1}}(1+\boldsymbol{\pi}_{t+1})\boldsymbol{\pi}_{t+1}\right]\end{array}\right\}} \tag{3.49}$$

which boils down to the traditional mark up for $k = 0$ or in the absence of inflation.[43] Equilibrium real profits are:

$$\pi_t^M = \left(1 - \frac{1}{\mu_t^M} - k\boldsymbol{\pi}_t^2\right)\frac{C_t}{N_t\left(1-k\boldsymbol{\pi}_t^2\right)} \tag{3.50}$$

where we used the fact that only a fraction $1-k\boldsymbol{\pi}_t^2$ of total demand represents private consumption C_t, the rest being spent in adjustment costs. The model is closed by the usual equation of motion for the number of firms and by the asset pricing equation under endogenous entry, where the real interest rate is now decomposed in $1 + r_t = (1 + i_{t-1}) / \left(1 + \boldsymbol{\pi}_t^C\right)$, with i_t nominal interest rate and $\boldsymbol{\pi}_t^C = P_t/P_{t-1} - 1$ that can be defined as the rate of consumer price inflation.

Here, inflation acts as a distortionary tax on firms profits and biases the allocation of resources between production of goods and of new firms against the latter. As we have seen in the previous section, it is always optimal to avoid mark up non-synchronization, which is possible here by implementing a policy of producer price stability (we will find a similar result in Chapter 5 when studying monetary policy in an endogenous growth model of creative destruction):

PROPOSITION 3.10. **In the presence of price stickiness, the optimal monetary policy requires zero producer price inflation.**

Notice that the optimal consumer price inflation satisfies $1 + \boldsymbol{\pi}_t^C = (N_t/N_{t-1})^{1/(\theta-1)}$, and moves freely to accommodate changes in the number of firms. Therefore, the Central Bank should target producer price inflation rather than the (mismeasured) consumer price inflation. Notice that Bilbiie, Ghironi and Melitz (2008,b) have obtained this result under the assumption of monopolistic behavior, which (as we have seen in Chapter 2) avoids inefficiencies in the endogenous number of firms (and generates efficient steady state EMSs). Introducing strategic interactions that lead to inefficient EMSs, the result holds again if fiscal policy can take care of these inefficiencies. When this is not the case, however, non-zero inflation becomes optimal, for instance

[43] Notice that in a stationary environment there is a negative relation between inflation and mark ups at least for zero inflation or low inflation. Contrary to this, in the growth model of Section 5.6 we will show the possibility of a positive relation between inflation and mark ups.

to reduce entry of firms when this is excessive in equilibrium (as under competition in quantities). A related result emerges in the model of growth with EMSs in the competition for the market of Etro (2007,c), where inflationary distortions affect firms' profits and their incentives to invest, with negative consequences on the growth rate. Nevertheless, those distortions can be useful in the absence of fiscal instruments to correct the inefficient endogenous structure of the market for innovations (see Chapter 5).

Log-linearizing (3.49) around the steady state with zero inflation, Bilbiie, Ghironi and Melitz (2008,b) obtain a new version of the New-Keynesian Phillips curve:

$$\boldsymbol{\pi}_t = \beta(1-\delta_N)\boldsymbol{\pi}^e_{t+1} + \frac{(\theta-1)}{2k}\left(\hat{w}_t - \hat{A}_t\right) - \frac{\hat{N}_t}{2k} \tag{3.51}$$

where a hat denotes percent deviations from steady state levels.

Current inflation depends on expected inflation and on two additional terms. The first one is standard and depends on the variations of the real marginal cost of production. Moreover, log-linearizing the endogenous entry condition $V_t = \eta w_t/A_t$ to obtain $\hat{V}_t = \hat{w}_t - \hat{A}_t$, it emerges that the equation ties inflation dynamics to the relative price of investment in new firms: for given expected inflation and number of firms, inflation is positively related to equity prices. The real novel term compared to the standard New-Keynesian Phillips curve with an exogenous number of firms - for instance in (1.76) - is the last one, which links the predetermined number of firms to inflation, introducing a new degree of endogenous inflation persistence.

The behavior of the equilibrium system augmented with an interest rate rule has been analyzed by Bilbiie, Ghironi and Melitz (2008,b), who confirm the validity of the Taylor principle under weak conditions, and simulate the impact of a monetary shock. An exogenous increase in the interest rate reduces output and inflation and increases the real interest rate. However, the latter effect has a positive impact on the expected return on equity (due to the no-arbitrage condition between the two forms of investment), which induces a counterempirical entry of firms; this is associated with the reduction in entry costs due to the reduction of wages. This phenomenon disappears (and a contractionary policy induces exit of firms) when the cost of entry is not in units of labor, but in units of consumption.[44] Lewis (2007,a) has augmented the model in a number of directions, showing that also sticky wages, costs of entry with congestion externalities and an endogenous exit rate can improve substantially its performance in matching the VAR evidence on the impact of monetary shocks. This class of models could be useful to study optimal monetary policy in the absence of optimal fiscal tools: during recessions monetary policy could stimulate business creation through interest rate reductions aimed at increasing the (stock market) value of firms and promoting

[44] See also Rotemberg (2008).

investments in the creation of new firms and products, and during booms it could limit excessive investments with a tight monetary policy.

Strategic complementarities and monetary policy. An important task for future research remains the introduction of strategic interactions to verify their effects of monetary policy. As suggested more generally by Ball and Romer (1990), real rigidities associated with these strategic effects may generate interesting interactions with the nominal rigidities.[45]

We can provide a simple insight on this point. Let us reconsider the static competition in prices between firms with the profit function (3.14), and let us assume that a fraction λ of the N_t firms cannot adjust the nominal price and maintains the pre-determined price level $\bar{p}_{t-1}$, while the fraction $1-\lambda$ can reoptimize. The Bertrand equilibrium is characterized by new nominal prices reset at a level:

$$p_t = \mu^P(\theta, N, \lambda, \bar{p}_{t-1}) \left(\frac{W_t}{A_t} \right)$$

where the equilibrium mark up is:

$$\mu_t^P = \frac{\theta N_t \Gamma_t - \theta + 1}{(\theta - 1)\left[N_t \Gamma_t - 1\right]} \quad \text{with } \Gamma_t = \lambda \left(\frac{p_t}{\bar{p}_{t-1}} \right)^{\theta-1} + 1 - \lambda$$

which shows that the presence of firms that do not adjust their prices leads also the optimizing firms to adjust less their own prices. This is a consequence of the strategic complementarity between price setters (at the microeconomic level), which is usually ignored in the New-Keynesian analysis, even in the literature focusing on strategic complementarities at the aggregate level (Cooper, 1999), which emphasizes mainly equilibrium multiplicity. In concentrated markets, when firms increase their prices other firms are induced to increase their prices as well, but when only some firms increase their prices, the other firms increase their prices *by less*. As long as the price level is increasing, in the sense that $p_t > \bar{p}_{t-1}$, the prices are reset at a new level which is decreasing in the fraction of firms that do not adjust.[46]

This suggests that small nominal rigidities can have larger real effects in the presence of strategic interactions. In a dynamic context, where firms adjust their nominal prices to maximize the discounted value of future profits, these forms of strategic interactions would affect the New Keynesian Phillips curve as well.

---

[45] See Romer (1996) for an overview.

[46] In a related model with small menu costs of price adjustment and quadratic costs of non adjustment, Ball and Romer (1991) endogenize the degree of nominal rigidity λ.

3.8 Labor and Credit Market Imperfections

A large part of the recent macroeconomic literature has been dedicated to the analysis of imperfections in the labor market leading to real wage rigidities and involuntary unemployment[47] and in the credit market, leading to problems of credit rationing and credit cycles.[48] For the sake of simplicity, most of this literature has focused on these distortions neglecting those in the goods markets, which kept being characterized by perfect competition and market clearing. For this reason, the application of the analysis of input markets imperfections to macroeconomic models with EMSs is an important task for future research.

Unemployment and the Blanchard-Giavazzi model. Labor market imperfections typically emerge because of bargaining between firms and unions leading to suboptimal employment and wages above the market clearing level, because of minimum wage laws binding on low productivity workers, or because of informational asymmetries between firms and workers leading to "efficiency wages" above the market clearing level.[49] All these imperfections generate involuntary unemployment. Moreover, they affect the EMSs in non-trivial ways and can generate interesting modifications of the aggregate behavior of the economy. In our basic model with perfect competition in the labor market, we obtained a counterfactual large volatility of the real wages, and the introduction of the above rigidities may help to improve the performance of the model in this sense. Moreover, one can study the interdependence between imperfections in the goods and labor market to investigate product and labor market reforms and the role of labor unions as non-atomistic wage setters.[50]

Blanchard and Giavazzi (2003) have introduced (efficient) wage bargaining *à la* McDonald and Solow (1981) in a simple model with monopolistic behavior of the firms and a reservation wage for the unemployed labor force which is decreasing in the unemployment rate.[51] They characterize the equilibrium with an exogenous number of firms, which is associated with the

[47] See Shapiro and Stiglitz (1984) or Mortensen and Pissarides (1994). See also the related literature on the macroeconomics of specificity (Caballero and Hammour, 1998).

[48] See Bernanke and Gertler (1989), Greenwald and Stiglitz (1993), Kaplan and Zingales (1997), Azariadis and Smith (1998), and Kiyotaki and Moore (1997).

[49] For different explanations of efficiency wages see Solow (1979), Weiss (1980), Malcomson (1981), Salop (1979), Shapiro and Stiglitz (1984) and Summers (1988).

[50] Contrary to what happens in a basic New-Keynesian model, Di Bartolomeo, Tirelli and Acocella (2008) have shown that, in the presence of predetermined nominal wages, positive inflation is optimal to discipline an exogenous number of strategic wage setters. It would be interesting to apply this analysis of strategic interactions in the labor market to a model with EMSs.

[51] This is what happens when the unemployment benefits are financed by the workers. Bokan and Hughes Hallett (2008) have extended the model with an endogenous unemployment benefit and distorsive taxation without altering most of the results.

short run, and the equilibrium with endogenous entry, which is interpreted as the steady state equilibrium. The Blanchard-Giavazzi model employs *ad hoc* preferences leading to mark ups decreasing in the number of firms, but it could be easily extended with genuine strategic interactions to obtain mark ups endogenously decreasing in the number of firms.[52]

One of the most important results is that the real wage is increasing in the bargaining power of the union in the short run, but it is independent from it in the long run. The intuition for the second result is interesting: in the long run, more bargaining power of the unions reduces the rents of the firms and, for a given entry cost, it reduces their number so as to relax competition and increase the mark ups and counterbalance the short run effect (of the higher bargaining power of the union) on real wages. This reduces the reservation wage, which has a positive impact on the unemployment rate.

There is another crucial implication: labor market deregulation reduces wages and increases the profit rate in the short run, but in the long run this leads to entry of firms, strengthens competition and reduces the mark ups, so as to lower the unemployment rate and increase the real wage. For these reasons, labor market deregulation can be opposed by myopic unions or by unions with a bias toward currently employed workers. The introduction of similar mechanisms and other distorsive rents in a dynamic model as the one of this chapter may provide important insights in the study of goods and labor market dynamics.

The dynamics of unemployment can be also endogenized if we introduce a mechanism of entry and exit in the tradition of the job matching models.[53] For instance in case of exogenous labor supply, unemployment must follow a process as $u_{t+1} = (1 - m)u_t + \delta_N(1 - u_t)$, where m is the rate of matching between vacancies (in the production sector and in the business creation sector) and unemployed workers (which should be increasing in labor demand and decreasing in the unemployment force), while δ_N is the fraction of firms that shut down and therefore of the workers that are fired in each period. In steady state, the unemployment rate is $\tilde{u} = \delta_N/(m + \delta_N)$, which must be increasing in the rate of business destruction and decreasing in all the factors that improve the matching technology. This creates a further channel through which structural parameters affect the labor force and, through that, the EMSs and the macroeconomy.

Banking inefficiencies and stock market shocks. Credit market imperfections typically derive from problems of asymmetric information or hold up between savers and investors leading to suboptimal credit for the households or the firms, but they may also derive from forms of limited rationality (as in the

[52] Strategic interactions between firms could be easily introduced in the version of the model where unions choose the wage first and the firms choose their prices afterward (this is the inefficient bargaining model known as the Right to Manage model).

[53] See Pissarides (1990) or Mortensen and Pissarides (1994).

empirically relevant case of the "rule of thumb" consumers, who tend to consume all their current income without optimizing their savings decisions). These imperfections crucially affect the process of business creation which is at the basis of the EMSs approach to macroeconomics. In our economy with labor as the single factor of production, we could examine the impact of exogenous frictions in the matching of savings by households and investments in business creation. As long as these frictions delay the entry process, the impact of a temporary shock can be fundamentally affected, and if they alter the cost of entry in the long run, the result could be steady state EMSs with a lower number of firms and therefore higher mark ups. Ghironi and Stebunovs (2007) have extended a related model to an imperfectly competitive banking sector.

When access to credit for business creation is positively correlated with the current profitability or with the stock market value because of endogenous constraints on loans (that require collaterals in terms of current profits or stocks), mechanisms of propagations of the shocks could be magnified. In a boom, larger profits induce a wider access to credit to create new firms, which strengthens competition, reduces the prices and boosts consumption, so as to accelerate the positive path of the economy. On the contrary, in a recession banks extend less credit to the business creation activity, which contributes to relax competition and to increase mark ups and prices, so as to limit consumption and magnify the contraction of the economy.[54]

While our baseline model does not incorporate an explicit credit channel, it can account for real effects of stock market shocks. For instance, we may introduce unexpected but temporary and persistent shocks to the aggregate stock market value and verify their impact on the economy. A stock market crash would reduce the incentives to create new business, which would reduce the number of firms and the degree of competition, raising mark ups and depressing labor income. The stock market shock would then affect the real economy with a negative impact on consumption and output, which would feedback into worse stock market evaluations. At least in part, the stock market shock becomes self-fulfilling. On the other side, positive random shocks to the firms evaluation can feed a boom of the real economy through business creation which ends up justifying part of the initial irrational exuberance. This may contribute to explain how stock market bubbles can affect the real economy in otherwise efficient financial markets.

Finally, one could introduce EMSs in the credit market as well.[55] In particular, in case of fixed costs of entry in the banking sector and strategic interactions leading to a positive spread between the interest rates for lending and borrowing, new mechanisms of propagation of the shocks would emerge.

[54] On the impact of credit constraints on investments in innovation see Galor and Zeira (1993) and Aghion and Howitt (2009, Ch. 14).

[55] For an interesting analysis of competition and moral hazard in the banking sector see Hellmann, Murdock and Stiglitz (2000).

A boom would be associated with lower spreads and a more efficient financial market leading to additional business creation, while a recession would be associated with larger spreads and endogenous limitations of the entry process. The impact on policy could be interesting. On one side, notice that the Ricardian Equivalence would not hold, because changes in savings would affect entry and therefore the endogenous spread between borrowing and lending rates: fiscal policy should take into account this impact. On the other side, monetary policy would have an additional channel, through its impact on the interest rate spread, to affect the real economy.[56]

3.9 Conclusions

In this chapter we have studied a dynamic stochastic general equilibrium model with flexible prices where the structure of the markets is endogenous and accounts for strategic interactions of different kinds. The model provides some improvement in the explanation of the business cycle compared to the standard RBC framework. Our characterization of the market structure allows us to explain the procyclical variability of the profits and entry together with the countercyclical variability of the mark ups. Nevertheless, we have emphasized a mark up and profit volatility puzzle: further examinations of alternative market structures should be aimed at matching the high levels of volatility that emerge from the empirical investigation of U.S. mark ups and profits.

We want to end this chapter with a limited but hopefully fertile conclusion: endogenous market structures do seem to matter for macroeconomic issues. While most of the recent approach to the study of business cycles has been based either on perfect competition, price-taking behavior and zero mark ups or on monopolistic behavior, exogenous entry and positive and constant mark ups, we have shown that strategic interactions leading to a link between entry, mark ups and prices can substantially affect the way an economy reacts to shocks.

[56] On the credit channel of monetary policy see Bernanke and Blinder (1988).

4. Endogenous Market Structures and International Trade and Finance

> I am so busy doing nothing that the idea of doing anything—which as you know, always leads to something—cuts into the nothing and then forces me to have to drop everything. **Jerry Seinfeld**

The gains from trade are traditionally associated with international specialization of production activities following the comparative advantage of nations within the neoclassical tradition started by Heckscher (1929) and Ohlin (1933), and with the availability of new varieties of goods produced and exchanged in intra-industry trade within the new trade approach, pioneered by Krugman (1980) and summarized by Helpman and Krugman (1985).[1] The latter approach has introduced monopolistic behavior of firms *à la* Dixit and Stiglitz (1977) in the trade literature emphasizing the endogenous increase in the number of products that can be bought (at the same price as before) after a country opens up to trade.

The normative theory of trade has mainly emphasized the optimality of import tariffs and export taxes either to improve the terms of trade of a large open economy within the neoclassical tradition (Lerner, 1936),[2] or for profit shifting reasons within the modern analysis of strategic trade policy, exemplified by Brander and Spencer (1985) and Eaton and Grossman (1986),[3] and surveyed by Helpman and Krugman (1989). The latter approach has introduced strategic interactions between international firms to analyze unilateral optimal trade policies, emphasizing the need for passive protectionist policies as tariffs on importers and taxes on price-setting exporters.

All these approaches neglect a fundamental role of trade, the competitive impact of international entry on the strategic interactions between firms and on the resulting international prices. This role leads to a new and separate source of gains from trade, which partially crowds out the traditional ones, and to different, and sometimes opposite, conclusions on the optimal trade policy.

---

[1] See also Dixit and Norman (1980).

[2] See Bhagwati, Panagariya and Srinivasan (1998) for a wide discussion.

[3] See also Dixit (1984).

F. Etro., *Endogenous Market Structures and the Macroeconomy*,
DOI: 10.1007/978-3-540-87427-0_4, © Springer-Verlag Berlin Heidelberg 2009

The analysis of international EMSs emphasizes that the gains from trade are mainly associated with a reduction in the prices due to international competition. This effect is of course absent in the standard neoclassical framework, since perfect competition implies that all goods are sold at the marginal cost both before and after countries open up to trade (comparative advantage only depends on factor endowments when the same technologies are available everywhere). Moreover, in the new trade literature which adopts a framework of monopolistic behavior by firms producing different varieties, trade enlarges the amount of produced varieties keeping mark ups and prices unchanged: it is only the increase in the number of varieties available to consumers that creates gains from trade. In this chapter we show that when strategic interactions occur and entry affects them, globalization (i.e.: opening up to trade) enhances international competition, reduces the equilibrium mark ups and induces a generalized reduction of the prices. This creates a negative feedback effect on profitability which crowds out the impact on the number of produced varieties and leads to business destruction at the local level: therefore the gains from trade come mainly from lower prices and not from more varieties. These implications have stimulated a growing empirical literature on the relation between market size and the endogenous elements of the market structure, number of firms, mark ups and production per firm. Building on the initial contributions by Bresnahan and Reiss (1987) and Berry (1992), recent works by Manuszak (2002), Campbell and Hopenhayn (2005) and others have provided empirical evidence which is in direct support of the implications of the EMSs approach. Novel evidence will be analyzed below.

At the theoretical level, the static analysis of intra-industry trade between countries can be translated in a dynamic general equilibrium model extending the one introduced in Chapter 3, and we will illustrate recent attempts of the literature in this direction, mainly built around the pathbreaking works of Melitz (2003) and Ghironi and Melitz (2005), whose intuitions have also started a novel interest in the relation between industrial organization and international trade theory.

The analysis of international EMSs affects the normative predictions as well. This does not lead to radical changes of the traditional results for the optimal unilateral policy for the domestic markets, since we will show that moderate import tariffs keep being optimal (but lead to excessive protectionism from a global perspective). However, the case of trade policy for strategic export promotion reveals a new general result in the presence of EMSs. We will show that the optimal unilateral trade policy for exporting firms is the opposite of the traditional one: countries should always subsidize exports rather than tax them (moreover, this remains the optimal policy even when more countries can choose independently their policies, and in such a case the global equilibrium is not characterized by excessive protectionism). These results suggest the necessity of a sort of active protectionism based on strategic export promotion which incentivates trade (and does not affect foreign countries in a negative way), as opposed to the passive protectionism based on import tariffs and export taxes which limits trade volumes.[4]

The investigation of EMSs in international trade theory has a number of precursors. The so-called "reciprocal dumping" model of Brander and Krugman (1983) is the pathbreaking work in the analysis of strategic interactions in intra-industry trade, but it ignores fixed costs of entry in the foreign market and is limited to partial equilibrium. The well-known model of Krugman (1979a, 1980, 1981) of monopolistic competition adopts a general equilibrium approach, but it neglects the strategic interactions.[5] An early attempt to endogenize market structures in the analysis of intra-industry trade and strategic trade policy is due to Horstmann and Markusen (1992), but they focus on the entry decisions of just two multinationals firms in two segmented national markets with competition in quantities and fixed costs of entry at the firm and plant level (equivalent to fixed export costs), characterizing how technological conditions and trade policy induce switching between equilibria with a single firm or with two firms (and with plants in a single country or in both countries). An important work by Lahiri and Ono (1995) has analyzed a 2x2x2 model with Cournot competition and endogenous entry in one sector, limiting the analysis to the case of homogenous goods. More recently, foreign direct investment in the presence of EMSs has been studied by De Santis and Stahler (2004) who endogenize the number of national and multinational firms active in two countries. Finally, Etro (2007,b) has analyzed strategic interactions of different kinds (Cournot and Bertrand competition with product differentiation) in the Krugman model emphasizing the gains from trade associated with the competition effect and the reduction of mark ups. A similar point has been independently made by Ottaviano and Melitz (2008), who study monopolistic behavior of a continuum of firms, but under quadratic preferences which deliver linear demand functions.

[4] In a historical perspective, merchantilism has promoted export subsidies, as noticed by Smith (1776, Book IV, Ch. 5, Of Bounties): "*Bounties upon exportation are, in Great Britain, frequently petitioned for, and sometimes granted, to the produce of particular branches of domestic industry. By means of them, our merchants and manufacturers, it is pretended, will be enabled to sell their goods as cheap or cheaper than their rivals in the foreign market. A greater quantity, it is said, will thus be exported, and the balance of trade consequently turned more in favour of our own country... We cannot force foreigners to buy their goods, as we have done our own countrymen. The next best expedient, it has been thought, therefore, is to pay them for buying. It is in this manner that the mercantile system proposes to enrich the whole country.*" Notice, however, that classical economists, including Smith, opposed these policies and were in favor of free trade.

[5] The Krugman model was however extended to the case of a preference parameter exogenously depending on the number of firms, so as to lead to mark ups decreasing in the number of firms under monopolistic behavior. This exogenous and demand side mechanism is able to obtain similar results to those of the endogenous and supply side mechanism of the EMSs approach to trade.

Applications of the literature on endogenous entry and trade to dynamic macroeconomic models have emerged more recently. Melitz (2003) has extended a dynamic version of the basic Krugman model to fixed costs of export and heterogenous productivity between firms to endogenize the fraction of domestic firms that do export and to explain their higher average productivity through a "selection effect". This framework has been applied by Ghironi and Melitz (2005) to a DSGE model of a two-country world with trade and endogenous entry of firms, but without considering strategic interactions between them and the impact of entry on these interactions.[6] Atkeson and Burnstein (2008) have introduced strategic interactions in a related model, reproducing deviations from relative purchasing power parity similar to those observed in the data thanks to the fact that firms "price-to-market".

The introduction of fully fledged EMSs in the literature on trade policy has been rather slow, since most of the many initial contributions have been constantly focused on duopoly cases. Only in the works of Venables (1986), Horstmann and Markusen (1986) and Markusen and Venables (1988)[7] there are preliminary studies of examples of strategic trade policy in a reciprocal dumping model and in a model of integrated markets with free entry. Bhattacharjea (1995) has analyzed the effects of changing the number of firms on the optimal domestic trade policy, but without fully endogenizing entry. More recently, Etro (2007,b) has derived the optimal domestic trade policy in a simple example with EMSs. Finally, Etro (2010) has provided the first analysis of strategic export promotion with EMSs characterizing the optimal unilateral export subsidies for a large class of models of competition in quantities and prices, and applying the same principles to the analysis of R&D subsidies in the competition for the market and competitive devaluations.

The chapter is organized as follows. Section 4.1 discusses empirical evidence on crucial predictions of the EMSs approach that are relevant for trade issues. Section 4.2 introduces EMSs in the Krugman model of international trade deriving a new source of gains from trade. Section 4.3 discusses the Ghironi-Melitz model, which has introduced intra-industry trade in a dynamic two-country framework. Section 4.4 is about tariff policy for the domestic market. Section 4.5 is about export promotion policy for foreign markets and Sections 4.6 employs it to fully characterize the optimal unilateral trade policy for exporting firms. Sections 4.7 evaluates a number of key issues in trade, including different forms of protectionism, lobbying and trade wars. Section 4.8 evaluates exchange rate policy and the effects of exchange rate depreciation in the short run under different forms of competition in the international markets. Section 4.9 analyzes the optimal unilateral R&D policy in partial equilibrium. Section 4.10 concludes.

---

[6] See also Devereux and Lee (2001)

[7] More recently see De Santis and Stahler (2001).

4.1 Empirical Evidence on EMSs and Market Size Effects

In this section we provide some microeconometric evidence in support of the EMSs approach, and we argue that this evidence is in line with one of its main implications for international trade theory, the effect of opening up to trade with new markets on the production structure, and in particular on the number and size of firms.

The empirical literature on EMSs started with the pathbreaking works by Bresnahan and Reiss (1987, 1990) and Berry (1992). These and subsequent works moved beyond the naive view for which lower mark ups should be associated with stronger competition and a larger number of firms. Such a simple correlation should hold in the presence of exogenous market structures, but when entry is endogenous there is an opposite mechanism at work: higher mark ups attract a larger number of firms. Therefore, empirical analysis should take seriously into account this problem of reverse causality.

A correct test of the predictions of the EMSs approach should be based on the impact that exogenous variables exert on the endogenous variables, as the mark ups and the number of firms in different markets, after controlling for a number of market specific characteristics. For instance, in a mark up regression, this requires to introduce a correction term derived from the model of EMSs (as in standard selection models *à la* Heckman, 1979) to account for possible correlation between market structure and the error term in the mark up equation.[8]

Moreover, a useful empirical analysis should be able to discriminate between the implications of the EMSs approach and those of alternative approaches: since the neoclassical framework leads to indeterminate market structures, the main possible comparison is with the market structure emerging from a model of monopolistic behavior with endogenous entry *à la* Dixit and Stiglitz (1977) and Krugman (1980). Between the exogenous variables affecting the market structure, the fixed costs of entry, the degree of substitutability between goods and the marginal costs of production are often hard to measure, while the size of the market can be easily measured with good proxies. Between the endogenous variables, mark ups are often difficult to measure as well, while data on the number of firms and the production per firm are easily accessible. Therefore, a large part of the empirical literature has been focused on these last two endogenous variables as dependent variables and on the size of the market as a key independent variable.

Let us look at the number of firms first. Both the Dixit-Stiglitz approach and the EMSs approach generate a positive correlation between market size and the number of firms, however the relation is linear in the former case and concave in the latter. This has a crucial implication. According to the Dixit-

[8] See Cohen and Manuszak (2005) and Manuszak and Moul (2008) for a discussion and interesting applications.

Stiglitz framework with monopolistic behavior, it takes a market of double size to double the number of firms. In the EMSs approach this result is substantially confirmed under competition in prices, but not under competition in quantities: in this case, the size of the market must be more than double for the number of firms to be twice as much.

The modern empirical literature on EMSs has investigated the relation between market size and entry. The pathbreaking work in the field by Bresnahan and Reiss (1987) has studied professional and retail markets in small and isolated U.S. towns providing a certain support for the EMSs approach. More recently, an interesting work by Manuszak (2002) has examined the XIX century brewing industry in American frontier towns, which was extremely fragmented (due to the lack of a well-integrated transportation system and the perishability of beer). His calculations suggest that a town with an effective population of between 850 and 950 people could support a single firm, between 2200 and 2600 inhabitants were required for a second firm to enter, while a population of 4300-4500 people could support a third firm. According to the author, "a single brewer in a western town would have had substantial monopoly power due to his position as the sole provider of beer in the market. However, as the results indicate, it is reasonable to expect that entry of a second firm would have significantly compromised the initial brewer's advantageous position. Similarly, entry of a third firm continued to erode the market power possessed by firms." While these results are interesting, we need more systematic evidence on the relation between market size and market structure. A recent and growing literature is exactly looking at this relation through the analysis of cross-sectional and panel data at the industry level.

4.1.1 Panel data on the number of firms

Systematic evidence on the relation between market size and the number of firms can be obtained from a panel study of different sectors.[9] Following a preliminary work by Czarnitzki and Etro (2009), we have estimated the following structural relation between number of firms N_{jt} in sector j at time t and the market size S_{jt} of the same sector at the same time:

$$\ln N_{jt} = \beta_{0j} + \beta_1 \ln S_{jt} + \varepsilon_{jt} \tag{4.1}$$

where ε_{jt} is an error term. The Dixit-Stiglitz approach implies $\beta_1 = 1$, as can be verified taking logs of (2.26). Bertrand competition with endogenous entry leads to approximately the same results, as one can check from (2.40). Instead, Cournot competition with endogenous entry implies $\beta_1 \in [0.5, 1)$, with a coefficient that should decrease with the degree of substitutability between products. In particular, in case of homogenous goods we have $\beta_1 = 0.5$, as it

[9] I am extremely thankful to Dirk Czarnitzki for help with this analysis.

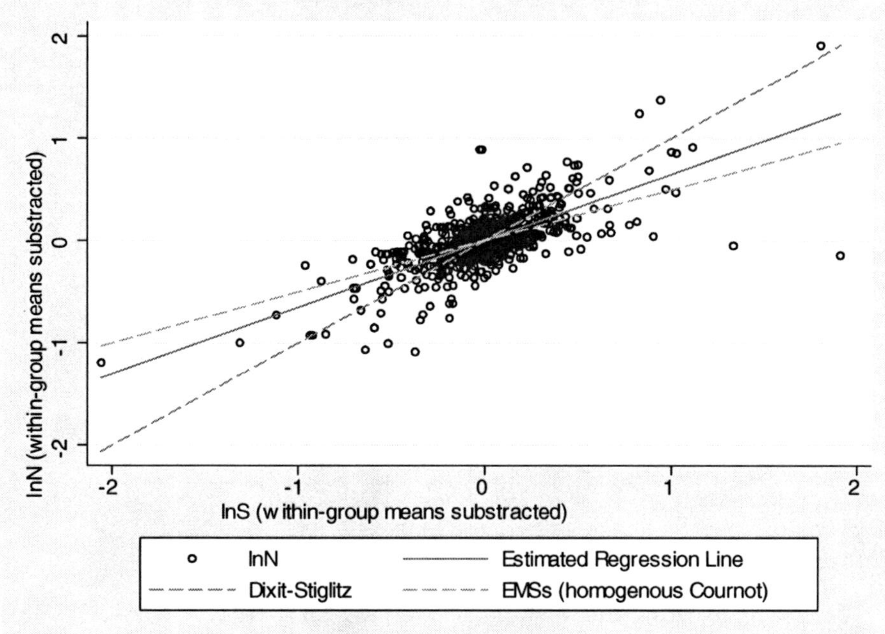

Fig. 4.1. Number of firms and Market Size for German industries (from regressions). NACE 3-digit level. Source: Czarnitzki and Etro (2009)

can be verified taking logs of (2.17).[10] Notice that the same size effects can also be obtained from the long run EMSs in general equilibrium: for instance, see (3.37) and (3.34) from the steady state of our DSGE model in Chapter 3. On this basis we can test the Dixit-Stiglitz hypothesis and compare it with the predictions of the EMSs approach.

One can also check the relation between concentration and market size. Common indexes of market concentration can be highly criticized as a measure of market power, nevertheless we used the Herfindahl-Hirschman Index of concentration for robustness analysis. This index corresponds to the sum of the squared market shares, and in the case of symmetric competition it becomes $HHI_{jt} = (1/N_{jt}) \cdot 1000$, therefore we would expect an inverse relation compared to that found for the number of firms.

[10] Our structural equation can be derived also from (4.6), (4.15) and (4.10) in the trade model described below. Notice that a strategic model in which mark ups depend on the number of firms (negatively) and on the market size (positively) in more general ways could rationalize even $\beta_1 < 0.5$. Moroever, one could interpret market structure indeterminacy associated with perfect competition with $\beta_1 = 0$ (no relation between size and number of firms).

Our preliminary investigation of the relation between market size and number of firms is based on a panel of industry level data for the German manufacturing sector. The data stem from the bi-annual official publications of the German Monopolies Commission, that is obligated by law to publish regularly concentration statistics for competition policy purposes. This data source contains the number of firms, total industry sales, which we adopt as a measure of market size, and the HHI index for German manufacturing for different aggregations of the European standard industry classification called NACE. In the following, we utilize data at the NACE 3-digit level, which results in 102 different manufacturing industries. We have bi-annual data between 1991 and 2003. Table 4.1 shows the summary statistics of the number of firms, N_{jt}, the total industry sales, S_{jt}, that we use as market size variable, and the Herfindahl Index, HHI_{jt}.

Table 4.1. Descriptive Statistics (711 observations, 102 industries)

	Mean	Std. dev.	Min.	Max.
N_{jt}	387	540	1	3851
S_{jt}	22456.38	34878.98	31.99	38569.31
HHI_{jt}	84.05	141.75	3.15	2431.14

We estimate three different panel models. First, we estimate a pooled cross-sectional OLS regression (POLS) where we implicitly assume that $\beta_{0j} = \beta_0$ for any j. This is then relaxed by estimating two fixed effects models. We allow β_{0j} to be industry specific by applying the well-known within panel regression (FE: Within) and by estimating a first-difference model (FE: FD).[11]

The results are displayed in Table 4.2. Standard F-tests confirm the presence of industry-specific β_{0j}, so that the POLS models are rejected. Consequently, we should turn our attention to the fixed effects models. In the regression concerning the number of firms, we find that the slope of market size is positive and significant in both the within and the FD regression. The Dixit-Stiglitz hypothesis that $\beta_1 = 1$ is clearly rejected in both models, with respectively $F = 22.9^{***}$ and 27.69^{***}.

Fig. 4.1 is representative of our results, and plots the (log) number of firms of the industries against the (log) sales per industry, here on the basis of the within regression. The 45°-dashed line represents the hypothetical relation between number of firms and size that emerges from the Dixit-Stiglitz approach ($\beta_1 = 1$), while the flatter dashed line is the relation that emerges from the EMSs approach in case of Cournot competition with homogenous goods ($\beta_1 = 0.5$). We expect that the data lie between these two extreme predictions, and this is exactly what we find out. The Dixit-Stiglitz approach is rejected, meaning that there is a robust positive but less than proportional relation between the size of the market and the number of firms. The estimates

[11] This methodology was introduced by Mundlak (1961). See Greene (2003) for a survey.

for β_1 in the within and first-difference model are respectively $\beta_1 = 0.650$ (the case of Fig. 4.1) and $\beta_1 = 0.507$ (which in Fig. 4.1 could not be distinguished from the flatter dashed line): this suggests that the EMSs model with Cournot competition and homogenous goods is a better approximation to the data (indeed, the hypothesis $\beta_1 = 0.5$ cannot be rejected in both models). The results are confirmed when we group firms in macrosectors (ICT, machinaries, chemicals, paper, metal, food- and textile-related industries) with the Dixit-Stiglitz hypothesis rejected always except for textile-related sectors. Finally, notice that even without controlling for other variables, our basic results are quite powerful, with R^2 around 50%.[12]

Table 4.2. Regression Results

	POLS		FE: Within		FE: FD	
	Coef.	(Std. err.)	Coef.	(Std. err.)	Coef.	(Std. err.)
	Regression of $\ln N_{jt}$ on $\ln S_{jt}$					
β_1	0.611	(0.07)***	0.650	(0.07)***	0.507	(0.09)***
Obs.	711		711		609	
R^2	0.45		0.56		0.43	

Note: Each regression includes a full set of time dummies. They are always significant at the 5% level. *** (**, *) indicate a significance level of 1% (5%, 10%).

The regressions on the HHI show a weaker relation with market size, but they nevertheless identify a negative slope much lower than one, again in contradiction with the Dixit-Stiglitz hypothesis.

We have also verified the robustness of our results with different datasets, always for the German manufacturing sector, obtaining similar patterns. Fig. 4.2 reports the basic within panel regression based on data at the NACE 4-digit level from the annual "Kostenstrukturerhebung" published by the German Federal Statistical Office (235 groups bewteen 1995 and 2006 for a total of 2695 observations). The results are in line with the earlier ones, except for one difference: when we look at more disaggregated data, the estimated coefficient β_1 decreases, here to $\beta_1 = 0.40$. This may support another implication of the EMSs approach: a higher degree of substitutability, that we find at more disaggregated levels in the data, changes the relation between market size and number of firms and reduces the associated coefficient. Further research should try to verify whether the results of Czarnitzki and Etro (2009) are supported in other countries or datasets.

4.1.2 Cross-sectional data on firms' production

Let us finally focus on the relation between market size and equilibrium production of each firm. In this case, the Dixit-Stiglitz approach and the EMSs approach provide radically different results. In the absence of strategic

[12] The results are not driven by the dynamic framework: similar coefficients emerge in cross-sectional regressions based on each single year.

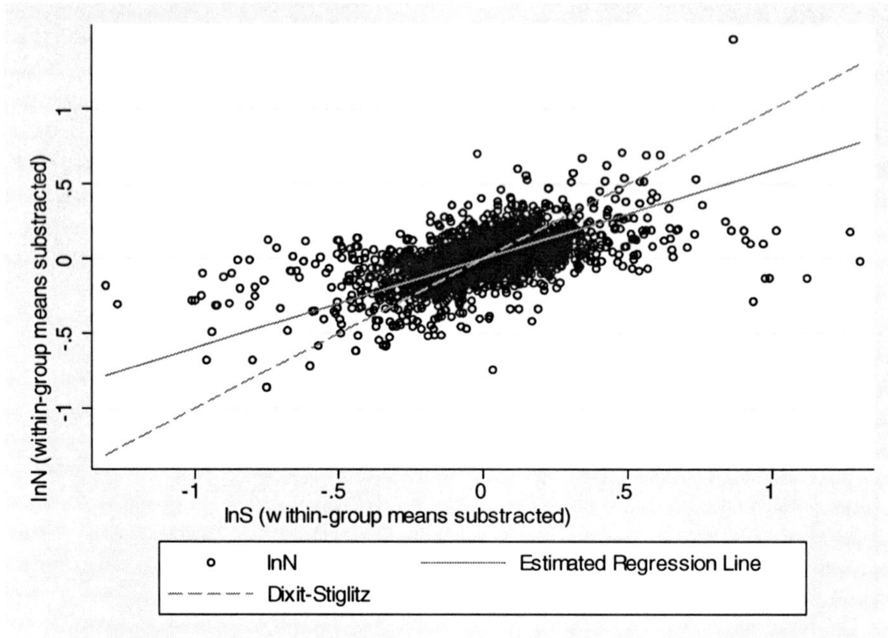

Fig. 4.2. Number of firms and Market Size for German industries (from regressions). NACE 4-digit level. Source: Czarnitzki and Etro (2009)

interactions the endogenous production of each firm is independent from the size of the market - see (2.27), while the EMSs approach generates a positive correlation between market size and production: larger markets attract entry which strengthens competition, reduces the mark ups and requires larger output per firm to cover the entry costs - see (2.37) under competition in quantities and (2.42) under competition in prices.

An important work by Campbell and Hopenhayn (2005) has tested this hypothesis examining the relation between the size distribution of establishments for thirteen retail trade industries across 225 U.S. cities and the size of these cities (each of which was identified with a separate market), after controlling for a number of economic and demographic variables as retail wages, commercial rents, advertising costs, income, and percentages of black and college educated population.

Fig. 4.3 is representative of the results of Campbell and Hopenhayn (2005) and plots the (log) average sales per establishment against the (log) population of the cities for Women's Clothing and Specialty Stores (SIC 526,3). All the variables are defined as residuals from regressions against the control variables. The coefficient of the regression line for average sales is positive

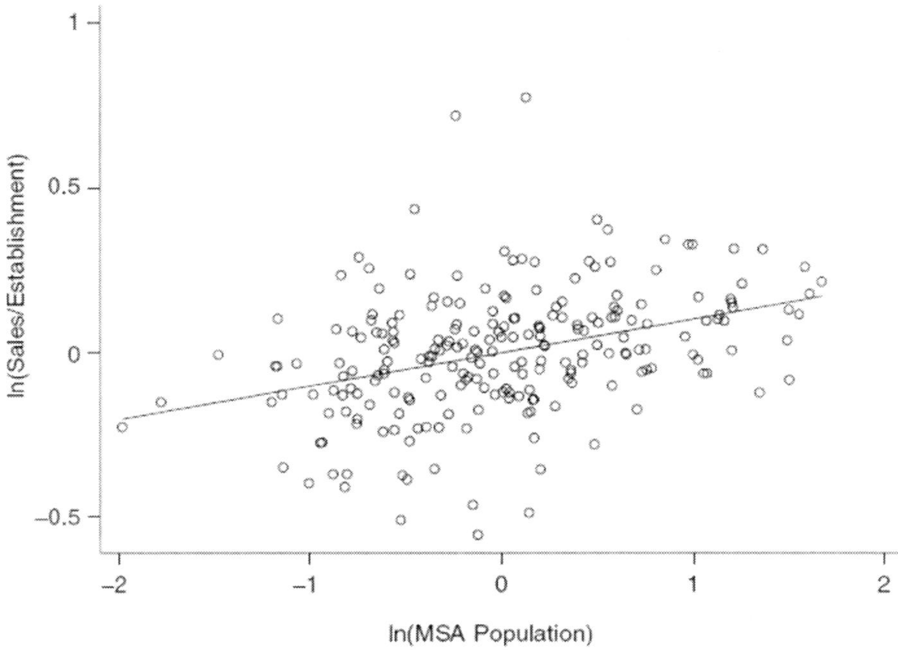

Fig. 4.3. Average Sales and Population of U.S. Metropolitan Statistical Areas for Women's Clothing and Specialty Stores (from regressions). Source: Campbell and Hopenhayn (2005)

and equal to 0.1, and it is significantly different from zero at the 1 % level.[13] Similar relations can be found in other industries (in particular car dealers, gasoline service stations, furniture stores, restaurants, TV/PC stores), and they are robust to the use of average employment as a measure of the size of the firms and to the use of density population or value of industry sales as measures of the market size.

The positive correlation between the size of a market and the production of each firm is economically interesting as well. In the example above, doubling the size of a city increases the average sales of its firms by 7.1 % and increases their average employment by 4.4 %, a result that is due to the reduction in mark ups associated with the larger number of firms competing in a larger city. More generally, this reflects the idea that an increase in the size of a market, due for instance to an increase in the population or to the integration with a formerly separate market, leads to more firms, lower mark ups and larger production for each firm.

[13] Notice that under Cournot competition this coefficient should be at most 0.5, which is its predicted value in case of homogenous goods.

This result is particularly interesting when studying trade because one of the main aspects of opening up to trade is exactly the increase in the size of the market where domestic and foreign firms interact. As we will see in the next section, the EMSs suggests that globalization leads firms to increase their production as a consequence of a more intense competition, which in turn reduces the prices and the number of firms at the local level. The microeconometric evidence presented in this section provides some support for this macroeconomic phenomenon.

4.2 A Model of Trade with EMSs

To start our exploration of an open economy context where market structures are truly endogenous, we re-examine the classic Krugman (1980) model and extend it to take into account different forms of strategic interactions between the firms. These lead to an impact of market size and international trade on the price level which motivates a novel source of gains from trade: gains from competition beyond the traditional gains from variety pointed out in the Kruman model. More recently, Krugman (1991,a) has applied his basic model to the study of economic geography and urban economics,[14] but also this application has ignored the strategic effects of the EMSs approach, which could provide a novel explanation for the concentration of population in metropolitan areas: lower prices (compared to the wages) due to stronger competition. Here, however, we focus on the international trade interpretation of the Krugman model.

Consider a market with L agents consuming multiple goods to maximize utility from our usual consumption index:

$$U = \left[\sum_{j=1}^{N} C(j)^{\frac{\theta-1}{\theta}}\right]^{\frac{\theta}{\theta-1}} \tag{4.2}$$

where N is the number of goods and $\theta > 1$ is the elasticity of substitution between the goods. Each consumer maximizes utility under the budget constraint $\sum_{j=1}^{N} p(j)C(j) = Wl$ and we normalize the nominal wage W to unity. Individual labor supply l is exogenous and normalized to unity as well, so that total labor supply is L. Each good i is produced according to the linear function:

$$x(i) = Al_i \tag{4.3}$$

where total factor productivity A is constant and normalized to one, and l_i is the labor input used by firm i. As we have seen in Chapter 3 in detail, gross

[14] See Fujita, Krugman and Venables (1998) on the new economic geography and Fujita (1989) on the neoclassical approach to urban economics.

profits for each firm can be expressed in terms of production levels or prices as:

$$\Pi(i) = \frac{x(i)^{\frac{\theta-1}{\theta}} L}{\sum_{j=1}^{N} x(j)^{\frac{\theta-1}{\theta}}} - x(i) = \frac{[p(i)-1]\, p(i)^{-\theta} L}{\left[\sum_{j=1}^{N} p(j)^{-(\theta-1)}\right]} \tag{4.4}$$

and net profits must take into account the fixed cost of entry F. Competition in quantities or prices between firms determines the market structure. In a symmetric equilibrium, each one of the N goods is produced in quantity x, so that the utility of each agent is $U = (x/L)\, N^{\theta/(\theta-1)}$, which increases in the number of goods and in the consumption of the agent.

4.2.1 The Krugman model

Following Dixit and Stigliz (1977), the Krugman model and most of the subsequent literature[15] ignores the strategic interactions between firms, assuming monopolistic behavior and obtaining the constant mark up:

$$\mu^M = \frac{\theta}{\theta - 1} \tag{4.5}$$

In a symmetric equilibrium, $x = C(i)L$ is the production of each firm. Imposing the endogenous entry condition $\pi = \left(\mu^M - 1\right) x - F = 0$, the number of firms can be derived from the market clearing condition for the labor market, $L = N(F + x)$. Accordingly, the endogenous number of firms is:

$$N = \frac{L}{\theta F} \tag{4.6}$$

and the equilibrium production of each good is

$$x = \frac{F(\theta - 1)}{\theta} \tag{4.7}$$

Notice that mark ups and production per firm are independent from the size of the economy L, while the number of firms is directly proportional to it, results that were broadly rejected by the empirical evidence of Section 4.1.

The fundamental contribution of the Krugman model is in the analysis of market integration with other countries, which leads to intra-industry trade of the differentiated goods produced by each country, and to gains for the consumers associated with an increase in the number of varieties available. For instance, consider opening up to trade with other countries with identical preferences and total population L^*. The proportional increase in the total demand generates an equilibrium number of varieties:

[15] See Helpman and Krugman (1985) or Feenstra (2004) for a survey.

$$N + N^* = \frac{L + L^*}{\theta F}$$

while the price and the production of each firm remain the same as in the closed economy, $\mu^M = \theta / (\theta - 1)$ and $x = x^* = F(\theta - 1)/\theta$: globalization does not change the equilibrium production of the firms and their prices.

In other words, if the consumers of two identical countries have ten different varieties available in autarchy, after opening up to trade they can consume twenty different varieties in half quantity as before. The prices, however, remain the same. Since consumers love variety, trade increases utility.

The assumption of monopolistic behavior is meaningful if we interpret the economy as characterized by a single market with an infinity of differentiated goods. Nevertheless, if we look at economies as characterized by many sectors with a limited number of firms producing relatively similar goods in each sector, the results of the Krugman model can become misleading. In what follows, we take this alternative approach and study a generalized version of this model with a representative sector characterized by competition in quantities or in prices. The following analysis is based on Etro (2007,a).

4.2.2 Cournot competition

Consider a Cournot equilibrium between N firms that choose their production to maximize profits (4.4) taking as given the production of the other firms. As we have seen in Chapter 2, the equilibrium is characterized by the following price:

$$\mu^Q(\theta, N) = \frac{\theta N}{(\theta - 1)(N - 1)} \tag{4.8}$$

For the sake of simplicity, we start our investigation in the extreme case of homogenous goods ($\theta \to \infty$), in which the possibility of gains from variety is absent. Consider first the equilibrium of a closed economy with EMSs. The Cournot equilibrium with N firms is characterized by the mark up condition $\mu^Q(N) = N/(N - 1)$, but the endogenous entry and market clearing conditions lead to the following price for each variety:

$$\mu^Q = \frac{\sqrt{L}}{\sqrt{L} - \sqrt{F}} \tag{4.9}$$

and to the production level:

$$x^Q = \sqrt{F}(\sqrt{L} - \sqrt{F})$$

Contrary to what happens in the monopolistic framework, where price and production per firm are independent from the domestic labor force, now the price is decreasing in the labor force and the production of each firm is increasing in it to cover the fixed costs of entry (as we have seen in the previous section this is empirically plausible). This result is due to a competition effect associated with the positive impact of the size of the market on the equilibrium number of firms, which is now:

$$N^Q = \sqrt{\frac{L}{F}} \tag{4.10}$$

Notice that the number of firms is increasing and concave in the size of the economy. Taking logs, we obtain a simple relation $\ln N^Q = 0.5\,(\ln L - \ln F)$ that we have successfully tested in Section 4.1.

Consider opening up to trade with an other country characterized by identical agents and total population L^*. It is immediate to derive that the new equilibrium implies the price:

$$\mu^Q = \frac{\sqrt{L+L^*}}{\sqrt{L+L^*} - \sqrt{F}}$$

The total number of firms becomes:

$$N^Q + N^{*Q} = \sqrt{\frac{L+L^*}{F}}$$

which is lower than the number of firms active in autarchy, $\sqrt{L/F} + \sqrt{L^*/F}$. Nevertheless, the strengthening of competition in the integrated market leads to a lower international price level, and to the larger production for each firm $x^Q = x^{*Q} = \sqrt{F}\left(\sqrt{L+L^*} - \sqrt{F}\right)$. Notice that trade has reduced the total number of firms, inducing an increase in world market concentration: in our case of homogenous goods, this represents an example of beneficial concentration, because the price level has decreased because of the competition effect, and production has become more efficient thanks to the reduction of the aggregate spending in fixed costs.

To clarify the implications of the EMSs approach for trade, let us consider the earlier example of two identical countries opening up to trade. Suppose that ten firms producing an homogenous good are active in each country in autarchy. After the two countries open up to trade with each other, the EMSs approach implies that only fourteen firms, seven in each country, remain active. This implies that trade leads to the foreclosure of three firms in each country, which is the consequence of a reduction of the mark up by about 30%.[16] Nevertheless, it is clear that trade increases utility because the price of the homogenous good is reduced everywhere (and there is a more limited waste of labor resources in fixed costs). In this case, contrary to the Krugman model, the gains from trade derive entirely from the price reductions associated with strategic interactions.

[16] Formally, if $N = \sqrt{L/F} = 10$, we must have $N + N^* = \sqrt{2}N \simeq 14$. Moreover, if the autarchic price was $1/(1 - 1/10)$, the new one must be $1/(1 - 1/14)$, which is equivalent to a reduction of the mark up from 11.1% to 7.7%. This is a 30% reduction.

Of course, in case of imperfect substitutability between goods, product variety is beneficial and the impact of trade is more complex, but the competition effect of trade on prices and the reduction in the equilibrium number of varieties produced in each country persist. In particular, one can verify that a closed economy is characterized by EMSs with the following mark up and number of firms:

$$\mu^Q = \frac{\theta}{(\theta-1)}\left(1-\frac{2\theta F}{L+\sqrt{L^2+4\theta(\theta-1)FL}}\right)^{-1} \tag{4.11}$$

$$N^Q = \frac{L}{2\theta F}+\sqrt{\left(\frac{L}{2\theta F}\right)^2+\frac{(\theta-1)L}{\theta F}} \tag{4.12}$$

and an increase in the size of the economy leads to a reduction of the former and to a less than proportional increase of the latter. As before we can verify the impact of opening up to trade with similar countries. The equilibrium mark up decreases, which leads to business destruction at the local level and to a larger production by each surviving firm. The price reduction/business destruction effect is reduced when the degree of substitutability θ decreases.

4.2.3 Bertrand competition

Consider now competition in prices between N firms choosing their prices to maximize profits (4.4) taking as given the prices of the other firms. The Bertrand equilibrium price is given by:

$$\mu^P(\theta,N) = \frac{\theta(N-1)+1}{(\theta-1)(N-1)} \tag{4.13}$$

as we have seen in Chapter 2. Under endogenous entry of firms in autarchy, the equilibrium price goes down to:

$$\mu^P = \frac{\theta L}{(\theta-1)(L-F)} \tag{4.14}$$

which corresponds to the following number of goods/firms:

$$N^P = \frac{L}{\theta F}+\frac{\theta-1}{\theta} \tag{4.15}$$

and to the production level:

$$x^P = \frac{F(\theta-1)(L-F)}{L+(\theta-1)F}$$

In a larger economy there are more firms and the strengthening of competition between them reduces the mark ups and increases the production of each firm.

Consider now the opening up of this economy to trade. The new equilibrium is characterized by the following price:

$$\mu^P = \frac{\theta (L + L^*)}{(\theta - 1)(L + L^* - F)}$$

and the total number of varieties produced becomes:

$$N^P + N^{*P} = \frac{L + L^*}{\theta F} + 1 - \frac{1}{\theta}$$

with $x^P = x^{*P} = F(\theta - 1)(L + L^* - F) / [L + L^* + (\theta - 1)F]$.

In this case, opening up to trade leads to a lower international price level without reducing the number of varieties produced in each country (or reducing it by at most one unit if we take in consideration the integer constraint on the number of firms),[17] and increasing the production of each one. Once again, the beneficial impact of trade emerges quite clearly, because world prices diminish and the number of varieties produced does not decrease.

4.2.4 Gains from trade and business destruction

Our analysis of the international EMSs could be extended to two inputs, for instance introducing capital besides labor, or considering qualified and non-qualified labor. In an early investigation, Lahiri and Ono (1995) have studied a model with two inputs, two countries and two sectors: one is competitive and the other is characterized by homogenous goods and Cournot competition. They have shown that, contrary to the case with an exogenous number of firms in each country, analyzed by Markusen (1981), autarchy generates lower prices for larger countries, but free trade leads to factor price equalization and re-establishes the validity of the Heckscher-Ohlin theorem. Moreover, both countries enjoy gains from trade due to the price reduction induced by the stronger competition in the integrated market, even if trade does not occur when the countries are identical.[18] Introducing product differentiation would lead to intra-industry trade and would generate gains from trade derived from both the price effect and the variety effect.

Finally, the analysis could be extended to trade frictions without altering most of its basic insights, while showing that the larger economy attracts a larger number of firms.[19]

---

[17] A similar result emerges in a model of endogenous growth with EMSs and competition in prices in the goods market by Peretto (2003).

[18] More recently, Wälde and Weiss (2007) have derived a positive relation between opening up to trade and inequality between qualified and non-qualified labor inputs in a related model.

[19] Let us focus briefly on the case of homogenous goods and competition in quantities. Each firm can sell at home or abroad, but exports require transport costs in the form of iceberg costs $d \in (0, 1]$: selling one unit of goods abroad requires

Summing up, our findings on the effects of trade with international EMSs are the following:

PROPOSITION 4.1. **Under endogenous market structures, opening up to trade decreases the price level, increase the production of each firm and (weakly) decreases the number of firms in each country, so that the total number of consumed varieties increases less than proportionally, with a beneficial impact on consumer welfare.**

The empirical analysis of Section 4.1 provides support for these results against those of the Dixit-Stiglitz-Krugman approach, and therefore in favor of the gains from competition against the gains from variety as the main source of gains from trade. The relevance of the mechanism associated with the strategic interactions depends on the type of traded goods under consideration. At one extreme we have perfectly differentiated goods with competition in prices: for these goods, all the gains from opening up to trade derive from an increase in the number of consumed varieties and not from price changes, while business destruction is absent. At the other extreme we have homogenous goods with competition in quantities: for these, all the gains from trade derive from lower prices, but business destruction is heavy. Notice that in the absence of labor market imperfections, business destruction is inconsequential for the agents: they switch jobs reallocating their work between a smaller number of firms that have a larger share of the global market. Nevertheless, job destruction due to globalization can have dramatic effects in the presence of labor market rigidities, both in terms of unemployment and income inequalities. Our distinction between different cases above can help us to understand better the consequences of globalization. One could think of sectors producing highly differentiated goods with competition in prices as sectors where innovation and design are the fruit of skilled labor: in these sectors business destruction due to globalization is limited. On the

---

shipping $1/d$ units of good. As long as exports are profitable for all firms, the Cournot equilibrium is characterized by a mark up:

$$\mu(N, N^*) = \frac{dN + N^*}{d(N + N^* - 1)}$$

The mark up in the foreign market $\mu^*(N, N^*)$ is symmetric. Each domestic firm obtains profits from the home market and from the foreign one, summing up to:

$$\pi(N, N^*) = \left[\frac{(1-d)N^* + d}{dN + N^*}\right]^2 L + \left[\frac{1-(1-d)N^*}{dN^* + N}\right]^2 L^* - F$$

The profits of the foreign firms $\pi^*(N, N^*)$ are specular. If the number of firms was the same in both countries, the firms of the larger country would obtain larger profits. However, endogenous entry induces a larger number of firms in the larger country. Once again, opening up to trade induces entry of foreign firms in both markets strengthening competition and reducing the price level, which increases welfare.

other side, sectors with homogenous goods and competition in quantities can be seen as sectors characterized by standard production processes employing low-skilled workers: for these sectors, we have seen that business destruction due to globalization is radical. Further investigations in this direction may help the study of the impact of globalization on unemployment and wage inequality.[20]

In an important recent work, Sutton (2007,a, 2008) has augmented a related model with competition in quantities and with heterogeneity between firms both in costs and quality, and he has confirmed the positive effect of globalization on the international prices.[21] The model emphasizes how global markets are characterized by a lower bound to quality, below which firms cannot sell however low their local wages are in general equilibrium. The range of quality levels of the goods produced shifts upwards after opening up to trade. The key question is then how wage adjustments can compensate for low levels of productivity and quality, and under the assumption that material inputs are independently tradable, the Sutton model shows that productivity differences can be fully offset by wage differences, but differences in quality cannot. This implies that the initial impact of globalization may be associated with a welfare reduction in countries with intermediate levels of capability. Nevertheless, these countries may be the most important gainers as capabilities are transferred in subsequent phases.

Summing up, in the presence of EMSs, the gains from trade largely derive from a reduction in the mark ups associated with more competition, larger production per firm and partial business destruction at the local level. Empirical evidence in support of the double channel through which trade creates gains from trade (increased varieties and reduced prices) can be found in the recent analysis of U.S. imports by Broda and Weinstein (2006).

4.3 The Ghironi-Melitz Model

International market structures characterized by monopolistic behavior or by strategic interactions and intra-industry trade can be introduced in a fully dynamic environment as the one developed in Chapter 3. The pathbreaking work of Ghironi and Melitz (2005) has applied the Krugman framework with trade frictions and export costs to a similar DSGE model of a two-country

[20] Alternative explanations for the recent increase in wage inequality in the Western countries are technology based (see Acemoglu, 1999, Galor and Moav, 2000).

[21] The category of models used by Sutton can be seen as an extension of the model used in this book and based on the utility function $U = \sum_{j=1}^{N} q(j)C(j)$ where $q(j)$ is the quality of good j. This implies that consumers buy only the goods with the best quality-price ratio and competition in quantities between goods with a homogenous price/quality level takes place - there is a loose analogy with Kremer (1993). For an interesting extension see Symeonidis (2000).

economy with trade and endogenous entry of firms in the domestic and foreign markets.[22]

The Ghironi-Melitz model considers a domestic country with a representative agent endowed with a standard intertemporal utility as (3.1) depending on our usual consumption index C_{jt} as (3.5), for different sectors or a representative one, associated with a price index P_t as (3.8). However, the model is enriched with possibly different productivity levels following Melitz (2003). Each firm i in sector j produces:

$$x_{jt}(i) = \omega(i) A_t l_{ijt} \tag{4.16}$$

where l_{ij} is labor employed, A_t is the common TFP and $\omega(i)$ is a firm specific factor which is known by each firm only after entry and is randomly drawn from a given distribution. Exporting abroad requires an extra fixed cost in units of effective labor in every period and transport costs, in form of iceberg costs - selling one unit of goods abroad requires shipping $1/d > 1$ units of good because a fraction d melts on its way. In each period, entry of domestic firms occurs until the discounted value of expected profits equals the fixed costs of entry, and given the exogenous exit rate δ_N, the number of firms follows the same equation of motion as (3.21).

The foreign country is characterized in the same exact way, possibly with different average productivity, entry costs and export costs. Given the nominal exchange rate E_t (units of domestic currency per units of foreign currency) and the price indexes of the two countries P_t and P_t^*, the consumption-based real exchange rate (units of domestic consumption for units of foreign consumption) is defined as $Q_t = E_t P_t^* / P_t$. As we have seen in Section 1.3.1, in a neoclassical world where all goods are traded (as here in the absence of trade frictions) this should be always unitary, so as to satisfy the purchasing power parity (PPP). However, when only some goods are traded, more productive countries are characterized by higher wages and therefore higher prices for the non-tradable goods (*Harrod-Balassa-Samuelson effect*). In the Ghironi-Melitz model, the endogeneity of the number of goods produced, traded and consumed in each country implies further deviations from the PPP.

4.3.1 International trade and business cycle

Let us first characterize the EMSs in the absence of strategic interactions as in the original Ghironi-Melitz model. Each domestic firm chooses a price for domestic demand in home currency and, in case it is productive enough

[22] The core ideas are somewhat related to an earlier work by Baldwin and Krugman (1989), who developed a simple partial equilibrium example to show for the first time that the endogeneity of entry in foreign markets establishes a relation between shocks (including exchange rate shocks) and market structure (with possible forms of hysteresis in trade). I am thankful to Fabio Ghironi for pointing out this early work.

(relative to the export costs), a price for its exports in foreign currency: under monopolistic behavior, the optimal prices are given by a mark up $\mu^M = \theta/(\theta-1)$ over the marginal cost, which is simply $W_t/A_t\omega(i)$ for domestic production, and $(W_t/A_t\omega(i))/dQ_t$ for exports. Total profits are the sum of the profits from domestic and foreign sales. Notice that more productive firms set lower prices but earn higher profits, therefore, given the fixed cost of exports in every period, only the firms with productivity above a certain cut-off (depending on the fixed cost of export) do actually export their goods.[23] This "selection effect" implies that there are gains from trade derived from the greater efficiency of foreign exporters: beyond the gains from variety we have the gains from selection (Melitz, 2003).

Notice that the number of domestic firms N_t and, between them, the number of domestic exporters N_{Xt}, and also the average nominal price of domestic firms at home p_t and abroad p_{Xt} are all endogenous variables. This implies that the domestic price index is:

$$P_t = \left[N_t p_t^{1-\theta} + N_{Xt}^* p_{Xt}^{*1-\theta}\right]^{1/(1-\theta)} = (N_t + N_{Xt}^*)^{1/(1-\theta)} \tilde{P}_t \tag{4.17}$$

where we defined the average nominal price level as $\tilde{P}_t^{\theta-1} = s_t p_t^{1-\theta} + (1-s_t)p_{Xt}^{*1-\theta}$, with $s_t = N_t/(N_t + N_{Xt}^*)$ fraction of domestic firms on the total number of firms selling in the domestic country. Of course, similar formulas apply in the foreign country with:

$$P_t^* = \left[N_t^* p_t^{*1-\theta} + N_{Xt} p_{Xt}^{1-\theta}\right]^{1/(1-\theta)} = (N_t^* + N_{Xt})^{1/(1-\theta)} \tilde{P}_t^* \tag{4.18}$$

Assuming financial autarchy for both countries, the budget constraints correspond to those of a closed economy and usual Euler conditions for bonds and stocks as (3.19) and (3.20) must hold for both countries. Domestic exports can be expressed as:

$$EXP_t = Q_t N_{Xt} \left(\frac{p_{Xt}}{P_t^*}\right)^{1-\theta} C_t^*$$

and domestic imports as:

$$IMP_t = N_{Xt}^* \left(\frac{p_{Xt}^*}{P_t}\right)^{1-\theta} C_t$$

A similar result has been used by Helpman, Melitz and Rubinstein (2008) to generalize the so-called "gravity equation",[24] $IMP_t \propto C_t C_t^* \Upsilon_t(d)$, which

[23] The introduction of heterogeneity between firms and incomplete contracts is key to recent progress in the theory of international trade, outsorcing, foreign direct investments and the organization of multinationals. See also Grossman and Helpman (2005) and the survey by Helpman (2006).

[24] For the standard gravity equation based on the Krugman model see Anderson and van Wincoop (2003).

links bilateral trade flows to the product of the sizes of the two markets and to a factor Υ depending on the trade barriers (in theory related to the transport cost d and in empirical analysis typically associated with the distance between the two countries). They show that taking in consideration the selection effect (and in a multicountry setup of the same existence of trade between two countries) eliminate an important bias of the traditional estimates of trade flows due to the omission of the endogeneity of entry in the export markets.

The Ghironi-Melitz model is closed by the condition for balanced trade, which equates domestic exports, from which we can solve for the real exchange rate:

$$Q_t = \frac{N_{Xt}^*}{N_{Xt}} \left(\frac{P_t^*}{P_t} \frac{p_{Xt}^*}{p_{Xt}} \right)^{1-\theta} \frac{C_t}{C_t^*} \tag{4.19}$$

This guarantees labor market clearing by Walras' law. Notice that the theoretical counterpart to the empirical real exchange rate (which is calculated as the ratio between average nominal price abroad in domestic currency and average nominal price at home) can be derived as:

$$\tilde{Q}_t = \frac{E_t \tilde{P}_t^*}{\tilde{P}_t} = Q_t \left(\frac{N_t^* + N_{Xt}}{N_t + N_{Xt}^*} \right)^{\frac{1}{\theta-1}}$$

and our following discussion will refer to this definition of the real exchange rate.

Imagine that the domestic economy is hit by a temporary increase in aggregate productivity A_t. In a standard model with an exogenous number of firms, this would have the typical effect of increasing domestic demand for both national and foreign goods, exerting an upward pressure on wages. Since foreign productivity is not changed, this would lead to an increase in the relative prices abroad which is in contradiction with a wide empirical evidence. However, in the Ghironi-Melitz model where entry is endogenous, the opposite result occurs. The main reason is that the positive shock increases domestic consumption and profits, which attracts entry of new firms in the domestic economy. This increases labor demand at home compared to abroad, and exerts upward pressure on the domestic wages, which causes the relative price of non-traded goods at home to increase relative to foreign. Meanwhile, the positive impact on the profits of the foreign firms exerts an indirect expansionary effect abroad, which contributes to propagate the boom across countries.

Taking strategic interactions into account in the Ghironi-Melitz model would augment this mechanism with a competition effect analogous to the one emerging in the closed economy of the previous chapter. Entry in the domestic country would strengthen competition, reduce the mark ups and increase the real wages even more, dampening to some extent the increase in the number of new firms - once again, in the presence of full fledged EMSs, globalization brings more benefits through a tendency toward mark up reduction rather than through increased number of goods.[25] The entry process would increase labor demand at home, putting additional pressure for an increase in the domestic wage. Moreover, effects of intertemporal substitution similar to those studied in the previous chapter would strengthen the domestic boom and the associated entry mechanisms.

Another phenomenon emphasized by Ghironi and Melitz (2005) has to do with the endogenous number of non-tradable goods. The higher relative cost of labor at home reduces the fraction of domestic exporters and increases the fraction of foreign ones. Since only the most efficient firms export their goods, the average price of domestic imports increases (because of the new and more costly imported goods) and the average price of foreign imports decreases (because only less costly imported goods remain).

Finally, the relative increase in the number of domestic goods induces the domestic consumers to switch toward domestic production. As long as this is relatively more expensive than the imported production (which arrives from the most efficient foreign firms), there is third effect in favor of the appreciation of the real exchange rate. We can summarize as follows the main result of Ghironi and Melitz (2005):

PROPOSITION 4.2. **Under endogenous market structures in a dynamic two-country model, an increase in domestic productivity generates a real exchange rate appreciation due to higher prices of domestic non-tradables and imports, and to a larger demand for domestic goods, which leads to persistent deviations from Purchasing Power Parity.**

Ghironi and Melitz (2005) calibrate and simulate the model confirming these results, and showing that a positive technology shock at home increases the average domestic price level compared to the foreign one so as to appreciate the empirical real exchange rate ($\tilde{Q}_t$ decreases). Nevertheless, the shock increases enough the number of goods consumed at home that the consumption-based real exchange rate depreciates (Q_t increases): as intuitive, the shock is such that the consumer derives higher utility from spending the same amount in the home market with higher prices but also more goods available.

4.3.2 International finance and macroeconomic policy

Similar mechanisms to those found under financial autarchy would take place in case of international trade in bonds issued by the two countries. In such a

[25] Therefore gains from trade would derive from three sources: 1) gains from variety *à la* Krugman, 2) gains from lower mark ups due to EMSs, and 3) gains from selection *à la* Melitz. The first extension of the Melitz (2003) model to EMSs with competition in quantities and homogenous goods is due to Long, Raff and Stähler (2009).

set-up the domestic current account reflects the changes in aggregate holdings of domestic and foreign bonds, say $\mathbf{B}_t$ and $\mathbf{B}_{*t}$:

$$CA_t = \mathbf{B}_{t+1} - \mathbf{B}_t + Q_t \left(\mathbf{B}_{*t+1} - \mathbf{B}_{*t}\right) \tag{4.20}$$

which must match the opposite changes of the foreign country. Bond trading allows the domestic economy hit by a positive productivity shock to finance entry of new firms by running a CA deficit. Home households borrow from abroad to finance higher initial investments in new firms, while foreign households share the benefits of higher domestic productivity by lending. However, in case of a temporary shock, the domestic economy runs a CA deficit initially, but a surplus afterward.

The model can be also simulated to verify its cyclical properties. In contrast to standard international business cycle models (Backus, Kehoe and Kydland, 1992), this framework generates positive GDP correlation across countries, it does not automatically produce high correlation between relative consumption and the real exchange rate, and it substantially reduces the so-called "consumption-output anomaly" (the counterfactual prediction that consumption is more strongly correlated across countries than output).

Finally, the Ghironi-Melitz model can be used to verify the reaction of the economy to domestic deregulation, defined as a reduction in the costs of entry (this attracts investment in the creation of new firms and puts upward pressure on the relative cost of labor at home), or to a reduction in the trade frictions (which again induces an increase in the number of varieties available for the consumers). Ghironi and Stebunovs (2007) have also extended the model with an imperfectly competitive banking sector to study financial deregulation (as a reduction in the market power of financial intermediaries): this enhances investments in business creation, and induces a CA deficit of the country which deregulates, with an expansion in both countries.

The framework analyzed in this section could be extended with a more accurate consideration of the forms of competition between firms. For instance, an interesting work by Atkeson and Burnstein (2008) has introduced strategic interactions in a related model, managing to produce deviations from relative PPP similar to those observed in the data (mainly because firms choose to price-to-market).

Trade policy can be studied in a similar dynamic framework with EMSs. For instance, a generalized reduction in the trade barriers induces an increase in the number of exporters and in the average productivity (as a consequence of the selection effect), a smaller reduction in the number of domestic firms for both countries, and an increase in welfare. Richer consequences would emerge in the presence of strategic interactions between firms. In such a case, unilateral trade policy could be used for profit shifting reasons and to provide a strategic advantage to the domestic firms, at least in the short run. However, simpler static models as those of Section 4.2 (focusing on the domestic or the foreign market) can provide better insights on the role of trade policy in international markets with EMSs, as we will see in the next sections.

Fiscal policy and monetary and exchange rate policy could be studied within the Ghironi-Melitz model as well. In a preliminary work, Cavallari (2007, 2008) has augmented a related model in the tradition of the New open economy macroeconomics and studied macroeconomic policy.[26] In the presence of price rigidities, monetary expansions (in case of a flexible exchange rate regime) and devaluations (in case of a fixed exchange rate regime) would induce real effects on the economy in the short run, and in particular they would affect the endogenous number of goods produced, traded and consumed in each country. Moreover, these policies could provide temporary strategic advantages to the domestic firms that cannot emerge in a neoclassical framework with perfect competition.

After working out such a complex dynamic model (which will certainly be extremely useful for future research), we need to remind the reader that interesting insights on policy issues can be obtained from simpler and static models with EMSs as those of Section 4.2. To these models we return now for the rest of the chapter, examining trade policy first and then exchange rate policy and R&D policy.

4.4 Trade Policy and Import Tariffs

In this section we study optimal trade policy for a domestic market in the simplest static framework. As we know from Chapter 1, the neoclassical theory of trade policy suggests that a small open economy should endorse a free trade policy, and only a large country able to affect its terms of trade should adopt a positive tariff on imports so as to improve its terms of trade. Moreover, such a policy is equivalent to an export tax, therefore the above result implies the optimality of a positive export tax for large countries - a result on which we will return in the next section.

In the presence of imperfect competition, the incentives to adopt tariffs depend on profit shifting reasons as well. Once again, the entry conditions are crucial. Endogenous entry has been considered in the literature only to study special cases and, in particular, the case of monopolistic behavior in a model with two countries: in this setup Helpman and Krugman (1989) have confirmed the optimality of a positive tariff for a large country. A general treatment taking fully fledged EMSs into account is much more complicated: a system of tariffs on imports and subsidies to domestic production should balance the effects on consumer surplus, on the profits of the domestic firm and on the net tariff revenue taking into account both strategic interactions

[26] The fundamental application of the New-Keynesian framework to the study of international finance solving for both the current account and the exchange rate is due to Obstfeld and Rogoff (1995, 1996, 2000).

and entry of foreign firms. We cannot provide such a general analysis (the same concept of domestic welfare is model-dependent), but we can focus on a simple example which shows that the traditional protectionist outcome is substantially robust to the introduction of EMSs.

For the sake of simplicity, we will consider a model of competition in quantities for the domestic market where the national government chooses a specific tariff t on the imports of all the foreign firms and possibly also a specific production subsidy s on the sales of a single national firm. Our aim is to compare the optimal trade policy when entry is exogenous (the typical case analyzed in the literature) and when it is endogenous.

Assume a linear inverse demand $p = a - X$ where X is total production in the domestic market. The constant marginal cost of production is c and the fixed cost is F. The profit function of the domestic firm is:

$$\pi = (a - X + s)x - cx - F \tag{4.21}$$

where x is its own production. The profit function of a representative foreign firm is:

$$\pi^* = (a - X - t)x^* - cx^* - F \tag{4.22}$$

where x^* is its production. Notice that we could think of the fixed cost F that constraints entry as the fixed cost of export that determines which international firms are present in the domestic market.

In Cournot equilibrium with N firms, the production levels are respectively:

$$x(s,t) = \frac{a - c + (N-1)t + Ns}{1+N} \quad \text{and} \quad x^*(s,t) = \frac{a - c - 2t - s}{1+N}$$

and domestic welfare, defined as the sum of consumer surplus, domestic profits and tariff revenue net of the cost of subsidies, is:

$$W(s,t) = \underbrace{\frac{[x(s,t) + (N-1)x^*(s,t)]^2}{2}}_{\text{Consumer surplus}} + \underbrace{x(s,t)^2 - F}_{\text{Domestic profits}}$$
$$+ \underbrace{t(N-1)x^*(s,t)}_{\text{Tariff revenue}} - \underbrace{sx(s,t)}_{\text{Subsidy cost}} \tag{4.23}$$

In case the government can only use the tariff (and $s = 0$), its welfare maximizing level can be derived as:[27]

$$t = \frac{3(a-c)}{7+N} \tag{4.24}$$

[27] This optimal tariff is a particular case of the one derived by Bhattacharjea (1995) for the case of multiple domestic firms.

which is associated with entry of some foreign firms in the domestic market (as long as the fixed cost is low enough). However, if both instruments are available, welfare is maximized by $\tilde{t} = 0$ and $\tilde{s} = a - c$, that is by setting the price equal to the marginal cost and driving out of the market all foreign firms. This entry deterrence result (obtained through a positive subsidy that equates the price to the marginal cost) emerges with more general demand functions as long as the goods are homogenous.

Now imagine that entry of foreign firms in the domestic market is endogenous. As long as there are foreign firms in the market and $t + s > 0$, the zero profit condition must be binding on them, which implies:

$$N = \frac{a - c - 2t - s}{\sqrt{F}} - 1$$

with $x^* = \sqrt{F}$, $x = \sqrt{F} + s + t$ and total production:

$$X = a - c - t - \sqrt{F}$$

This equilibrium is consistent with $N \geq 2$ for $t \leq (a - c - s - 3\sqrt{F})/2$. If this is the case, welfare becomes:

$$W(s,t) = \underbrace{\frac{\left(a - c - t - \sqrt{F}\right)^2}{2}}_{\text{Consumer surplus}} + \underbrace{\left(\sqrt{F} + s + t\right)^2 - F}_{\text{Domestic profits}} + \underbrace{+t\left(a - 2t - s - \sqrt{F}\right)}_{\text{Tariff revenue}} - \underbrace{s\left(\sqrt{F} + s + t\right)}_{\text{Subsidy cost}} \tag{4.25}$$

For any given small subsidy, and in particular in the case of no subsidy (when the government can just use the tariff and $s = 0$), the optimal unilateral tariff can be derived as:

$$t = \sqrt{F} \tag{4.26}$$

which is consistent with entry of some foreign firms as long as the fixed cost is small enough (precisely if $F < (a - c)^2/25$). Otherwise, it is optimal to set the prohibitive tariff:

$$\bar{t} = \frac{a - c - 3\sqrt{F}}{2}$$

However, notice that welfare turns out to be a linearly increasing function of the subsidy. Again, if both the instruments are available, the first best can be obtained by setting $\tilde{t} = 0$ and $\tilde{s} = a - c$.

These results can be generalized to the case of imperfect substitutability between products and competition in quantities, for instance with inverse demand functions derived from the utility (4.2). In such a case it would be optimal to adopt non-prohibitive tariffs on imports of some foreign differentiated products, because consumer welfare is enhanced by variety, but the domestic firm would keep producing more than any other foreign firm for the domestic market.

In a model of competition in prices with product differentiation, as that derived from the utility (4.2), it is easy to show that domestic welfare is maximized when the tariff and the subsidy maximize the sum of tariff revenue and the profits of the national firm net of its subsidies, taking into account that the number of foreign firms is endogenous and negatively related to the tariff. Despite a closed form solution is not available, in this case it is still optimal to choose a non-prohibitive tariff on imports and (possibly) a positive subsidy to domestic production.

In general, we can state the following principle:

PROPOSITION 4.3. **Under endogenous market structures in the domestic market, the optimal trade policy implies a positive import tariff and (possibly) a production subsidy for the domestic firm, leaving positive profits to the latter.**

It is important to remind the reader that we are referring to optimal unilateral policies. As well known, also in this contest retaliation by foreign countries by imposing similar import tariffs would lead to an inefficient Nash equilibrium trade policy. Moreover, keeping as neutral the policies of the foreign countries for their exporting firms has implied a substantial loss of generality: these countries may have strong incentives to support their firms with export taxes or, more realistically, with export subsidies, with consequences for the domestic market as well. We now turn to the complementary issue of the optimal policies that promote exports, which will deliver new policy recommendations compared to the traditional outcomes.

4.5 Strategic Export Promotion

Common wisdom on the benefits of export subsidization largely departs from the implications of trade theory. While export promotion is often seen as welfare enhancing at least in the short run and often supported by governments, theory is hardly in favor of its direct or indirect implementation. The classical theory since Smith (1776) has been strongly against forms of protectionism including export premiums, or bounties, and tariff rebates - which, nevertheless, were used in England at that time.[28] The neoclassical theory based on perfect competition (and analyzed in Section 1.3) associates the scope of

[28] See Irwin (1991) for an interesting application to strategic trade policy in the Netherlands and England engaged in trade in East India in the XVII century. In that case, Dutch supremacy was not reached through subsidies but through managerial incentives to shipping companies.

trade policy with improvements in the terms of trade, that is the price of exports relative to the price of imports: as long as a country is large enough to affect the terms of trade, the optimal unilateral policy is to set a positive tax on exports to increase the relative price of exports (see Helpman and Krugman, 1989). The new trade theory based on imperfect competition associates the scope of strategic trade policy with profit shifting toward the domestic firms (Brander and Spencer, 1985). Eaton and Grossman (1986) have shown that the optimal unilateral policy is an export tax under price competition, or whenever strategic complementarity holds. Under quantity competition, an export subsidy can be optimal only under certain conditions (Dixit, 1984). Moreover, the recent literature has developed many other arguments against export subsidies, as in case of passive foreign competitors (Rodrik, 1989; De Santis and Stahler, 2001) or in case of asymmetric information between firms and government. Finally, trade wars lead to inefficient Nash equilibrium trade policy, and this result supports once again the optimality of a commitment to free trade. Such a view is broadly accepted by the World Trade Organization which coordinates international trade policies and forbids export subsidies, except for agricultural products.[29]

Nevertheless, different forms of direct or indirect export subsidies are widespread in the industrialized and developing world. It appears quite surprising that, in front of this, trade economists do not have clear and unambiguous arguments to explain why export subsidies could be the optimal unilateral trade policy. Etro (2010) provides such an argument, studying a model of trade policy for a foreign market with EMSs.[30] Under endogenous entry, export subsidization is always the best unilateral policy both under quantity and price competition, or, more generally, under strategic substitutability and strategic complementarity. The intuition is simple. While firms are playing some kind of Nash competition in the foreign market, a government can give a strategic advantage to its domestic firm with an appropriate trade policy. When entry is free, an incentive to be accommodating is always counterproductive, because it just promotes entry by other foreign firms and shifts profits away from the domestic one. It is rather optimal to provide an incentive to be aggressive, that is to expand production or (equivalently) lower the price, since this behavior limits entry increasing the market share of the domestic firm. This is only possible by subsidizing its exports.[31]

4.5.1 A model of international competition

In this section we adopt the general model of market structures introduced in Section 2.1 to study international competition and to evaluate the impact of

[29] See Bhagwati (1988) for a discussion.

[30] Notice that free entry is a realistic assumption since a foreign country without a domestic firm in the market can only gain from allowing free entry of international firms.

[31] See Boone, Ionescu and Zigic (2006) for a related analysis.

strategic export promotion. Consider an international market where N firms from different countries are competing in Nash strategies. Assume that each firm chooses a strategic variable $x(i)$ with $i = 1, 2, ..., N$ which delivers the net profit function:

$$\pi_i = \Pi\left[x(i), \beta_i, s_i\right] - F \tag{4.27}$$

where F is a fixed cost of production that, given the issue at hand, can be interpreted as the fixed cost of export that is key to determine which international firms are active in the foreign market. The spillovers from the strategies of the other firms are summarized by $\beta_i = \sum_{j \neq i} h(x(j))$ for a positive and increasing $h(x)$ function. In this case, s_i is the export policy chosen by the government of country i: in our main application, this is an export subsidy, but we will take in consideration also other policies. Assuming without loss of generality that an increase in the policy raises profits, $\Pi_3 > 0$, we define s_i as an export promotion policy for country i. We allow Π_{13} to be positive or negative: in the first case, the policy increases marginal profitability. All forms of trade subsidies under quantity and price competition imply $\Pi_{13} > 0$, but other indirect forms of export promotion can be characterized by $\Pi_{13} < 0$.

The welfare of country i, $W(s_i)$, depends positively on the profits of the domestic firm and negatively on the cost of its policy. In case of export subsidization, the cost of trade policy is the collection of tax revenue, but this may imply tax distortions or other kinds of costs due to general equilibrium or political considerations. Moreover, in case of lobbying activity, the weight given by the politicians to the social costs of the policy may be smaller. Finally, other forms of export promotion can have different costs for national welfare. Nevertheless, our focus will be on the strategic incentive to export promotion, which will be defined as the indirect marginal benefit of an increase in s_i on the profit, $SI = \Pi_2\left(x(i), \beta_i, s_i\right) \partial\beta_i/\partial s_i$. As long as this is positive, the government of country i has a strategic reason to promote exports beyond any direct reason which depends on the first order impact of the policy on welfare.

4.5.2 Optimal trade policy *à la* Brander-Spencer

Let us briefly generalize the standard results (Brander and Spencer, 1985, Eaton and Grossman, 1986) on the optimal unilateral trade policy for a foreign market with a fixed number of firms. More specifically, assume that $s_i = 0$ for all firms except the domestic one, whose policy s is chosen by the government of its home country at an initial stage.

Consider the second stage after a policy s has been chosen. We assume that an equilibrium exists with the same strategy for each foreign firm, say x^*, and a different strategy for the domestic one, say x, depending on the policy s. The first order equilibrium conditions are:

$$\Pi_1^*\left[x^*, (N-2)h(x^*) + h(x), 0\right] = 0 \tag{4.28}$$

$$\Pi_1\left[x,(N-1)h(x^*),s\right]=0 \tag{4.29}$$

Let us totally differentiate the system under the stability assumption, which requires a positive determinant Δ. The equilibrium strategies $x^* = x^*(s)$ and $x = x(s)$ are two functions with $dx^*/ds \gtrless 0$ if $\Pi^*_{12}\Pi_{13} \gtrless 0$ and $dx/ds \gtrless 0$ if $\Pi_{13} \gtrless 0$.

In the initial stage the government chooses a policy to maximize welfare $W(s)$. Using the envelope theorem and the previous results, we obtain the strategic incentive to export promotion as:

$$SI = \frac{(N-1)h'(x^*)h'(x)\Pi_2\Pi^*_{12}\Pi_{13}}{\Delta} \tag{4.30}$$

When $\Pi_{13} > 0$ this effect is positive under strategic substitutability ($\Pi^*_{12} < 0$) and negative under strategic complementarity ($\Pi^*_{12} > 0$), while the result is inverted when $\Pi_{13} < 0$. It is now immediate to conclude with:

PROPOSITION 4.4. **With an exogenous number of firms in the international market, there is (not) a strategic incentive to adopt a policy that increases the marginal profitability of the domestic firm if strategic substitutability (complementarity) holds, and *vice versa* for a policy that decreases the marginal profitability of the domestic firm.**

The optimal policy implies an aggressive firm under strategic substitutability (as in Brander and Spencer, 1985) and an accommodating firm under strategic complementarity (as in Eaton and Grossman, 1986). For instance, assuming $\Pi_{13} > 0$, a strategic incentive to promote exports emerges in models of competition in quantities but not in models of competition in prices. This is the kind of ambiguity in the normative predictions that has largely limited the utility of the modern literature on strategic trade policy.

Notice that the above result is not affected by the (exogenous) number of international firms, but it is sensitive to the number of domestic firms. In particular, when the number of domestic firms is high there may not be a strategic incentive to export promotion even in models of competition in quantities. As Dixit (1984) has shown this is due to a terms of trade effect: subsidies increase production and reduce the equilibrium price. As in the neoclassical theory of trade policy, this terms of trade effect supports the optimality of export taxes.

4.5.3 Optimal trade policy with EMSs

Following Etro (2010), we now examine the general nature of the optimal policy when the structure of the international market is endogenous. In the next section we will characterize the optimal policy for a number of examples.

Let us assume that the number of potential entrants is great enough that a zero profit condition pins down the effective number of firms competing in the foreign market. The equilibrium conditions are the two first order conditions, (4.28) and (4.29), and the zero profit condition which binds on the international firms (since these do not profit from trade policy):

$$\Pi^* \left[x^*, (N-2)h(x^*) + h(x), 0\right] = F \tag{4.31}$$

Totally differentiating the system we can derive a fundamental result for what follows:

LEMMA 4.1. **Under endogenous entry in the international market, a change in the domestic policy does not affect the equilibrium strategy of all the other firms but only their equilibrium number.**

When the domestic policy is changed, the marginal profitability of the strategy of the domestic firm changes and its optimal strategy changes as well. Nevertheless, the policy change does not affect the marginal profitability for the other firms, and any impact on the competition emerges through an impact on the number of competitors.

More specifically, notice that optimization by the foreign firms and the free entry condition constraining their number pin down both the strategy of each firm and the level of spillovers that each firm receives from the strategies of the other international firms and the domestic firm. In other words, the equilibrium determines x^* and β^* independently from s. Since the domestic policy affects the strategy of the domestic firm but not the spillover $\beta^* = (N-2)h(x^*)+h\left[x(s)\right]$, it follows that the number of firms must be influenced by the domestic policy. In particular, we have:

$$\frac{dN}{ds} = \frac{h'(x)\Pi_{13}/h(x^*)}{\Pi_{11} - h'(x)\Pi_{12}} \lesseqgtr 0 \text{ and } \frac{dx}{ds} = -\frac{\Pi_{13}}{\Pi_{11} - h'(x)\Pi_{12}} \gtreqless 0 \text{ if } \Pi_{13} \gtreqless 0$$

while $dx^*/ds = d\beta^*/s = 0$. In the initial stage, the government chooses the policy to maximize welfare. Using the envelope theorem and the previous results, we obtain the strategic incentive to promote exports:

$$SI = \frac{h'(x)\Pi_2\Pi_{13}}{\Pi_{11} - h'(x)\Pi_{12}} \tag{4.32}$$

whose sign is the sign of Π_{13}. Therefore, we have:

PROPOSITION 4.5. **Under endogenous entry in the international market, there is (not) a strategic incentive to adopt a policy that increases (decreases) the marginal profitability of the domestic firm.**

Notice that the result would not change in the presence of more than one domestic firm, as long as some entry of foreign firms takes place in equilibrium. Actually, it is immediate to verify that with N_H domestic firms, the equilibrium strategy of each firm would not change and the strategic incentive to promote exports would be $SI(N_H) = N_H SI(1)$. Contrary to the case of exogenous entry, here there is not the traditional terms of trade effect induced by an export promoting policy, because each domestic firm crowds out foreign firms accumulating positive profits.

4.6 Trade Policy and Export Subsidies

In this section we apply the previous results to the theory of strategic trade policy, analyzing both models with competition in quantities and prices and deriving the optimal unilateral trade policy in both cases. The focus is going to be on specific subsidies, but similar results could be obtained with *ad valorem* subsidies.

4.6.1 Optimal trade policy with Cournot competition

Consider a general model of competition in quantities with a specific subsidy. The general profit function of a firm i obtaining a subsidy s_i is:

$$\pi_i = x(i)[p(x(i), \beta_i) + s_i] - c(x(i)) - F \tag{4.33}$$

where $x(i)$ is its production and $\beta_i = \sum_{j \neq i} h(x(j))$, while $c(x)$ is a positive and increasing cost function. One can easily verify that in this case the export subsidy increases the marginal profitability of the subsidized firm:

$$\Pi_{13} = 1 > 0 \tag{4.34}$$

Given the subsidy for the domestic firm, the equilibrium conditions in the competition stage are:

$$p(x^*, \beta^*) + x^* p_1(x^*, \beta^*) = c'(x^*), \qquad s + p(x, \beta) + z p_1(x, \beta) = c'(x)$$

where $\beta^* = (N-2)h(x^*) + h(x)$ is the spillover received by an international firm from the strategies of all the other firms in the market and $\beta = (N-1)h(x^*)$ is the spillover for the domestic firm. Under barriers to entry the two conditions above define $x^*(s)$ and $x(s)$. Comparative statics analysis shows that $x^{*\prime}(s) < 0$ as long as strategic substitutability holds, and $x'(s) < 0$ always. Assuming that the welfare cost of subsidization is just the necessary tax revenue, the welfare function of the home country can be expressed as follows:

$$W(s) = x(s)\left\{p\left[z(s), (N-1)h(x^*(s)\right] - c[x(s)]\right\}$$

which is maximized by a subsidy $s_H(N)$ implicitly defined as:

$$s_H(N) = \frac{x(\cdot)(N-1)h'[x^*(\cdot)]x^{*\prime}(\cdot)p_2}{x'(\cdot)}$$

Therefore, the optimal export subsidy is positive under strategic substitutability ($x'(s) < 0$) and negative under strategic complementarity ($x^{*\prime}(s) > 0$). For instance, in case of perfectly substitutable goods we have:

$$s_H(N) = \frac{p}{\varepsilon\left\{1 + 1/\left[N(1-\Sigma)\right]\right\}}$$

where $\varepsilon \equiv -p/xp'$ is the elasticity of demand (with respect to domestic production) and $\Sigma \equiv -zp''/p'$ is the elasticity of the slope of the inverse demand which represents the degree of convexity of demand. More specifically, consider the case of linear demand $p = a - X$. Assuming a constant marginal cost c, the optimal subsidy can be derived as:

$$s_H(N) = \frac{(a-c)(N-1)}{2N} > 0 \tag{4.35}$$

which is increasing in the number of international firms.

Consider now the case of EMSs in the international market. When entry is endogenous, the productive stage is characterized not only by the optimality conditions for the domestic firm and for the representative foreign firm, but also from the free entry condition:

$$x^* p(x^*, \beta^*) - c(x^*) = F$$

The equilibrium system defines output of each firm and the number of firms as functions of the subsidy s. We know from Lemma 4.1 that the production of the foreign firms x^* and their spillovers β^* do not depend on the subsidy, while $x(s)$ and $\beta(s)$ depend on it. Therefore, we can write the welfare of the domestic country as follows:

$$\begin{aligned} W(s) &= x(s)p\left(x(s), \beta(s)\right) - c(x) - F = \\ &= x(s)p\left[x(s), \beta^* + h(x^*) - h(x)\right] - c(x) - F \end{aligned}$$

Welfare maximization entails an interior solution for the optimal subsidy if goods are poor substitutes or if marginal costs are increasing enough; if this is not the case, a prohibitive subsidy is optimal. When an interior solution exists, it must satisfy the first order condition:

$$p\left(x(s), \beta\right) + x(s)\left[p_1\left(x(s), \beta\right) - p_2\left(x(s), \beta\right) h'(x)\right] = c'(x)$$

which is a complex implicit expression. Nevertheless, if we substitute this in the equilibrium first order condition for the domestic firm, we can derive a neat expression for the optimal export subsidy (Etro, 2010):

$$s_H = \left[-p_2\left(x, \beta\right) h'(x)\right] x \tag{4.36}$$

whose sign is always positive. We now verify this in a number of examples.

The case of isoelastic sub-utility. Let us consider the usual isoelastic preferences (4.2), which lead to the inverse demand:

$$p\left(x, \beta\right) = \frac{x^{-\frac{1}{\theta}} L}{x^{1-\frac{1}{\theta}} + \beta}$$

with $h(x) = x^{1-1/\theta}$. Applying our formula we obtain the optimal subsidy $s_H = (\theta - 1)p^2 x/\theta L$, which leads the domestic firm to sell its good at a price $p = c\theta/(\theta - 1)$ lower than the other firms. This allows us to rewrite the optimal export subsidy as:

$$s_H = \frac{cx}{L} p > 0 \tag{4.37}$$

which is decreasing in the size of the international market L. In other words, large countries exporting in small markets (with small L) should adopt large subsidies for their exporting firms, while small open economies (exporting to large markets) should tend to commit to free trade.

The case of homogenous goods. In general, the upper bound for the optimal subsidy emerges in the case of homogenous goods, which is consistent with a non-prohibitive subsidy if the marginal costs are increasing enough. Using the previous definition of the elasticity of demand $\varepsilon \equiv -p/xp'$, the optimal export subsidy (4.36) simplifies to (Etro, 2010):

$$s_H = \frac{p}{\varepsilon} > 0 \tag{4.38}$$

Under perfect substitutability, the optimal subsidy as a percentage of the price of the domestic firm is equal to the inverse of the elasticity of the international demand to the price: the more elastic is demand the smaller is the optimal subsidy, and this tends to zero when demand becomes infinitely elastic. This is exactly the opposite of the optimal trade policy in the neoclassical international trade model (Lerner, 1934), where the optimal policy was an export tax equal to p/ε because this would have optimized the terms of trade.

Notice that the optimal subsidy implies that domestic firms produce until their marginal cost equates the equilibrium price ($p = c'(x)$) and enjoy positive profits because of decreasing returns to scale. Finally, we should remark that the optimal subsidy would be the same in the presence of other domestic firms: there is not a terms of trade effect here because the equilibrium price is independent from the subsidy, while domestic firms crowd out foreign ones.

The case of linear demand and quadratic costs. As a further example of the homogenous goods case, consider the a linear demand function $p = a - X$ together with a quadratic cost function $c(x) = x^2/2$. Looking at the Cournot equilibrium between N firms for a given subsidy s of the domestic firm, and imposing the free entry condition, we obtain the equilibrium production for each international firm $x^* = \sqrt{2F/3}$ and the number of firms $N = (a - s/2)\sqrt{3/2F} - 2$, which imply total production $X = a - \sqrt{8F/3}$. Consistently with Lemma 4.1, the subsidy does not affect the individual production of the other firms, but decreases their number. The equilibrium production of the subsidized firm is $x(s) = \sqrt{2F/3} + s/2$, which generates profits $\Pi = (3/8)\left(s + \sqrt{8F/3}\right)^2 - F$. The government maximizes profits net of the tax revenue necessary to finance the subsidies:

$$W(s) = x(s)\sqrt{\frac{8F}{3}} - \frac{x(s)^2}{2} - F$$

This welfare function is maximized by:

$$s_H = \sqrt{\frac{8F}{3}} > 0 \tag{4.39}$$

which implies that the domestic firm produces the double than any other international firm. Its net profits are $\pi_H = 3F$ and domestic welfare is $W = F/3$.[32]

The case for a prohibitive subsidy. When there is low substitutability between goods and the marginal costs are constant or decreasing (or even not too much increasing), it becomes optimal to set a prohibitive subsidy, that is a subsidy large enough that entry of international firms is deterred. Such a subsidy has to satisfy the free entry condition for N slightly smaller than 2, which implies that just one firm (the domestic firm) can profitably remain in the market:

$$x^* p\left[x^*, x(s_H)\right] - c(x^*) = F$$

For instance, consider the linear demand case with constant marginal cost c. Imagining that there is entry in equilibrium and imposing the free entry condition for a given subsidy s, we obtain the equilibrium production for each international firm $x^* = \sqrt{F}$ and the number of firms $N = (a - c - s)/\sqrt{F} - 1$, which imply total production $X = a - c - \sqrt{F}$. Once again, the subsidy does not affect the level of production of the other firms but decreases their number. The equilibrium production of the subsidized firm is instead $x(s) = \sqrt{F} + s$, which generates profits $\Pi = (\sqrt{F} + s)^2$. The government maximizes these profits net of the tax revenue necessary to finance the subsidies:

$$W(s) = (\sqrt{F} + s)^2 - s\left(\sqrt{F} + s\right) - F \tag{4.40}$$

Since this is a linearly increasing function of s, it is optimal to increase subsidization as long as there is entry. But entry is deterred at:

$$s_H = a - c - 3\sqrt{F} > 0 \tag{4.41}$$

which is the optimal subsidy. The domestic firm obtains positive profits $\Pi = (a - c - 2\sqrt{F})^2$ which are larger than the cost of the subsidy $s_H x(s_H)$. The intuition for this result is the following. Free entry pins down the equilibrium price level of the foreign firms as long as some of them enter. This implies that the choice of the subsidy does not affect the equilibrium price at

[32] Notice that when the fixed cost of entry decreases, the level of concentration in the market is reduced and the optimal subsidy goes down: in the limit case of perfect competition (zero fixed costs) we obtain the traditional result for which free trade is optimal.

which the domestic firm sells its production, but simply increases its market share. Since there are positive fixed costs of production, an increase in the market share reduces the average costs and, therefore, increases the net profits. Consequently it is optimal to raise the market share as much as possible, which amounts to full entry deterrence.

We can summarize our results as follows:

PROPOSITION 4.6. **Under quantity competition and endogenous entry in the international market, a positive export subsidy is always optimal, since it helps the domestic firm to produce more than the international firms in the foreign market.**

4.6.2 Optimal trade policy with Bertrand competition

Consider a general model of price competition with a specific subsidy such that the profit function of firm i with a subsidy s_i is:

$$\pi_i = (p_i - c + s_i) D (p_i, \beta_i) - F \tag{4.42}$$

where c is the constant marginal cost and the demand function is decreasing in both arguments with $\beta_i = \sum_{j \neq i} g(p_j)$ for positive and decreasing functions $g(p)$: this implies that the demand for the domestic good is decreasing in its price and increasing in all the other prices. Notice that by setting $x = 1/p$ and $h(x) = g(1/x)$, the profit function is nested in the general model.

The gross profits of the domestic firm are:

$$\Pi = (p - c + s) D (p, \beta) \tag{4.43}$$

and we have:

$$\Pi_{13} = -\frac{D_1}{x^2} > 0 \tag{4.44}$$

Let us start from the case of an exogenous number of firms N. The first order equilibrium conditions in the second stage for the foreign and domestic firms are:

$$(p^* - c)D_1(p^*, \beta) + D(p^*, \beta) = 0, \quad (p - c + s)D_1(p, \beta) + D(p, \beta) = 0$$

where $\beta^* = (N-2)g(p^*) + g(p)$ is the spillover received by an international firm from the strategies of all the other firms in the market and $\beta = (N-1)g(p^*)$ is the spillover for the domestic firm. This system provides the prices and the number of firms as functions of the subsidy s, with $p'(s) < 0$ and $p^{*\prime}(s) < 0$. Domestic welfare is:

$$W(s) = [p(s) - c] D \{p(s), (N-1)g[p^*(s)]\}$$

which is maximized by:

$$s_H(N) = \frac{(p-c)D_2 g'(p^*)p^{*\prime}(N-1)}{D_1 p'} < 0$$

Even if this is an implicit expression, it shows that the optimal export policy under price competition requires an export tax, as first suggested by Eaton and Grossman (1986).

Let us move on to the analysis of EMSs. When entry is endogenous, beyond the first order conditions above, we need to take into account the free entry condition:

$$(p^* - c)D(p^*, \beta^*) = F$$

Lemma 4.1 tells us that the price of the foreign firms p^* and their spillover β^* do not change with the subsidy, while the price of the domestic firm $p(s)$ and its spillover $\beta(s)$ depend on the subsidy. Therefore, we can write the welfare of the domestic country as:

$$\begin{aligned} W(s) &= [p(s) - c]\, D\,[p(s), \beta(s)] - F = \\ &= [p(s) - c]\, D\,[p(s), \beta^* + g(p^*) - g(p(s))] - F \end{aligned}$$

which is maximized by a subsidy satisfying the first order condition:

$$D\,(p, \beta) + (p - c)\,[D_1\,(p, \beta) - D_2\,(p, \beta)\, g'(p)] = 0 \tag{4.45}$$

If we now substitute this in the equilibrium first order condition for the domestic firm, we can derive a neater expression for the optimal export subsidy (Etro, 2010):

$$s_H = \frac{(p-c)D_2\,(p,\beta)\, g'(p)}{[-D_1\,(p,\beta)]} > 0$$

Of course this is an implicit expression, since on the right hand side p depends on the optimal subsidy. However, this expression makes clear our main result, which is in contradiction with the celebrated result by Eaton and Grossman (1986): the optimal export subsidy is always positive. Summarizing, we have:

PROPOSITION 4.7. **Under price competition and endogenous entry in the international market, an export subsidy is always optimal, since it helps the domestic firm to lower its price in the foreign market.**

We can also notice that the optimal subsidy as a percentage of the price of the domestic firm must be smaller than the inverse of the demand elasticity, which was the optimal subsidy under competition in quantities: in other words, the inverse of the demand elasticity represents an upper bound for the size of the optimal export subsidy.

This result overturns common wisdom for models with strategic complementarity and barriers to entry. Inducing an accommodating behavior of the domestic firm is not optimal anymore because it would just attract entry of new international firms. The only chance for the government to increase welfare is to induce an aggressive behavior. Then, the domestic firm undercuts the competitors gaining a larger market share, and spreads a lower mark up over a large portion of the market, leaving the few remaining firms with zero profits.

The case of isoelastic sub-utility. The direct demand function derived from our usual isoelastic preferences (4.2) is

$$D(p,\beta) = \frac{Lp^{-\theta}}{p^{1-\theta}+\beta} \tag{4.46}$$

with $g(p) = p^{1-\theta}$. Under the optimal policy, the equilibrium price of the domestic firm must be $p(s_H) = \theta c/(\theta-1)$ - exactly as in the case of competition in quantities - and the foreign firms must sell at the higher price $p^* = \theta cL/(\theta-1)(L-F)$. The associated optimal export subsidy can be derived as:

$$s_H = \frac{c}{(\theta-1)\left\{(1-F/L)^{\frac{1}{\theta}}\left[L/\theta F+(\theta-1)/\theta\right]-1\right\}} > 0 \tag{4.47}$$

which is again decreasing in the size of the international market.

The case of Logit demand. Another explicit result can be obtained in the case of a Logit demand for L consumers:

$$D_i = \frac{Le^{-\lambda p_i}}{\sum_{j=1}^{N} e^{-\lambda p_j}} \tag{4.48}$$

with $\lambda > 0$. In this case, the international firms choose the price $p^* = c + F/L + 1/\lambda$, and it is easy to derive that the optimal subsidy must induce a price for the domestic firm equal to $p(s_H) = c + 1/\lambda$, which requires a very simple expression for the optimal export subsidy:

$$s_H = \frac{F}{L} \tag{4.49}$$

Again, this implies that a relatively large country should heavily subsidize its exports toward a smaller country.

4.7 Protectionism, Lobbying and Trade Wars

Our model of strategic trade policy can be applied and extended in different directions. Here we will discuss further applications to indirect forms of export promotion (as demand enhancing or cost reducing policies) which represent positive forms of active protectionism, a theory of lobbying for export subsidies, and an analysis of equilibrium policy between multiple countries.

4.7.1 Indirect export promotion and infant industries

The general principles on optimal export promotion can be applied to other policies which increase demand for the domestic product in an indirect way (Spencer and Brander, 1983). In particular, when export promotion increases demand for the domestic goods (without making it too rigid), one can show that under Cournot competition, overinvestment in export promotion is optimal a) only under strategic substitutability when the number of firms is exogenous, b) always under endogenous entry. moreover, under Bertrand competition, a) overinvestment in export promotion is optimal when the number of firms is exogenous, and b) underinvestment is optimal under endogenous entry.

Other forms of indirect export promotion can affect the supply side. In a context of international trade, transport costs are crucial since the marginal cost of exports depend on them. The government can implement policies to reduce transport costs for all exporting firms. A main example is given by investments in infrastructures for international communication, but more indirect examples include the establishment of easier business connections with other countries, the reduction of bureaucracy for export duties and even the development of trade unions and currency unions to reduce import tariffs and uncertainty costs related with the exchange rate. Applying the usual principles, one can show that there may or may not be a strategic incentive to reduce transport costs when there are barriers to entry abroad, but under endogenous entry this incentive always exists.

Finally, even forms of the infant industry argument can be defended (only) under special circumstances on the basis of our theory: if initial overproduction can reduce production costs and create a comparative advantage in this sense, it may be optimal to close the domestic sector from to the competition of international firms so as to develop a cost effective domestic industry able to compete in the international market in the future.

4.7.2 The political economy of export promotion

It is widely claimed that trade policy is often determined by lobbying activity of special interest groups rather than by welfare maximizing politicians (Grossman and Helpman, 2002). In this section we study the way firms that are targets for export promotion policies can affect trade policy through rent-seeking activities.

Define $C(s) = C[z(s), \beta(s), s]$ as a reduced form for the social cost of the policy, which we assume increasing and convex: for instance, in case of subsidies, this is the tax revenue necessary to finance them. The optimal policy of export promotion studied in the previous sections is the one that maximizes $W(s) = \pi - C(s)$, and satisfies the first order condition $SI + \Pi_3 = C'(s_H)$.

Here we want to extend this basic model to take into account the lobbying activity of the exporting firm (exporters represent a minoritarian but well organized part of the society). Following Grossman and Helpman (1994) and Goldberg and Maggi (1999),[33] the easiest way to endogenize lobbying is to imagine that the government chooses its policy to maximize a weighted average of welfare and firm's profits:

$$s = \arg\max\left[(1-\lambda)W(s) + \lambda\pi\right] = \arg\max\left[\pi - (1-\lambda)C(s)\right]$$

where λ is the weight given to the profits of the firm, which will be endogenized later. Given λ, the equilibrium policy will satisfy the first order condition $SI + \Pi_3 = (1-\lambda)C'(s)$, which delivers a policy $s_H(\lambda)$ increasing in λ, that is in the weight given by the politician to the firm's profits in its objective function, and equal to the optimal one if this weight is zero ($s_H(0) = s_H$). Notice that the strategic incentive to adopt export promoting policies is the same as before, just the costs of the policy changes. If we define $C[s_H(\lambda)]$ as the social cost of lobbying, this is clearly increasing in the lobbying activity.

Let us look at lobbying for export subsidies in the model with Cournot competition and homogeneous goods. When there is an exogenous number of firms, the derivation of the equilibrium subsidy is quite complex.[34] Under endogenous entry, assuming that the elasticity of the marginal cost $\sigma = c''(x)x/c'(x) > 0$ is high enough to have an interior solution, the politico-economic equilibrium subsidy can be derived as:

$$s_H(\lambda) = s_H\left[\frac{1+\lambda+\lambda\sigma(1-\varepsilon)/\varepsilon}{1-\lambda-\lambda\sigma}\right] \tag{4.50}$$

which is clearly increasing and even convex in the weight that politicians give to the profits of the exporting firm λ. Moreover, the equilibrium subsidy is unambiguously increasing in σ but, contrary to the optimal subsidy, it is decreasing in the elasticity of demand ε for λ or σ large enough.

Finally, the lobbying activity by the exporting firm determines λ, and this framework provides a simple way to understand the benefits of lobbying. If the cost of the lobbying activity to obtain a weight λ in the objective function of the politician is $L(\lambda)$, which is assumed increasing and convex, the investment in lobbying will select λ to maximize $\Pi - L(\lambda)$. The first order condition is $(SI + \Pi_3)\, s'_H(\tilde{\lambda}) = L'(\tilde{\lambda})$, and it allows us to derive a more informative expression for the equilibrium lobbying:

---

[33] Endogenous entry in the lobbying activity has been first characterized by Mitra (1999).

[34] In the particular case of linear demand and cost we can derive:

$$s_H(N,\lambda) = \frac{(a-c)[N-1+\lambda(N+1)]}{2N[1-\lambda(N+1)]}$$

which is increasing and convex in λ.

$$\tilde{\lambda} = 1 - \frac{L'(\tilde{\lambda})}{C'[s_H(\tilde{\lambda})]s'_H(\tilde{\lambda})} \tag{4.51}$$

The right hand side contains the ratio between the marginal cost of lobbying for the exporting firm, and the product of the marginal cost of subsidization with the derivative of the policy with respect to λ, which is just the social marginal cost of lobbying.

The bottom line is that even if there is a strategic incentive to export promotion, lobbying activity induces excessive export promotion.[35] If this distortion is strong, a commitment to free trade may be still optimal for domestic welfare.[36]

4.7.3 Nash equilibrium trade policy

The spirit of our results holds also in an equilibrium analysis where all the countries can choose optimally their trade policy. This situation was studied first by Brander and Spencer (1985) in a model of Cournot competition between two firms of different countries. The Nash equilibrium was characterized by excessive export subsidization: although export subsidies are unilaterally optimal, they are jointly suboptimal (for the countries involved) because they reduce prices without increasing profits. Analogously, in case of strategic complementarity, the Nash equilibrium is characterized by suboptimal export taxation by both countries.

Etro (2010) extends the model of equilibrium trade policy between multiple countries to the case of endogenous entry of international firms. In equilibrium there is an endogenous number of countries adopting a positive export subsidy, while all the others commit to free trade. Moreover, the equilibrium subsidies are the same as the optimal unilateral subsidies. The welfare gain is identical for all the countries that actively subsidize their firms, and the same as in the case of a unilateral optimal policy - indeed, even through coordination those countries could not reach a better outcome. In other words, in the presence of EMSs, strategic trade policy is not a beggar-thy-neighbour policy in the traditional sense. Nevertheless, only a limited number of countries can exploit the benefits of this policy: the adoption of export subsidies by some countries induces the exit of some international firms compared to the free trade equilibrium.

35 However, notice that, if other groups are lobbying the equilibrium may imply a policy closer to the optimal one, since its costs are born by the whole society (see Dixit, Grossman and Helpman, 1997).

36 On the political economy of international unions under lobbying activity see Brou and Ruta (2006) and Cheickbossian (2008).

4.8 Exchange Rate Policy

In this section we apply the general results on strategic export promotion to exchange rate policy. We think of a fixed exchange rate regime where the government can adjust the exchange rate within certain limits, but our arguments on the advantages of a devaluation equally apply to a flexible exchange rate regime where a depreciation takes place for different reasons (including a monetary expansion).

Governments often undertake competitive devaluations with the specific aim of supporting exporters and improving the trade balance. In spite of this, economic theory is uncertain on the merits of these policies. The traditional Mundell-Fleming model emphasizes the expansionary effects of unilateral devaluations: as long a devaluation improves the trade balance,[37] it boosts aggregate demand with an expansionary effect on the domestic country and possibly a negative effect on the trading partners.[38] However, the recent New Open-Economy Macroeconomics has shown that these devaluations can be even beggar-thy-self policies (Corsetti and Pesenti, 2001): in the presence of imperfectly competitive markets and sticky prices, they lower the purchasing power of domestic agents (due to the reduction of real wages under nominal rigidities) and this negative terms of trade externality can more than offset the positive espansionary effect.[39] The IMF broadly accepts this negative view on competitive devaluations and tends to oppose them unless a fixed exchange rate clearly appears unsustainable.[40]

---

[37] Devaluations tend to increase exports and reduce imports, which tends to improve the trade balance, and they raise the price of imports, which works in the opposite direction but does not change the net effect under plausible conditions: the so-called *Marshall-Lerner condition* guarantees that the overall effect is positive. Empirically this is true in the long run but not in the short run (after a depreciation the trade balance deteriorates on impact because imports and exports do not react instantaneously and then it gradually improves).

[38] See Fleming (1962) and Mundell (1968). Consider a two country model with trade balance of the domestic country $TB = vQ + mY^*$, where $Q = EP/P^*$ is the real exchange rate, m the marginal propensity to import and Y^* is income in the foreign country, whose trade balance is $TB^* = v/Q + mY$ with Y domestic income. Under perfect capital mobility the interest rate is r^*. Under fixed unitary prices, the IS and LM relations are $Y = (G+\bar{I}-br^*+TB)/s$ and $M/P = kY - hr$ for the domestic country and $Y^* = (G^* + \bar{I}^* - br^* + TB^*)/s^*$ and $M^*/P^* = k^*Y^* - h^*r^*$ for the foreign one. Assume $s^*s > m^*m$. Under fixed exchange rates a devaluation can be beggar-thy-neighbour, while under flexible exchange rates monetary policy induces a depreciation and is always beggar-thy-neighbour. Fiscal policy is always prosper-thy-neighbour.

[39] Moreover, economists have emphasized the perverse consequences that competitive devaluations have in terms of inflationary bias and creation of self-fulfilling financial crises and bank runs (Obstfeld, 1996, Chiang and Velasco, 2001), which may have a recessionary impact on the real economy. For a survey see Obstfeld and Rogoff (1996).

[40] On the political economy of IMF policy see Marchesi and Sabani (2007).

In front of this theoretical ambiguity it is difficult to make sense of the common wisdom according to which unilateral devaluations provide a positive strategic advantage in the international markets. Following the pioneeristic work by Dornbusch (1987) and our previous analysis, in this section we evaluate the strategic incentives to implement devaluations in a model where the shift of exchange rate variations on prices is endogenous. Our analysis shows that under barriers to entry competitive devaluations may be a bad idea to provide a strategic advantage to domestic exporters (especially under price competition), but under free entry there is always a strategic incentive to depreciate the currency to promote exports.

Our model of Section 4.5 with imperfect competition between international firms for a foreign market is particularly useful to study exchange rate policy because it is consistent with international market segmentation, which allows firms to choose different prices for different markets (*pricing to market*: the price of a good in domestic currency does not need to be the same in the domestic and the foreign market).[41] Therefore, following Dornbusch (1987), we can endogenize the effects of variations in the nominal exchange rate on prices. The effects of exchange rate policy for exporting firms crucially depend on the location of production, on whether local currency pricing or producer currency pricing holds[42] and on the strategic reaction of firms to the policy. In our partial equilibrium context, we will focus on the strategic effects of a devaluation on the domestic firm. Clearly, a devaluation has other consequences in general equilibrium, but the point here is just to understand whether the usual claim that a devaluation gives a strategic advantage to exporting firms is correct.

Imagine first that all international firms produce and compete abroad with independent production units. This is typical of multinationals which are directly active in other countries where they sell their products. Under price competition, this case of *local currency pricing* with market power implies no pass-through of the nominal exchange rate on prices. In this situation, a devaluation is not going to affect the equilibrium in the foreign market. All firms would choose the same prices in foreign currency after a devaluation, but the profits of the domestic firm would be artificially increased in the domestic currency. The same would happen under quantity competition, since production decisions abroad would be again independent from the exchange rate, but profits in domestic currency would be inflated by a devaluation. It is clear that such a gain in profits should be compared with the losses for the society in terms of higher prices of the imports. However, this is not our focus; what matters for our purposes is that in such a context there is not a strategic incentive to implement a competitive devaluation. This policy does

[41] In other words, the law of one price does not hold.

[42] See Engel (2000) and Betts and Devereux (2000) for theoretical discussions on local currency pricing versus producer currency pricing respectively in a standard Mundell-Fleming model and in a New Open Economy Macroeconomics model.

not give a real strategic advantage to the domestic firm in the foreign market but just artificially increases its profits.

The opposite situation happens when all firms produce in their domestic country, bear production costs in domestic currency, choose their strategy taking into account the exchange rate and then export abroad. Under price competition this corresponds to the case of *producer currency pricing.* Such a case is typical of SMEs which are active at a national level, often producing typical domestic products and exporting some of them abroad, but also of larger firms which are not directly active in the foreign market under consideration but sell their goods to distributors of this market. This is the situation we will focus on, but we will have again to divide the discussion between the cases of quantity competition and price competition. The bottom line will be that competitive devaluations are always desirable to provide a strategic advantage to domestic firms when foreign markets are characterized by EMSs.

Clearly, one could employ this framework to derive an optimal competitive devaluation comparing its benefits on the export side with its costs on the import side. However, one must always keep in mind that this partial equilibrium analysis is relevant in the short run. In the long run with perfectly flexible prices and no frictions, markets equilibrate in such a way that nominal variables as the nominal exchange rate are irrelevant.

4.8.1 Competitive devaluations with Cournot competition

Let us start our exploration of the strategic effects of competitive devaluations in the case of Cournot competition taking place in the foreign market.

Foreign demand for good i produced by the firm of country i is $p_i = p\left[x(i), \sum_{j\neq i} h(x(j))\right]$ where $x(j)$ is production of firm j, but revenues of firm i in domestic currency are $E_i x(i) p_i$ where E_i is the price of the foreign currency in terms of the currency of country i, that is the nominal exchange rate of this country. For expository purposes, imagine an initial situation where, without loss of generality, all the exchange rates (with the foreign country where firms compete) are unitary. If the domestic country can adopt a competitive devaluation and rise the exchange rate to the level $E > 1$, the profits of the domestic firm become:

$$\Pi = Exp\,(x, \beta) - c(x) \tag{4.52}$$

which can be rewritten in our framework as $\Pi\,(x, \beta, s)$ where $s = E - 1$, implying:

$$\Pi_{13} = p + xp_1 = \frac{c'(x)}{E} > 0 \tag{4.53}$$

Accordingly, Prop. 4.4 and 4.5 apply. After a devaluation the domestic firm increases its production level and the consequences depend on the entry conditions. Under barriers to entry, as long as strategic substitutability holds, the other firms will decrease production so that the market share of the domestic firm increases (as it was shown by Dornbusch, 1987). This creates a strategic incentive to devaluate, despite it also creates a negative terms of trade effect (which can eliminate the strategic incentive to devaluate if there are many domestic firms). Also under endogenous entry the domestic firm expands its market share, but the other firms produce the same as before the devaluation, and some of them exit from the market. The strategic incentive to devaluate is always present (and there is not even a terms of trade effect in the presence of more domestic firms). Summarizing, in case of a single domestic firm we have:

PROPOSITION 4.8. **Under quantity competition in the international market, a) when the number of firms is exogenous there is a strategic incentive to adopt a competitive devaluation if strategic substitutability holds and b) when entry is endogenous there is always a strategic incentive to adopt a competitive devaluation.**

A devaluation always increases exports. When entry is free and goods are perfect substitutes, the elasticity of domestic production with respect to the exchange rate is simply $(\sigma+\mu)^{-1}$, that is decreasing in the elasticity of the marginal cost $\sigma = c''(x)x/c'(x)$ and in the mark up $\mu = [Ep - c(x)]/c(x)$. Since the devaluation does not affect the equilibrium price, the elasticity of exports (Exp) to the exchange rate is just $1 + (\sigma+\mu)^{-1} > 1$.

4.8.2 Competitive devaluations with Bertrand competition

The case of price competition is quite interesting because it is the usual case under study in macroeconomic models of the exchange rate, and it is probably the most realistic case for our purposes.

If the exchange rate between country i and the foreign importing country is E_i, demand for firm i in that country can be written as $D_i = D\left[p_i/E_i, \sum_{j\neq i} g(p_j^*)\right]$ where p_i is the price of firm i in the currency of its country and p_j^* are the prices of the international firms in the foreign currency. In general, if p_i^F is the price of firm i in foreign currency, the price of the same firm in domestic currency is $p_i = E_i p_i^F$. If the latter was constant, a devaluation (an increase in E_i) would reduce the price in foreign currency, and an appreciation of the exchange rate would increase it. However, prices in domestic currency for foreign segmented markets can be changed after a devaluation and our purpose is exactly to check how they are changed.

Imagine a situation in which the domestic exchange rate E is initially unitary and is increased with a devaluation, while all the other exchange rates with the foreign currency are and remain unitary. Since production takes place at home and demand depends on prices in foreign currency, the relevant profit function for the domestic firm is:

$$\Pi = (p-c)\,D\left(\frac{p}{E},\beta\right) = \left(Ep^F - c\right)D\left(p^F,\beta\right) \tag{4.54}$$

where we used the relation $p = Ep^F$ to express the strategic variable of the domestic firm as its price in foreign currency (even if what matters in the profit function is the value in terms of the domestic currency).[43] Notice that, given our assumptions, the price of all the other international firms is the same in foreign and domestic currency, p_j^*.

Defining with $\varepsilon_i = -D_1(p_i, \beta_i) p_i / D(p_i, \beta_i) > 1$ the perceived price elasticity of demand for firm i, we can express the equilibrium first order condition for the Bertrand equilibrium as:

$$p_i = \frac{\varepsilon_i}{\varepsilon_i - 1} \frac{c}{E_i} \tag{4.55}$$

while the second order condition requires $\epsilon_i < 2\varepsilon_i$ where ϵ_i is the price elasticity of the slope of demand and represents the degree of convexity of the demand function. This shows that a unilateral devaluation reduces the mark up of the domestic firm compared to the mark up of the international firms in foreign currency terms. In other words, a devaluation induces an aggressive behavior of the domestic firm.

We can now evaluate the incentives to implement a devaluation. First of all, notice that a devaluation has always a direct positive effect on the profit of the domestic firm, since it increases revenues in domestic currency (at the same time, there are direct costs from a devaluation, for instance in terms of higher prices of imports). However, these are not the effects we are interested in, since the case for a strategic advantage for the domestic firm depends on the indirect effect on the equilibrium strategies.

As usual, the incentives to change strategy for the domestic firm depend on the cross effect:

$$\Pi_{13} = -\left(p^F\right)^2 \left[D + p^F D_1\right] \tag{4.56}$$

which is positive in equilibrium. Therefore, the price of the domestic firm in foreign currency p^F is always decreasing in the exchange rate, as we have seen from (4.55). In general, Prop. 4.4 implies that a competitive devaluation is not desirable under barriers to entry. Such a policy forces the domestic firm to decrease its price in foreign currency, which induces also the other firms to do the same, reducing profits for all the firms in the market. There is instead a strategic incentive to appreciate the currency, which induces the domestic firm to increase its own price in foreign currency and the other firms to do the same. Clearly, this is the strategic incentive for the government: an appreciation would also have a negative direct effect on profits, reducing the mark-up of the domestic firm, and finally, it would induce other effects like a reduction in the price of imports. The comparison between these direct effects and the strategic effect provides the optimal unilateral policy, but the crucial

[43] Notice that the profit function can be rewritten in our general framework with $x = 1/p^F$ and $s = E - 1$.

point, here, is that there is not a strategic incentive to implement competitive devaluations when domestic firms export in markets with barriers to entry.

When entry is endogenous, the domestic firm does not obtain a strategic advantage when induced to increase its own price because this would promote entry in the foreign market. According to Prop. 4.5, there is instead a strategic incentive to devaluate the nominal exchange rate. This would reduce the price of the domestic firm in the foreign currency. International firms would not change their own prices, but fewer would enter in the market so that the market share of the domestic firm would expand. In this case, a devaluation has also a direct beneficial effect, since it increases revenues of the domestic firm in domestic currency. The positive direct and strategic effects of a devaluation should be compared with the costs in terms of a higher price of imports, which is beyond the scope of this discussion. What matters here, is that the usual claim that devaluations give a strategic advantage to exporting firms is correct only for competitive foreign markets. Summarizing:

PROPOSITION 4.9. **Under price competition in the international market, a) when the number of firms is exogenous, there is a strategic incentive to appreciate the domestic currency, but b) when entry is endogenous there is a strategic incentive to adopt a competitive devaluation.**

The bottom line is quite intuitive. Devaluations can be deleterious for exporting firms when they induce a war between international firms to reduce prices in foreign currency, and this happens when there are clear barriers to entry. However, when entry is endogenous, international firms cannot undertake such a price war and the domestic firm can unilaterally decrease its price in foreign currency expanding its market share: only in this case there is a strategic incentive toward competitive devaluations.

We can understand better the implications of a devaluation if we look at the change in the foreign price of the domestic good after the devaluation.[44] Focusing on the case with endogenous entry, if we define $\phi \equiv -\left(\partial p^F/\partial E\right)\left(E/p^F\right)$ as the elasticity of this price with respect to the exchange rate, we can derive a simple expression for it:

$$\phi = \frac{\varepsilon - 1}{\varepsilon(2+\kappa) - \epsilon} > 0 \tag{4.57}$$

where $\kappa \equiv -\left[D_2 + \varepsilon p^F D_{12}\right] g'(p^F)/D_1 > 0$. The percentage reduction in the foreign price after a devaluation is smaller when demand is highly elastic (ε is large) and when it is not too convex (ϵ is small). We can also have a

[44] In general, one defines the degree of pass-through from the exchange rate to import price (of the foreign country in our context) as the percentage by which the import price changes when the exchange rate varies by 1%. Here we are interested in something different, the change of p^F when the nominal exchange rate between our domestic economy and the foreign country changes.

clue about the size of this elasticity. For instance, if demand is linear in the price, ϕ is always below 50%.[45] However, when demand is isoelastic we have $\epsilon = 1 + \varepsilon$, which implies levels of ϕ always above 50%.

Let us now look at the equilibrium effect of a devaluation on exports, which in this model correspond to the value of the demand for domestic goods in domestic currency, $pD\left(p^F, \beta\right)$. The corresponding elasticity is $[1 - \phi(\varepsilon - 1)] + ED_2 (\partial\beta/\partial E)/D$. The first term is due to the increase in the value of exports given the prices of the other firms, and it must be positive as long as $\phi < 1$. The last term is due to the strategic effect on the other firms. Under barriers to entry this is negative because the other international firms respond to the devaluation by decreasing their prices, which reduces demand for domestic goods. However, under endogenous entry, the last term is positive since international firms do not change their prices after the devaluation, but their number shrinks, which increases demand for the domestic product.

Finally, we can look at welfare in the foreign country. Under barriers to entry prices decrease after a devaluation, which unambiguously improves welfare since the number of firms is fixed, and therefore foreign consumers can have more of each good at a lower price. Under free entry, just the price of the domestic good decreases, while the others remain at the same level and some international firms exit from the market. However, this is not likely to reduce foreign welfare. For instance, under isoelastic utility the price index remains the same before and after the devaluation, therefore welfare does not change abroad.

4.9 R&D Policy and IPRs Protection

Another related case of strategic export promotion is that of R&D policy, which is quite relevant for high-tech industries: its main aspect involves R&D subsidies, but there are other forms of R&D promotion. These include public investment in complementary R&D or the strengthening of the protection of intellectual property rights of the domestic firms.

One can analyze the role of unilateral R&D policy focusing on the competition *for* international markets rather than the competition *in* international markets. As we have seen in Chapter 2, traditional models of patent races are nested in our general framework and can be used to study R&D policy for firms investing in some forms of innovation to conquer foreign markets. For instance, consider a standard international patent race where each firm

[45] This suggests that the domestic firm is going to increase its price in domestic currency in a strong way after a devaluation, but less than the change in the exchange rate (notice that the elasticity of the price in domestic currency to the exchange rate is $1 - \phi$): this increases demand for its domestic product and expands profits, while crowding out some international firms from the foreign market.

i invests a flow of resources $x(i)$ in continuous time. This investment delivers innovations according to a Poisson process with an instantaneous arrival rate $h(x(i))$, which is a positive and concave function. When one of the firms innovates, it obtains a rent V^M and the race is over. The R&D subsidy s_i is assumed to be proportional to the investment flow. Given a constant interest rate r, the expected profit function for firm i can be expressed as:

$$\Pi(x(i), \beta_i, s_i) = \frac{h(x(i))V^M - x(i)(1 - s_i)}{r + h(x(i)) + \beta_i} \tag{4.58}$$

which is clearly nested in our general functional form. Notice that $\Pi_{12} > 0$ and:

$$\Pi_{13} = \frac{r + h(x) + \beta - h'(x)x}{[r + h(x) + \beta]^2} > 0 \tag{4.59}$$

therefore in case of a fixed number of international firms, it would be optimal to tax domestic R&D (to slow down the aggregate investment rate), while under the assumption of endogenous entry in the international competition for the market a positive R&D subsidy is always optimal. Following the usual method, it is immediate to verify that the optimal unilateral R&D subsidy satisfies:

$$s_H = \frac{1}{1 + \frac{V^M(r+\beta)}{h(x)V^M - x}} \in (0, 1) \tag{4.60}$$

where the investments of the domestic firm, x, and of the foreign firms, x^*, satisfy $h(x)V^M = h(x^*)(V^M - F) = 1$. Once again, the subsidy allows the domestic firm to commit to a more aggressive strategy, which is now represented by a larger investment flow. Summarizing:

PROPOSITION 4.10. **In an international patent race a) when the number of firms is exogenous it is optimal to set a positive R&D tax, but b) when entry of international firms is endogenous the optimal unilateral R&D policy requires always to set a positive R&D subsidy.**

As we noticed, the same point can be made for other aspects of R&D policy. For instance, a strengthening of the domestic protection of IPRs can provide domestic firms with larger incentives to invest in the competition for international markets with endogenous entry. More formally, imagine that domestic IPRs protection can increase the value of innovation as follows:

$$\Pi(x(i), \beta_i, s_i) = \frac{h(x(i))(1 + s_i)V^M - x(i)}{r + h(x(i)) + \beta_i} \tag{4.61}$$

In this case, the national firm will have an advantage in the international patent race because:

$$\Pi_{13} = \frac{h'(x)(r+\beta)V^M}{[r+h(x)+\beta]^2} > 0 \tag{4.62}$$

and the investment of the firms is the same as before.

The strong American position in favor of IPRs protection (and against compulsory licensing of IPRs for antitrust purposes) could be interpreted in this sense: it provides U.S. high-tech firms with a strategic advantage in the international competition. Until now, Europe has followed a different path, but a strengthening of IPRs protection at the E.U. level would have positive effects on the incentives to invest in R&D for the most dynamic European firms. As we will see in the next Chapter, the results on the optimality of R&D subsidies are strengthened in a dynamic growth context.

4.10 Conclusions

In this chapter we studied international economic theory and policy in the presence of EMSs. The theoretical analysis of closed and open economies with competition in quantities or prices and endogenous entry has emphasized that, contrary to what happens in the Krugman (1980) model, globalization leads to lower prices and to fewer, but larger, firms. These result are also supported by preliminary empirical evidence on the relation between markets' size and endogenous structure.

Similar trade models can be introduced in dynamic two-country frameworks, as the Ghironi-Melitz model, which can be extremely useful to explain the reaction to shocks in a global world, the propagation mechanism of the international business cycle, and the reasons behind systematic deviations from purchasing power parity.

We have also applied basic models of trade to the study of optimal trade policy, showing that in the presence of EMSs, the traditional bias in favor of a positive import tariff holds for a large country, but the traditional bias in favor of export taxes collapses. In particular, we have shown the general optimality of the unilateral adoption of export proming policies, including export subsidies, competitive devaluations and R&D subsidies. In the next chapter, we will return to some of these policy results in a growth framework.

5. Endogenous Market Structures and Growth

You can measure distance by time. "How far away is it?" "Oh about 20 minutes." But it doesn't work the other way. "When do you get off work?" "Around 3 miles." **Jerry Seinfeld**

According to Schumpeter (1942), market power provides the incentives to invest and innovate and so it stimulates growth. The recent revival of this important observation, started with the work by Romer (1990) and Aghion and Howitt (1992), has formalized this idea of creative destruction.[1]

However, the market for innovations has been usually described in a very simple and empirically arguable way: the production function of new ideas is characterized by CRS without fixed costs of R&D, and a simple no-arbitrage condition pins down the aggregate investment in R&D and the growth rate, as in Aghion and Howitt (1992). This approach to the analysis of markets is still in line with the neoclassical tradition, delivers indeterminate market structures in the innovation sector and does not allow one to open the "black box" of the engine of growth: it says nothing about how many firms invest in R&D and how much each one invests. A more recent alternative approach, exemplified by Aghion and Howitt (2009), describes the market for innovations with a production function of new ideas characterized more realistically by decreasing marginal productivity but used by monopolistic investors.[2] This approach delivers exogenous market structures in the innovation sector and again does not open the black box of the innovative activity: it explains how much a single firm would like to invest, but it completely neglects the incentives to invest of the other firms. The objective of this chapter is to open the black box of the engine of growth, and we will do it by characterizing the EMSs of the competition for the market. Therefore, we will derive the

[1] See Aghion and Howitt (2009) and Acemoglu (2009) for the most updated surveys.

[2] The assumption of Aghion and Howitt (2009) is dramatically simple: "In each period there is one person (the "entrepreneur") who has an opportunity to attempt an innovation" (p. 87). They extend the analysis to the case of two entrepreneurs (one of which may be the incumbent monopolist, see Ch. 12), but they never take into account the endogenous incentives that other entrepreneurs may have to invest in R&D.

F. Etro., *Endogenous Market Structures and the Macroeconomy*,
DOI: 10.1007/978-3-540-87427-0_5,

main determinants of the investment in R&D, and we will endogenize both the number of firms investing in R&D and their investment levels.

As noticed by Kortum (1993), Cohen and Klepper (1996) and others, investments in R&D are characterized by relevant fixed costs, decreasing marginal productivity,[3] and substantial duplications of efforts between firms. The first works to introduce some of these elements in an endogenous growth model with expanding varieties of intermediate goods *à la* Romer (1990) are due to Peretto (1996, 1999). However, the deep nature of the competition for the market can be realistically described through patent races where few firms invest in R&D taking the investment of each other as given, and the probability of being the first one to innovate depends on the relative investment. Patent races with this kind of strategic interactions have been proposed by Loury (1979) in partial equilibrium, and they have been introduced in a general equilibrium framework of Schumpeterian growth by Etro (2004,a).

Another important stylized fact about R&D investment is that a large part of it is done by incumbent monopolists with patents on the leading edge technologies. Blundell, Griffith and Van Reenen (1999) have provided wide evidence based on the British manufacturing sector for which market leaders have greater incentives to commit to high investment in innovation. Czarnitzki, Etro and Kraft (2008) have provided related evidence based on the German manufacturing sector, but they confirm the result only for leaders facing a strong entry pressure. Nevertheless, this important aspect has been neglected by most of the theoretical literature on Schumpeterian growth, in which incumbent monopolists do not invest at all in R&D because they have lower gains from replacing their own technology with a new one (compared to the gains from innovation of the outsiders). This traditional result, due to Arrow (1962), leads to a systematic change of leadership that is largely in contradiction with the Schumpeterian idea and with the empirical evidence on the persistence of leadership.[4] The EMSs approach allows us to obtain a more realistic description of the competition for the market, and also to solve the puzzle concerning the incentives of the leaders to invest in innovation. Etro (2004,a) has shown that incumbent leaders have more incentives than the other firms to invest in R&D when they face endogenous entry of innovative outsiders. In such a case the Arrow effect disappears because an accommodating strategy of the leaders (through low investments) would

[3] One of the stilyzed facts pointed out by Cohen and Klepper (1996) is that the number of patents and innovations per dollar of R&D decreases with the level of R&D. From a theoretical point of view, notice that, while in most of the productive sectors there are strong reasons to believe that doubling the amount of input total production will double, there are no reasons to believe that doubling the amount of input in the R&D activity will deliver a double expected amount of innovations. As a matter of fact, decreasing marginal productivity (in the single input) is the standard assumption in the industrial organization literature on patent races.

[4] See Sutton (2007,b), Kato and Honjo (2007), and especially Doi (2008) on the Japanese experience.

attract innovative entry of outsiders and would reduce the length of the leadership, while an aggressive strategy (through high investments) would limit entry and improve the relative chances to retain the leadership.[5]

Keeping in mind the above stylized facts and the related empirical evidence, this chapter extends our dynamic general equilibrium model with differentiated intermediate inputs to characterize endogenous growth due either to the creation of new intermediate goods (as in Romer, 1990) or to the improvement of the existing ones (as in the Schumpeterian theory). This allows us to describe the EMSs of the competition *in* the market (between producers of inputs) and *for* the market (between firms investing in R&D to improve existing inputs). The latter is the main interest of this chapter. A crucial outcome of the basic model is that, even if the decentralized equilibrium can be characterized by a growth rate below or above the optimal one, the EMS is always biased toward too small firms, that is firms that underinvest in R&D compared to what would be optimal for the society. This brings to a form of dynamic inefficiency that is similar to the one found in Chapter 3, but that is more disruptive: the inefficient organization of resources in the market for innovations leads to permanent consequences on growth. The basic model can be extended with a leadership of the incumbent patentholders in the competition for the market. In such a case, we show that the engine of growth is more powerful, in the sense that the value of innovations is enhanced, the aggregate incentives to invest are increased and this speeds up the growth process in a Schumpeterian sense.

Our framework can be employed to study the optimal fiscal and monetary policy for a growing economy. The former has to restore the efficient market structure with R&D subsidies and proper capital income taxation (or profit taxation). The latter should be aimed at minimizing the distortions induced by inflation on the investment in R&D, which requires price stabilization only when the optimal fiscal policy is available. In this way we generalize some of the results of Chapter 3 on fiscal and monetary policy to a different framework.

Finally, we extend the model to a multicountry version with trade of intermediate goods across countries, and we analyze global growth due to the creation of new or better intermediate goods. This allows us to study the relation between openness and growth, and to characterize the optimal unilateral and coordinated R&D policy, which requires heavy R&D subsidization and protection of IPRs.

The chapter is organized as follows. Section 5.1 provides new empirical evidence on the determinants of the R&D activity on the basis of a unique dataset on the German manufacturing sector. Section 5.2 describes a basic endogenous growth model with a number of extensions. Section 5.3 develops

[5] A similar result has been subsequently pointed out by Aghion *et al.* (2009), but with an exogenous entry pressure. They also provide additional empirical evidence on the impact of entry on incumbents' investments. On innovation by leaders see also Weinschenk (2008) and Dumav (2008).

a Schumpeterian growth model with EMSs in the competition for the market. Section 5.4 endogenizes the persistence of technological leadership. Sections 5.5 and 5.6 discuss optimal fiscal and monetary policy in the Schumpeterian model. Section 5.7 extends the analysis to a multicountry world. Section 5.8 concludes.

5.1 Empirical Evidence on EMSs and Innovation by Leaders

There is a lot of debate on the role of market leaders and followers in investing in R&D and promoting technological progress. A commonly held view is that firms invest more in a competitive market where the entry pressure is stronger, and that incumbents tend to be less innovative than their followers, so that the persistence of market dominance is typically the signal of market power and lack of entry pressure. This view is often associated with Arrow (1962), who has shown that incumbents have lower incentives to invest in R&D than the outsiders, and that in case of perfect competition for the market they do not invest at all, leaving the innovative activity to the outsiders.

As we have noticed in the introduction and we will see formally in the rest of the chapter, the EMSs approach challenges this view, showing the crucial role of the entry pressure on the different behavior of incumbents and outsiders. It is true that, when the number of firms investing in innovation is exogenous, the Arrow effect induces a lower investment by the incumbents, however, in markets where entry can be regarded as endogenous the Arrow effect disappears (Etro, 2004,a). In such a case, we obtain the exact opposite result of the common view associated with Arrow: entry pressure leads the outsiders to invest less and the incumbent to invest more. Ultimately, this leads to a surprising association between entry pressure and persistence of dominance through innovations. These theoretical results are robust to substantially different model specifications, in particular they hold in general patent races (as those presented in this chapter), and in models of preliminary investment in R&D as a strategic commitment.

In this section, based on Czarnitzki, Etro and Kraft (2008), we perform a simple empirical test on whether actual firm-level investment data support the basic hypotheses emerging from the EMSs approach on the determinants of R&D intensity, defined as the R&D-sales ratio. A first hypothesis is that the R&D intensity of the average firm is lower when there is an endogenous entry threat.[6] The second and most important empirical prediction can be

[6] This result mirrors what happens in models of competition in quantities and in prices, where output per firm decreases with the number of firms and therefore is minimal when entry is endogenous. It holds in patent races with strategic substitutability.

stated as follows: *the R&D intensity of the incumbent leader is larger than the one of the average firm when there is an endogenous entry threat.*[7]

5.1.1 Data Sources on R&D, Entry and Leadership

For our empirical investigation of the determinants of innovation, we use data from the 2005 Mannheim Innovation Panel (MIP). This innovation survey has been conducted by the Centre for European Economic Research (ZEW) in Mannheim, and covers a representative sample of the German manufacturing sector.[8] The 2005 spell of the MIP included some unique questions allowing to model entry threats and to identify market leaders in a novel way.[9]

The dependent variable of our analysis is the R&D intensity in the year 2004 at the firm level. The intensity for firm i is defined as $RDINT_i = R\&D_i/SALES_i \times 100$.

The most important right-hand side variables are aimed at identifying the endogeneity of entry in the market where each firm is active and the leadership position. An innovative aspect of our empirical approach is given by the fact that the same firms provide a subjective view on these two factors: rather than assigning a degree of entry intensity in a discretionary way or assigning a status of leadership on the basis of predetermined variables, we allow the firms to identify the existence of an endogenous threat of entry in the market and to identify who is the leader in the market.

The survey asked for several characteristics about the competitive situation in the firms' main product markets in 2002–2004. In particular, firms where asked to indicate if a list of six statements about the firms competitive environment apply to their situation or not. The response was based on a

---

7 Czarnitzki, Etro and Kraft (2008) provide theoretical background for these hypotheses and for a last one concerning a positive and concave relation between employment and investment.

8 The ZEW conducts the survey since 1992, and this represents the German part of the EU-harmonized Community Innovation Survey (CIS). It follows the Eurostat-OECD guidelines for collecting innovation data which are documented in the so-called Oslo Manual (see OECD and Eurostat, 1997, *Oslo Manual - Proposed guidelines for collecting and interpreting technological innovation data*, Paris)

9 The database has a cross-sectional structure, but the questionnaire collects information generally for the years 2002 to 2004. The quantitative variables, such as R&D investment, capital intensity, employment, sales etc., are surveyed for a certain year. For instance, R&D investment is only collected for the year 2004. Other information that we use as controls are, however, collected for the two years 2003 and 2004, so that we can make use of lagged controls to avoid direct simultaneity bias in the regressions. Qualitative information, such as the competitive situation in a firm's main market, the firm's competitive position etc., are collected through one question each referring to the time period 2002–2004. We use the qualitative information to construct variables on leadership and entry threats during this period, and argue that the situation between 2002 and 2004 will have an impact on strategic investment behavior in 2004.

4-point Likert scale, from "entirely applies" to "does not apply at all". Thus, our variable of entry threat, $ENTRY_i$, is an ordinal variable taking values from 0 to 3, where 3 indicates that the respondent firm strongly agreed to the statement that its market position is highly threatened by entry. When this is the case, we conjecture that entry in the industry where the firm is active can be regarded as endogenous; when the firm does not consider the threat of entry as present in its industry, this is regarded as one with an exogenous number of firms. According to our first hypothesis, we expect a negative sign of $ENTRY_i$ in the regressions for the average R&D intensity.

The theoretical definition of a market leader is associated with a strategic first mover advantage, but a more general definition can be based on the leading position of the firm compared to its main competitors. Therefore, our crucial variable is defined through a question on a firms' position compared to its main competitors. The respondents indicated if their competitors are larger, smaller, of a similar size, or larger and smaller than their firm. Consequently, an incumbent leader in our analysis is identified by an indicator variable, $LEADER_i$, describing a firm that is self-defined as larger than the competitors in its main product market. According to our main testable hypothesis, we expect that the incumbent leaders choose to invest more than other contestants if and only if their market is threatened by entry. We capture this by an interaction term of leadership and entry ($LEADER_i \times ENTRY_i$).

It is desirable to control for employment and capital requirement. We include firms' past employment (EMP_i) as well as past capital intensity ($KAPINT_i$) in the empirical model to account for such impacts on the investment decision. For the size of the employment we expect a positive (and concave) relation. Concerning the role of capital intensity, theoretical results are model-specific, thus we do not have strong priors on the sign of its coefficient. We also control for the Herfindahl-Hirschman Index of concentration (HHI_i) of the industry where the firm is active (multiplied by 1000).

The main general determinant of the investment in R&D is the degree of protection of the intellectual property rights (IPRs) associated with the innovations that each firm can obtain. It is difficult to measure the degree of protection of the IPRs at the firm level, but we can proxy this with a measure of the stock of patents at the firm level. In particular, the differences between firms in the size of the patent portfolio can be associated with the differences in the degree of expected protection of the innovations of the firms, therefore we expect a positive correlation between R&D intensity and the patent stock. Moreover, the introduction of this variable as a control variable for our main regressors allows us to obtain a robustness check that might account for unobserved heterogeneity even in the absence of panel data.

Our measure of the patent stock at the firm level accounts for all patent applications from 1978 onwards. In particular, we compute the patent stock using the perpetual inventory method for each firm. The survey data has been merged with the database from the German Patent Office which covers all patents filed at both the German and the European Patent Office since 1978. We follow the common practice in the literature and impose a rate of obsolescence δ_P of 15% per year when computing the patent stock.[10] Including such a rate of obsolescence implies, quite realistically, that knowledge loses its relevance as capital depreciates over time. The variable $PSTOCK$ is given by:

$$PSTOCK_{it} = (1 - \delta_P)PSTOCK_{i,t-1} + PA_{it},$$

where $\delta_P = 0.15$ and PA_{it} denotes patent applications of firm i in year t. We set the initial patent stock in year 1978 to zero for all firms. Since we use data from 2002-2004 in our regressions, the bias arising from a zero starting value will have disappeared due to the included depreciation rate δ_P.

Finally, we use twelve industry dummies to control for unobserved heterogeneity in investment across industries. The industries are: Food, Textiles, Paper/Publishing, Chemicals, Rubber, Glass/Ceramics, Metal, Machinery, Electronics, Information & Communication Technology, Instruments/ Optics and Vehicles.

Table 5.1 shows the descriptive statistics of core variables used in the upcoming regression analysis. In total, we can use 1,857 observations for the empirical study. The average R&D intensity of firms is about 2.3 % and average firms size amounts to 307 employees in the sample; 8% of all firms are classified as leaders.

Table 5.1. Descriptive statistics (1,857 observations)

Variable	Mean	Std. dev.	Min.	Max.
$RDINT_{it}$	2.271	5.112	0	38.914
$EMP_{i,t-1}/1000$	0.307	1.356	0.001	36.761
$KAPINT_{i,t-1}$	0.078	0.090	0.001	0.861
$LEADER_{it}$	0.080	0.271	0	1
$ENTRY_{it}$	1.531	0.851	0	3
$HHI_{i,t-1}$	36.778	61.022	3.15	650.17
$PSTOCK_{i,t-1}/(EMP_{i,t-1}/1000)$	8.864	26.906	0	222.447
IV candidates				
MES_{t-1}	0.079	0.166	0.009	2.102
$ADVERT_{it}$	2.219	1.428	1	6
$SUBSTITUTE_{it}$	1.874	0.840	0	3

5.1.2 Tobit Regressions

As not all firms invest in R&D, we estimate Tobit models that take account for the left censoring of the dependent variable (see Tobin, 1958). The Tobit

[10] See e.g. Griliches and Mairesse (1984).

model to be estimated can be written as:

$$RDINT_i^* = X_i'\beta + \varepsilon_i \tag{5.1}$$

where $RDINT_i^*$ is the unobserved latent variable. The observed dependent variable is equal to:

$$RDINT_i = \begin{cases} RDINT_i^* \text{ if } X_i'\beta + \varepsilon_i > 0 \\ 0 \text{ otherwise} \end{cases} \tag{5.2}$$

X_i represents the matrix of regressors, β the parameters to be estimated, and ε_i the random error term. In our basic specification, X_i includes $EMP_{i,t-1}$, $EMP_{i,t-1}^2$, $KAPINT_{i,t-1}$, $HHI_{i,t-1}$, $LEADER_{it}$, $ENTRY_{it}$ as well as 12 industry dummies (Model I). In further models, we add the interaction term $LEADER_{it} \times ENTRY_{it}$ (Model II) and the patent stock $PSTOCK_{it}$ to control for previous R&D (Model III).

We first consider homoscedastic regressions, and subsequently test for heteroscedasticity as coefficient estimates may be inconsistent if the assumption of homoscedasticity is violated in Tobit models. In order to estimate heteroscedastic Tobits, the homoscedastic variance σ is replaced with $\sigma_i = \sigma \exp(Z_i'\alpha)$ in the likelihood function (see Greene, 2003, pp. 768–9). We consider groupwise multiplicative heteroscedasticity by using a set of five size dummies (based on employment) and the industry dummies in the heteroscedasticity term.

Table 5.2 shows the regression results for homoscedastic models, and Table 5.3 for the heteroscedastic models. In the homoscedastic Tobit Model I, we find that R&D investment decreases as the threat of entry increases. The leaders' investment does not differ from that of the outsiders. When we add the interaction term of leadership and entry threat in Model II, however, interesting differences occur. While the leader dummy is still insignificant, we now find that leaders who are faced by potential entry invest more than the outsiders. The results remain robust when we control for prior R&D using the patent stock in Model III. The patent stock is highly significant and positive, confirming that firms receiving stronger protection of IPRs through patents tend to invest more - alternatively, firms that (successfully) conducted R&D in the past will also invest more in the current period.

With respect to the other covariates, we find a positive and concave relation with employment, [11] while capital intensity is positively significant in all models, and the Herfindahl index is always insignificant. Furthermore there are differences in R&D investment across industries. The industry dummies are always jointly different from zero in the regressions, and our results emphasize a high correlation of R&D spending with firms of the Information & Communication Technology.

[11] The inverted U curve peaks at about 20 thousand employees. As we have only a single observation that has more employees, we can basically conclude that R&D investment is increasing and concave in firm size.

Table 5.2. Homoscedastic Tobit models on *RDINT* (1,857 observations)

Variables	Model I	Model II	Model III
$EMP_{i,t-1}/1000$	0.840***	0.877***	0.803***
	(0.266)	(0.267)	(0.260)
$(EMP_{i,t-1}/1000)^2$	−0.021**	−0.022**	−0.019**
	(0.010)	(0.010)	(0.009)
$KAPINT_{i,t-1}$	4.126**	4.039**	3.621*
	(2.066)	(2.065)	(2.017)
$HHI_{i,t-1}$	0.001	0.001	0.001
	(0.004)	(0.004)	(0.004)
$PSTOCK_{i,t-1}$			0.050***
			(0.006)
$LEADER_{it}$	−0.099	−0.161	−0.298
	(0.676)	(0.676)	(0.660)
$ENTRY_{it}$	−0.598***	−0.853***	−0.727***
	(0.223)	(0.246)	(0.240)
$ENTRY_{it} * LEADER_{it}$		0.541***	0.488**
		(0.217)	(0.212)
Intercept	−4.788***	−4.844***	−4.816***
	(0.939)	(0.939)	(0.915)
Industry dummies $\chi^2(12)$	304.69***	298.33***	239.66***
Log–Likelihood	−3769.18	−3766.07	−3735.12

Notes: Standard errors in parentheses. *** (**, *) indicate a significance level of 1% (5%, 10%).

As Table 5.3 shows, the assumption of homoscedasticity is rejected for all models (see the Wald tests on heteroscedasticity). The industry and firm size dummies are always jointly significant in the variance equation. However, our main results are robust to the model modification. Leaders, in general, are not investing differently from the outsiders, and R&D investment is negatively affected by the entry variable. However, leaders that suffer from entry threat invest more than outsiders also in the heteroscedastic version. There are no major changes in the estimates of the other covariates. The patent stock is still highly positively significant, and the estimated employment effect remains stable. However, the positive relationship between R&D and capital investment becomes statistically insignificant once we correct for heteroscedasticity.

To sum up, our findings on entry are in line with our first testable hypothesis, for which investment decreases with the strength of entry threats. Furthermore, we find that incumbent leaders do not differ in their investment from other firms ($LEADER_i$ is insignificant), unless they are threatened by endogenous entry (see the positive signs of the interaction term $LEADER_i \times ENTRY_i$): in line with our main testable prediction, the competitive pressure of the potential entry of other firms induces the market leaders to invest in R&D more than any other firm.

Table 5.3. Heteroscedastic Tobit models on *RDINT* (1,857 observations)

Variables	Model I	Model II	Model III
$EMP_{i,t-1}/1000$	0.625***	0.640***	0.610***
	(0.112)	(0.111)	(0.112)
$(EMP_{i,t-1}/1000)^2$	−0.016***	−0.017***	−0.016***
	(0.003)	(0.003)	(0.003)
$KAPINT_{i,t-1}$	1.047	1.037	1.031
	(0.919)	(0.927)	(0.924)
$HHI_{i,t-1}$	0.001	0.001	0.001
	(0.002)	(0.002)	(0.002)
$PSTOCK_{i,t-1}$			0.032***
			(0.005)
$LEADER_{it}$	0.147	0.135	0.045
	(0.271)	(0.269)	(0.271)
$ENTRY_{it}$	−0.203*	−0.322**	−0.317**
	(0.120)	(0.130)	(0.128)
$ENTRY_{it} * LEADER_{it}$		0.302***	0.291**
		(0.115)	(0.114)
Intercept	−0.802**	−0.909***	−0.949***
	(0.331)	(0.334)	(0.338)
Industry dummies: $\chi^2(12)$	143.09***	142.86***	109.11***
Log–Likelihood	−3533.40	−3529.90	−3511.60
Wald Test on heteroscedasticity: $\chi^2(17)$	534.22***	530.71***	514.14***

Notes: Standard errors in parentheses. *** (**, *) indicate a significance level of 1% (5%, 10%).

In economic terms, the findings are also highly significant. Calculating the expected value of $RDINT_i$ for outsiders under no entry threat, yields:

$$E(RDINT_i|LEADER_i = 0, ENTRY_i = 0, \bar{X}_i) = 0.98,$$

where the covariates are taken at the average. In contrast, the investment intensity of outsiders under high entry threat amounts to:

$$E(RDINT_i|LEADER_i = 0, ENTRY_i = 3, \bar{X}_i) = 0.49,$$

which means R&D intensity is reduced by more than half, all else constant. If a leader suffers from high entry threat, however, we get:

$$E(RDINT_i|LEADER_i = 1, ENTRY_i = 3, \bar{X}_i) = 0.93,$$

which corresponds only to a 5% decrease due to entry threat.[12] Statistically, the leader's reduction due to entry is not different from zero (see Czarnitzki, Etro and Kraft, 2008, for further details).

[12] See Greene (2003, pp. 768-9) for the computation of the expected value in Tobit models. Calculations are based on the heteroscedastic estimation of Model III.

5.1.3 Instrumental variables for entry

In our empirical investigation we have proxied the endogeneity of entry in the market where each firm is active with the existence of a threat of entry perceived by the same firm. This short cut avoids the need of investigating what are the determinants of the fact that a market is characterized or not by endogenous entry as opposed to be limited to an exogenous number of firms.

A main concern with respect to the results presented so far is related to the independence of our entry variable from the dependent variable, R&D intensity. First of all, there may be a problem of reverse causality in the relation between R&D intensity and the entry threat. In principle, it is possible that successful innovative activity in the past leads to technological advantage of the firms which are presently active in an industry so that entry becomes difficult for the outsiders. Analogously, if the incumbents are not research active and neglect the development of new processes and products, entry may become relatively easy. Second, Sutton (1998, 2007,c) and the EMSs approach have characterized R&D as a strategic factor, which is sometimes used by some companies to determine the market structure, including the entry conditions. For these reasons, the possibility of a reverse relationship has to be investigated with the IV approach through the analysis of the determinants of the endogenous entry threat variable (which, of course, is also interesting in itself for the EMSs approach).[13]

To find instrumental variables that explain our entry variable but not the R&D intensity variable, we need to look at the key element determining entry pressure, the difference between the expected profits in the market and the fixed costs of entry. Many empirical studies have emphasized the role of profitability and market growth.[14] One would expect that entry occurs more frequently in markets where profitability is expected to be high, and less frequently when profitability is expected to be low. However, expected profitability is hard to measure, and the adoption of past profitability of the incumbent firms has often led to mixed results.[15]

We then look at the fixed costs of entry as a (negative) determinant of entry. The empirical studies on entry barriers address this issue by examining on one side natural barriers, like sunk costs of entry as determinants of scale economies or the importance of advertising as determinant of the demand function, and on the other side strategic barriers, for instance excess capacity, limit pricing, product differentiation and also innovative activity.

It is not simple to find a measure of the natural barriers to entry. Sutton (1998) uses the size of the median plant in an industry as a proxy for minimum

[13] The classic reference on instrumental variables is Sargan (1958).

[14] A recent example is Berger *et al.* (2004).

[15] For instance, Geroski (1995) points to empirical evidence from the U.K. that entry and exit rates are positively correlated, which is difficult to reconcile with the static profitability interpretation.

efficient scale, and therefore for the size of the costs of entry. In other studies variants of size measures are used, but most studies rely on observed size as it is very difficult to get information on the minimum efficient size required by the technology used.[16] We have information on total industry sales and the number of firms active in an industry. This information is taken from official statistics and measured at a detailed industry level (NACE 3-digit level).[17] The ratio, industry sales per firm, is applied as a proxy for minimum efficient scale and enters the regressions as a lagged value (MES_{it-1}).

Another factor that can affect profitability and entry is the importance of advertising in determining demand. For our purpose, it is not relevant whether advertising is informative or has a direct impact on preferences. In a sector in which advertising is an important competitive factor, entry could be easier because firms can gain market shares just by advertising their products. On the other side, when advertising investment in the industry is large, entry may be quite costly. In one way or the other, when advertising is perceived as important by the firms, it is likely to affect entry. Our survey collects information on the importance of advertising. Firms are asked to rank the importance of several characteristics of their competitive environment where they are active (product quality, technological advance, service, product variety, advertising and price). Consequently, we employ the variable $ADVERT_i$ which takes values between 1 and 6, where the largest value corresponds to the highest importance of advertising in the industry where the firm is active (and is not a measure of investment in advertising of the single firm).

Finally, the degree of substitutability between goods can heavily affect entry pressure, as Sutton (1998) has emphasized. If products are homogenous, an entrant offering a product with a higher quality, captures a relatively large market share as many consumers are interested in a superior product. In contrast, if products are distant substitutes a firm investing in improved product quality will only gain a small share of the industry sales as consumer preferences are very heterogenous. Hence, product substitutability is a determinant of entry barriers, with higher substitutability supporting entry. The 2005 MIP questionnaire also collects information on the relation between products. The respective question is "Please indicate to what extent the following characteristics describe the competitive environment in your main market." One characteristic is "Products of rivals are easily substitutable with ours." The evaluations are rated by use of a four point Likert scale ranging from "applies entirely"(3) to "does not apply at all"(0) and we build on this the variable $SUBSTITUTE_i$.

We first test for endogeneity of the entry variable in our regressions for R&D intensity using the instrumental variables described above. We apply a

[16] Lyons *et al.* (2001) use engineering estimates based on the firms' technologies employed in the production process.

[17] NACE is the European standard industry classification, and the firms in our sample are active in 96 different NACE 3–digit industries.

Table 5.4. IV relevance tests and endogeneity test of entry variable

Test	MES_{t-1}	MES_{t-1}, $ADVERT_t$	MES_{t-1}, $ADVERT_t$, $SUBSTITUTE_t$
F-Test on IV significance in 1st stage regression	$F = 14.33$***	$F = 14.47$***	$F = 21.41$***
Blundell/Smith endogeneity test[a]	−0.53	−0.12	1.11
Hansen J–test[b]	—	0.028	7.704**

Notes: *** (**,*) indicate a significance level of 1% (5%, 10%).
[a] Based on heteroscedastic model I. t–statistics of first stage residuals are displayed.
[b] Based on 2SLS regressions as test is not available for Tobit.

regression based test following Hausman (1978, 1983).[18] Suppose our equation for R&D intensity $RDINT_i$ is given by:

$$RDINT_i^* = x_i'\beta + \alpha ENTRY_i + u_i, \quad (5.3)$$

where the possibly endogenous regressor $ENTRY_i$ can be explained by the reduced form equation:

$$ENTRY_i = z_i'\gamma + v_i, \quad (5.4)$$

where z_i' contains the vector x and the other instrumental variables described above, and γ are the corresponding parameters. Once we estimate (5.4), we obtain $\hat{v}_i$ and we can estimate our R&D equation including the generated residuals from the first stage regression:

$$RDINT_i^* = x_i'\beta + \alpha ENTRY_i + \rho\hat{v}_i + e_i, \quad (5.5)$$

The usual t–statistic of $\hat{\rho}$ is a valid test on the endogeneity of the variable $ENTRY_i$ in the initial regression. If it is not rejected that $\hat{\rho} = 0$, we do not find that the variable $ENTRY_i$ is endogenous in the R&D regression, or in other words, there is not a problem of reverse causality.[19] In the Tobit framework, the corresponding test has been introduced by Smith and Blundell (1986): one simply estimates a standard Tobit instead of OLS to determine whether $\hat{\rho} = 0$. Furthermore, we compute the Hansen J-Test (the heteroscedasticity robust version of the Sargan test) on overidentifying restrictions, that is, we test if our instrumental variables are valid candidates.

[18] See also Wooldridge (2002, pp. 118–120).
[19] One should be careful here in recognizing the difference between the theoretical concept of "endogenous entry" and the econometric concept of endogeneity of the variable in a regression!

Table 5.4 reports the IV relevance tests from the first stage regression shown in Table 5.5 (partial F-statistics), and the Smith-Blundell test on reverse causality between entry and R&D based on the heteroscedastic regressions of Model I, since the homoscedastic version led to the same conclusions.

Table 5.5. IV first stage regressions on $ENTRY$ (1,857 observations)

Variables	Model I	Model II	Model III
$EMP_{i,t-1}/1000$	-0.057^{*}	-0.051	-0.065^{**}
	(0.033)	(0.032)	(0.033)
$(EMP_{i,t-1}/1000)^2$	0.0014^{*}	0.0013	0.002^{*}
	(0.0009)	(0.0009)	(0.0009)
$KAPINT_{i,t-1}$	0.087	0.153	0.061
	(0.241)	(0.239)	(0.243)
$HHI_{i,t-1}$	-0.00001	0.0001	0.0002
	(0.0004)	(0.0004)	(0.0004)
$LEADER_{it}$	-0.242^{***}	-0.227^{***}	-0.223^{***}
	(0.068)	(0.068)	(0.068)
$MES_{i,t-1}$	-0.330^{***}	-0.319^{***}	-0.340^{***}
	(0.087)	(0.083)	(0.014)
$ADVERT_{it}$		0.050^{***}	0.054^{***}
		(0.014)	(0.014)
$SUBSTITUTE_{it}$			0.142^{***}
			(0.025)
Intercept	1.711^{***}	1.559^{***}	1.269^{***}
	(0.086)	(0.097)	(0.110)
F–test: industry dummies	2.44^{***}	2.07^{**}	1.88^{**}
F–test: IVs	14.33^{***}	14.47^{***}	21.41^{***}

Notes: Standard errors in parentheses. *** (**, *) indicate a significance level of 1% (5%, 10%). The 'F–test: IVs' refers to a joint significance test of our instrumental variables, which are MES in model I, MES and $ADVERT$ in model II and MES, $ADVERT$ and $SUBSTITUTE$ in model III.

Staiger and Stock (1997) emphasize that the first-stage significance levels of the instrumental variables may be misleading, as they do not necessarily exclude a weak instrument problem, which would lead to considerable bias in IV regressions. Instead of interpreting the significance level, they argue that, as a rule of thumb, the partial F-statistic should exceed the value of 10 in the case of a single endogenous regressor to confidently rule out weak instruments. As can be seen in Table 5.4, all F values exceed the value of 10, and consequently we can reject a weak instrument bias.[20]

[20] More recently, Stock and Yogo (2005) derived new critical values for the weak instrument test on the basis of the rank test (see Kleibergen and Paap, 2006), and it would be desirable to rely on these. However, the critical values are only available for a minimum of three instrumental variables. Although our Model III employs three instruments, we document below that these are not valid as

Furthermore, we test whether the instrumental variables are uncorrelated with the error term in our structural equation. Only if we are confident of having no weak instrument problem, and if the instruments are not correlated with the error term in the R&D equation, we can rely on our IV results. The validity of the IV candidates is usually assessed using the Sargan test or Hansen's J-test for a heteroscedasticity-robust version. Unfortunately, these tests are based on standard 2SLS estimations, and they are not available for Tobit regressions. Therefore, we employ regular 2SLS ignoring the censoring of our dependent variable for the test.

The results are also shown in Table 5.4.[21] The set-up where we use MES_i and $ADVERT_i$ as instrumental variables passes the Hansen's J-test, but when we include $SUBSTITUTE_i$, the test rejects the validity of this combination of instruments. As a final step, we test for reverse causality from $RDINT_i^*$ to $ENTRY_i$ using the Smith-Blundell test. As the results in Table 5.4 show, the exogeneity of the variable $ENTRY_i$ with respect to $RDINT_i^*$ is not rejected.

Notice that we tested more combinations of our IV candidates than shown in Table 5.4, but the results never changed. We also tested other IVs that are not mentioned above, as the average profitability in the industry, the percentage of defaults out of the total number of firms in an industry as a variable standing for risk in an industry, and the ratio of capital depreciation and total assets at the industry level as a further proxy for sunk costs. None of these were significant in the first stage regression explaining entry, nor did the Smith-Blundell test reject exogeneity.

In summary, we find relevant instrumental variables, but the potential reverse causality has been rejected by the tests. Furthermore, we can also confirm the validity of instruments based on 2SLS regressions using the Hansen J-Test for several IV combinations. Given these results, we conclude that the results as presented in Table 5.3 still hold, and that our two main hypothesis are thus confirmed: R&D investment decreases with larger entry threats in general, but leaders invest more than outsiders when threatened by entry.[22]

The determinants of endogenous entry. The first stage regressions for $ENTRY_i$ shown in Table 5.5 provide, as a side product, an interesting analysis of the determinants of the endogeneity of entry. They relate the perceived

this set—p does not pass the Hansen J-test. Therefore, we cannot utilize the Stock-Yogo test statistics.

[21] Note that the Hansen J-test is only applicable in case of overidentification. Thus, we cannot calculate the test for Model I, where only one instrument is used.

[22] In addition to feedback effects from R&D to entry, one may be concerned about feedback from R&D to our variable $LEADER_i$. We simply checked if past R&D intensity (which we have for a subsample of about 1,000 companies) determines our leadership variable to a certain extent. For this, we simply regressed $LEADER_i$ on past R&D intensity, past sales and industry dummies. It turns out that past sales, and thus past firm size, dominate the relationship. There is no additional effect of past R&D beyond firm size.

threat of entry to a number of control variables. In particular, we propose three models, all of which include the size of the firm, its capital intensity, the Herfindahl-Hirschman Index, the incumbent status variable and the minimum efficient size (as in Model I), with the addition of the importance of advertising (as in Model II) and also of the perceived substitutability between products (as in Model III). In this last case, we can emphasize a number of significant results.

First, larger firms, both in terms of employment and of their own perception of relative size, are less likely to be active in markets where entry is endogenous, while capital intensity and the index of concentration in the market do not appear to affect the extent of entry pressure in the market. More interesting, a large minimum efficient scale is negatively correlated with the perceived entry threat: in other words, natural entry barriers make it less likely that entry is endogenous. The perceived importance of advertising in the market is positively correlated with endogenous entry: this may suggest that entry is perceived as easy when investments in advertising are crucial to increase market shares. Also the perceived degree of substitutability between goods is associated with endogenous entry: when goods are highly substitutable, it is easy to enter and increase the market share by offering the products at low enough prices, while differentiated goods reduce the relevance of entry pressure.

Of course, this is only a preliminary and incomplete investigation of the determinants of the endogeneity of market structures. Further work should uncover other explanatory variables and verify the possible links between them. A deeper analysis of entry conditions is highly needed not only for research purposes, but also for policy purposes, since free entry should be the aim of industrial policy for most markets and because only its absence could motivate antitrust intervention.

5.2 A Model of Endogenous Growth

On the basis of our general and empirical discussion on the determinants of innovation and of the structure of the market for innovations, we now develop a general equilibrium model of endogenous technological progress and we analyze the EMSs of the competition in the markets and for the markets. In this section we present a basic general framework, which differs from the dynamic models used until now in two dimensions: it is developed in the continuous time (to be consistent with the conventional choice of the growth literature focusing on long run properties) and it accounts for product differentiation between intermediate goods rather than consumption goods (which allows us to directly associate innovations with endogenous productivity growth).

In line with most of the endogenous growth literature,[23] we neglect the role of capital accumulation in the growth process, assuming that labor and intermediate goods are the only inputs in the production of final goods, and that new products can be created through investment of the final goods; similar results would emerge if (highly qualified) labor was employed to create new business (as in Chapter 3). This does not mean that capital accumulation is not important in the growth process, but that the neoclassical approach has well explained this process for industrializing economies, and that technological progress and growth in the New Economy can be better explained with the allocation of resources and labor in the innovative process that leads to the accumulation of new ideas.

Consider a closed economy where the final good is produced according to the following Cobb-Douglas function:

$$Y_t = \mathbf{X}_t^{\alpha} (AL)^{1-\alpha} \tag{5.6}$$

where A is labor productivity, L is the fixed labor force, and $\mathbf{X}_t$ is an index which summarizes the quantity and quality of all the other differentiated inputs:

$$\mathbf{X}_t = \left[\sum_{j=1}^{N_X} \left[q^{\kappa_j} X_t(\kappa_j) \right]^{\frac{\theta-1}{\theta}} \right]^{\frac{\theta}{\theta-1}} \tag{5.7}$$

In general, $X_t(\kappa_j)$ is the input j of quality $\kappa_j \geq 0$, N_X is the number of the differentiated inputs, and $q > 1$, so that the index q^{κ_j} summarizes the quality of input j in a direct way. It can be easily verified that the production function satisfies constant elasticity of substitution between the differentiated inputs, CRS and decreasing marginal productivity for each input. The parameter $\theta > 1$ is the elasticity of substitution between the differentiated inputs. Notice that this is the production function adopted in most studies of endogenous technological change to analyze innovation in terms of creation of new inputs, as we will do in this section, or in terms of better varieties of the existing inputs, as we will do in the following sections. Acemoglu (2002) has also used this production function to examine directed technological change toward different inputs.[24]

---

[23] See Romer (1990), Segerstrom, Anant and Dinopoulos (1990), Grossman and Helpman (1991) and Aghion and Howitt (1992).

[24] Acemoglu (2002) assumes $\alpha = 1$ and $N_X = 2$, and interprets $X_1(\kappa_1)$ and $X_2(\kappa_2)$ as skilled and unskilled labor inputs, whose relative marginal productivities are:

$$\varpi \equiv \frac{MP_1}{MP_2} = \left(q^{\kappa_1 - \kappa_2}\right)^{1-1/\theta} \left(\frac{X_1(\kappa_2)}{X_2(\kappa_2)} \right)^{-1/\theta}$$

For given technology (κ_j constant), the skill premium ϖ is decreasing in the relative quantity of skilled labor. However, a relative increase in skilled labor can induce technological change that is biased in favor of skilled labor,

In what follows, we interpret the differentiated inputs $X_t(\kappa_j)$ as intermediate inputs used to produce the final good. Therefore, output can be used for consumption, production of intermediate goods or investment in R&D. We also assume $0 < \alpha \leq 1 - 1/\theta \leq 1$. In the standard growth literature the assumption $\alpha = 1 - 1/\theta$ (see Barro and Sala-i-Martin, 1995) creates a link between the substitutability between different intermediate goods and that between labor and the index of intermediate goods, which eliminates any strategic interaction between the firms. However, it is important to keep these two parameters conceptually separate when we are interested in the EMS for the production of output - for instance because the number of producers of intermediate goods N_X is small due to substantial fixed costs.

We consider an infinite horizon representative agent with isoelastic utility:

$$U = \int_0^\infty \frac{C_t^{1-\gamma}}{1-\gamma} e^{-\rho t} dt \tag{5.8}$$

where $\rho > 0$ is the time preference rate and $\gamma > 0$ is the intertemporal elasticity of substitution. The agent can consume its income or save it to earn capital income at the interest rate r_t. As we have seen in Chapter 1, the standard optimality condition for the choice of savings (1.63) implies that consumption grows at the rate:

$$\frac{\dot{C}_t}{C_t} = \frac{r_t - \rho}{\gamma} \tag{5.9}$$

which is constant only when the interest rate is constant as well.

The markets for the final good, which is the *numeraire*, for labor and for credit (to firms investing in R&D) are perfectly competitive, while we introduce imperfect competition in the market for intermediate goods, where we assume that each innovator has a patent on the leading technology and is the only one able to use it. The outcome of this competition depends on many factors and leads to an equilibrium price given by a mark up on the marginal cost. Since the latter is unitary, the price is also equal to the gross mark up for each intermediate good j, $p_j = \mu_j$.

When innovations are non-drastic, innovators can adopt limit pricing strategy, setting the price at the highest level at which competitors with the immediately inferior technologies cannot profitably enter in the market,

---

if $\partial(\kappa_1 - \kappa_2)/\partial(X_1/X_2) > 0$ (weak equilibrium bias). Moreover, if this is the case and if the elasticity of substitution θ is large enough, a relative increase in skilled labor increases the skill premium (strong equilibrium bias), so that the endogenous technology relative demand curve becomes upward sloping, $\partial\varpi/\partial(X_1/X_2) > 0$. Acemoglu (2002) develops a model of endogenous growth in which the first effect occurs always because innovation is mainly attracted by inputs with a larger market. This may explain why technological progress has been unskilled biased in the XIX century and skilled biased in the last decades (when respectively unskilled and skilled labor were increasing faster). See Acemoglu (2009, Ch. 15) for further applications.

namely $\mu = q$. A more interesting case emerges when innovations are drastic. In such a case, each intermediate good is produced by the firm with the cutting edge technology, but we could still have different outcomes according to the form of competition between intermediate goods' producers.

5.2.1 Cournot competition

Consider first competition in quantities. The competitive sector for the production of final goods demands intermediate inputs to equalize their marginal productivity to their price. It is immediate to derive the inverse demand for each intermediate good, which must be decreasing in the quantity of any intermediate good. This inverse demand implies the following gross profits for the producer of intermediate good j:

$$\Pi(\kappa_j) = \frac{\alpha\,(AL)^{1-\alpha}\left[q^{\kappa_j}X(\kappa_j)\right]^{1-1/\theta}}{\left[\sum_{j=1}^{N_X}\left[q^{\kappa_j}X(\kappa_j)\right]^{1-1/\theta}\right]^{\frac{\theta(1-\alpha)-1}{\theta-1}}} - X(\kappa_j) \tag{5.10}$$

where the denominator of the revenue function disappears in the traditional case in which $\alpha = 1 - 1/\theta$ (see Barro and Sala-i-Martin, 1995). In such a case strategic interactions would disappear and each firm would become a pure monopolist in the production of its intermediate good. The monopolistic output would be $X(\kappa_j) = AL\left(\alpha^2 q^{\kappa_j \alpha}\right)^{1/(1-\alpha)}$, corresponding to the following monopolistic price for each firm:

$$\mu^M = \frac{\theta}{\theta - 1} \tag{5.11}$$

In the more general case with $\alpha < 1 - 1/\theta$, the strategic interactions become relevant. Under Cournot competition, all firms choose their production level $X(\kappa_j)$ taking as given the production of the others. In the absence of technological differences between sectors (that is, with the same κ_j for any j) the symmetric equilibrium price can be easily derived as:

$$\mu^Q(\theta, N_X) = \frac{\theta N_X}{(\theta - 1)(N_X - 1) + \theta\alpha} \tag{5.12}$$

which is decreasing in the number of firms producing intermediate goods N_X, in the degree of substitutability between goods θ, and also in the factor share of income from intermediate goods α.

When there are technological differences between alternative intermediate goods the equilibrium mark up is different for each good and dependent on the number of sectors and on their technological conditions - however, one can show that firms with better technologies (higher κ_j) will produce more in equilibrium. Things are again simpler when the number of sectors N_X is high enough and the strategic effects are negligible. In such a case, the equilibrium price converges to (5.11) for all goods.

5.2.2 Bertrand competition

Consider now the case of competition in prices. One can easily derive the direct demand from the final good sector. The demand for each good is decreasing in its price and increasing in the price of all the other intermediate goods. The profit function for the producer of the intermediate good j can be derived as:

$$\Pi(\kappa_j) = (p_j - 1)p_j^{-\theta} \frac{\alpha^\theta (AL)^{(1-\alpha)\theta} q^{\kappa_j/\theta}}{\left[\sum_{j=1}^{N_X} \left(\frac{q^{\kappa_j}}{p_j}\right)^\theta\right]^{\frac{\theta(1-\alpha)-1}{(\theta-1)(1-\alpha)}}} \tag{5.13}$$

since the constant marginal cost of production is normalized to 1.

In the absence of technological differences between sectors, the Bertrand equilibrium generates a common price:

$$\mu^P(\theta, N_X) = \frac{\theta}{\theta - \frac{N_X}{N_X - F(\theta,\alpha)}} \tag{5.14}$$

where $F(\theta,\alpha) \equiv [\theta(1-\alpha)-1]/(\theta-1)(1-\alpha)$ is decreasing in the degree of substitutability between goods θ and in the factor share of income from intermediate goods α. This implies that under competition in prices the mark up is again decreasing in N_X, in θ, and in α.[25]

When there are technological differences between producers of different intermediate goods, the calculation of the Bertrand equilibrium is again quite complex due to the strategic interactions between heterogenous firms, but one can show that firms with better technologies will be able to set a higher price. Things are simpler when the number of sectors N_X is high enough that strategic effects are negligible. In such a case, profits are proportional to $(p_j - 1)p_j^{-\theta}$, which is maximized again by the price (5.11). This situation is particularly realistic when the number of sectors increases indefinitely and generates growth, which happens when the firms endogenously develop new kinds of intermediate goods, as first shown by Romer (1990), whose model is presented next.

5.2.3 The Romer model

The situation in which firms develop new intermediate goods is exactly at the foundation of the Romer (1990)[26] model of endogenous growth through

[25] In a related work, Minniti (2008) has introduced strategic interactions in a model that captures both the intra-industry competition between firms operating within an industry and the inter-industry competition between firms in different industries.

[26] The model is also known as the lab-equipment model or the product-variety model of growth. See Acemoglu (2009, Ch. 13) for a review, and Matsuyama (1999) for an interesting OLG extension of the Romer model with capital accumulation and cycling dynamics.

horizontal innovations (also known as the *lab-equipment* or *product variety* model): when the development of new intermediate goods requires a fixed cost, it happens only gradually and drives economic growth. Consider the simplest case in which there are no technological differences between existing and new intermediate goods (say $\kappa_j = 0$) and $\alpha = 1 - 1/\theta$. In such a case each good is priced at a constant monopolistic mark up μ and the gross value of each firm is the discounted sum of the future flow of profits, that are $\Pi = (\mu - 1)\left[(\theta - 1)/\theta\mu\right]^{\theta} AL$.

The endogenous entry process induces constant growth. In particular, the equilibrium interest rate can be obtained by the endogenous entry condition that equates the gross value of a new (monopolistic) firm $V^M = \Pi/r$ to the fixed cost of entry F:

$$V^M = \frac{(\mu - 1)\left[(\theta - 1)/\theta\mu\right]^{\theta} AL}{r} = F \tag{5.15}$$

This condition implicitly defines the rate of creation of new firms, and consequently the growth rate. The total amount of savings available must match total investment in the economy, with the interest rate adjusting to clear the credit market. It is standard to verify that this requires that income must grow at the same rate as consumption, that is at the rate given by (5.9). Substituting the interest rate from (5.15) in (5.9), one obtains the growth rate of consumption and output as:[27]

$$\tilde{g} = \frac{(\mu - 1)\left(\frac{\theta-1}{\theta\mu}\right)^{\theta} \frac{AL}{F} - \rho}{\gamma} \tag{5.16}$$

Growth is driven by the creation of new products attracted by monopolistic profits: in a sense the stock market value of the patentholders is what directly determines the growth rate. Moreover, there is not adjustment dynamics: the economy grows always at the constant growth rate. Notice that scale effects are present: this generates the counterfactual prediction that a constant rate of growth of the population leads to explosive growth of income per capita. However, these scale effects would disappear simply if the fixed costs of business creation were proportional to the size of the economy.

5.2.4 The Peretto model with EMSs

If we re-introduce strategic interactions between producers of intermediate goods, we obtain mark ups and profits that are decreasing in the number of intermediate goods. In this case, product creation is stronger when there are few intermediate goods and decreases along the innovative process: the

[27] Here we have followed the adaptation of Barro and Sala-i-Martin (2004). The original Romer model assumed that skilled workers are employed in the R&D activity.

consequence is that the long run endogenous growth rate is reached through a path with decreasing growth, in line with the convergence hypothesis.[28] In a number of pioneering contributions Peretto (1996, 1999, 2003) has augmented such an analysis of EMSs in the competition in the market with a stylized form of technological progress to examine a number of issues.[29]

Peretto (1996) develops a model in which firms undertake cost-reducing R&D subject to a research technology with incomplete spillovers. Concentration of sales and R&D resources determines the optimal scale and the efficiency of firms' R&D operations and, thus, the rate of productivity growth. In addition, R&D expenditures are one component of the fixed costs of entry and determine the number of active firms in the equilibrium with endogenous entry. This feedback makes the price, investment, entry, and exit decisions interdependent, leading to the possibility of multiple equilibria where firms' expectations about rivalry determine the economy's performance.

Peretto (1999) emphasizes two aspects of the relation between EMSs and growth. First, a larger number of firms induces fragmentation of the market and dispersion of R&D resources, which prevents exploitation of scale economies internal to the firm and slows down growth. Second, there is a fundamental trade-off between growth and variety that produces interesting results. In particular, the scale effect is bounded from above and converges to zero when the number of firms is large: this result suggests that if the labor force grows at a constant rate, the oligopolistic economy converges to monopolistic competition where the scale effect is always zero. The number of firms then grows at the same rate as the labor force while the rate of growth does not change, therefore it is possible to introduce population growth in this class of models without the counterfactual prediction of explosive growth rates. Moreover, the market supplies too many varieties, a result similar to what we found in Chapter 3 and that we will encounter again in a different framework below. Oligopolistic firms face a price elasticity of demand that is lower than the elasticity of product substitution. As a consequence, the private rate of return to R&D is too low and firms spend too little on R&D. Since R&D spending represents the fixed cost of entry, this implies that the rate of return to entry is too high and that too many firms enter in the market. The size of this distortion is endogenous and depends on the number of firms: as this becomes very large, the market equilibrium approaches asymptotically the social optimum.

The analysis of EMSs in the competition in the market of the intermediate goods has been exploited also to study other macroeconomic issues. For instance, Jaimovich (2007) has studied EMSs similar to those characterized above assuming endogenous entry in each sector (without growth) and emphasizing how the interaction between firms'entry-and-exit decisions gives

[28] We obtained a similar results in Section 2.9.

[29] See also the independent contribution by De Klundert and Smulders (1997) and the more recent works by Le (2008) and Peretto (2008a,b).

rise to self-fulfilling expectation-driven fluctuations in aggregate economic activity and in measured total factor productivity.[30]

5.3 A Model of Schumpeterian Growth

In the rest of the chapter, we are interested in technological progress due to improvements in the productivity of the existing varieties of intermediate goods, therefore we neutralize the process of creation of new varieties in the model of the previous section and assume that the number of intermediate goods is high enough that the equilibrium mark up is constant and equal to μ. To conclude the characterization of the production process (before moving to the analysis of innovation), we need to derive the behavior of the aggregate production.

The aggregate quantity produced of any intermediate good j can be determined as:

$$X(\kappa_j) = \left(\frac{\alpha}{\mu}\right)^{\theta} ALq^{\kappa_j(\theta-1)}Q^{-\frac{\theta(1-\alpha)-1}{(\theta-1)(1-\alpha)}} \tag{5.17}$$

where we have introduced the Barro and Sala-i-Martin aggregate quality index $Q \equiv \sum_{j=1}^{N_X} q^{\kappa_j(\theta-1)}$, which is supposed to grow with technological progress. Substituting the quantity $X(\kappa_j)$ from (5.17) in the production function, we obtain the output of final goods:

$$Y = \left(\frac{\alpha}{\mu}\right)^{\frac{\alpha}{1-\alpha}} ALQ^{\frac{\alpha}{(\theta-1)(1-\alpha)}} \tag{5.18}$$

and the total amount of intermediate goods $X = \alpha Y/\mu$. Since TFP and labor force are constant, the growth rate of income must be:

$$g \equiv \frac{\dot{Y}}{Y} = \frac{\alpha}{(\theta-1)(1-\alpha)} E\left[\frac{\dot{Q}}{Q}\right] \tag{5.19}$$

If we define with $\delta_N(\kappa_j)$ the probabilty of innovation for the intermediate good j of quality κ_j, the expected growth rate of the quality index must be:

[30] The analysis is based on a dynamic general equilibrium model in which net business formation is endogenously procyclical and leads to endogenous countercyclical variations in markups. This interaction leads to indeterminacy in which economic fluctuations occur as a result of self-fulfilling shifts in the beliefs of rational forward looking agents (see Farmer, 1993). When calibrated with empirically plausible parameter values and driven solely by self-fulfilling shocks to expectations, the model can quantitatively account for the main empirical regularities characterizing postwar U.S. business cycles and for 65% of the fluctuations in measured TFP. Jaimovich and Floetotto (2008) and Brito, Costa and Dixon (2008) have also extended the model adopting the same approach of Etro and Colciago (2007) to Cournot competition.

$$E\left[\frac{\dot{Q}}{Q}\right] = \delta_N \left[q^{\theta-1} - 1\right] \approx \delta_N(\theta - 1)\ln q$$

with $\delta_N \equiv \left[\sum_{j=1}^{N_X} \delta_N(\kappa_j)q^{\kappa_j(\theta-1)}\right]/Q$ representing a weighted average of the probability of innovations.[31] Accordingly, the growth rate of income becomes:

$$g \approx \frac{\alpha}{(\theta-1)(1-\alpha)}\left[\delta_N(\theta-1)\ln q\right] = \frac{\delta_N \alpha \ln q}{1-\alpha} \tag{5.20}$$

which depends on the size of the innovations q, on the factor share of income from intermediate goods α, and on the average rate of innovation δ_N which will be endogenized below. As before, credit market clearing requires that income grows at the same rate as consumption, that is (5.9). Using these two expressions for the growth rate, one obtains:

$$\tilde{g} = \frac{r + \delta_N - \rho}{\gamma + (1-\alpha)/\alpha \ln q} \tag{5.21}$$

5.3.1 Patent races

It is now time to apply the EMSs approach to the study of competition for the market in general equilibrium. The markets for innovations are characterized by sequential patent races. First of all, the flow of profit for each intermediate good producer with the sector's highest quality rung κ_j is:

$$\Pi(\kappa_j) = (\mu - 1)X(\kappa_j) \tag{5.22}$$

To participate to the patent race for a particular innovation $\kappa_j + 1$, each firm i has to pay a fixed cost $F(\kappa_j)$ and spend a flow of resources $z_i(\kappa_j)$. Contrary to the usual Schumpeterian growth literature, which adopts the neoclassical assumption of constant marginal productivity in the R&D sector - equivalent to CRS since there is just one input, we follow the empirical evidence and assume decreasing marginal productivity in the production of new ideas. In particular, the investment of firm i gives birth to the innovation κ_j according to a Poisson process with arrival rate $h_i(\kappa_j)$ given by a concave function of $z_i(\kappa_j)$. Following Etro (2004,a) we assume the specification:

$$h_i(k_j) = [\phi(\kappa_j)z_i(\kappa_j)]^{\epsilon} \tag{5.23}$$

where the function $\phi(\kappa_j)$ expresses how difficult is to discover technology κ_j and $\epsilon \in (0,1]$ represents the degree of returns to scale in the innovation sectors or the elasticity of expected revenue with respect to the flow of investment. This parameter is unitary in the traditional versions of the quality-ladder

[31] Notice that, contrary to Chapter 2 to 4, here the probability of innovation δ_N is an instantaneous probability, ranging between 0 and ∞.

model, but empirical research, for instance by Cohen and Klepper (1996) and Kortum (1993) suggests an elasticity much smaller than 1.[32] The aggregate arrival rate of innovation κ_j which corresponds to the exit rate of the leading edge technology, is the sum of the individual innovation rates of the firms including the incumbent:

$$\delta_N(\kappa_j) = \sum_{j=1}^{N_j} [\phi(\kappa_j) z_i(\kappa_j)]^\epsilon$$

where N_j is the number of firms engaged in R&D activity in sector j. Therefore, there are decreasing returns both at the firm level and at the industry level, as suggested by the empirical evidence in this field. Using the properties of Poisson processes in a standard fashion, the expected gross value of entrant i in the patent race in sector j when the current quality is κ_j can be written as:

$$V^i(\kappa_j) = \frac{[\phi(\kappa_j) z_i(\kappa_j)]^\epsilon V^M(\kappa_j + 1) - z_i(\kappa_j)}{r + \delta_N(\kappa_j)} \tag{5.24}$$

where $r + \delta_N(\kappa_j)$ could be defined as the effective discount factor at the time of the innovation of vintage κ_j and $V^M(\kappa_j + 1)$ is the value of being monopolist with the next technology $k_j + 1$. It is immediate to verify that these functions of expected profit are nested in the general form (2.1), and the EMSs in the competition for the market inherits all their properties.

Since also the incumbent monopolist with technology κ_j can invest to innovate, we need to consider its objective function, which is given by:

$$V^M(\kappa_j) = \max_{z_M \geq 0} \left\{ \frac{[\phi(\kappa_j) z_M]^\epsilon V^M(\kappa_j + 1) + \Pi(\kappa_j) - z_M}{r + \delta_N(\kappa_j)} - F(\kappa_j) \right\} \tag{5.25}$$

where the fixed cost is paid only if $z_M > 0$. Deep down, this value of the innovation is the engine of growth, because what drives investment and growth is exactly the attempt to conquer this value.

We assume that new ideas are more difficult to obtain when the scale of the sector, as represented by expected production with the new technology, increases, so that:

$$\phi(\kappa_j) = \frac{1}{X(\kappa_j + 1)}$$

and that the fixed cost is a constant fraction of the expected cost of production with the new technology, that is:

[32] Kortum (1993) suggests a range between 0.1 and 0.6 for this elasticity. Segerstrom (2007) assumes that decreasing returns hold just for the incumbent monopolist, while constant returns to scale characterize all the other firms. He solves the model through simulations and assumes $\epsilon = 0.3$ as the average between the values proposed by Kortum.

$$F(\kappa_j) = \eta \int_0^\infty X(\kappa_j+1)e^{-[r+\delta_N(\kappa_j+1)]t}dt$$

with $\eta \in (0,\mu)$. With these assumptions, we want to capture the idea that the larger is the scale of expected production of a firm, the larger are the costs necessary to discover it, to develop the associated technology and the infrastructures needed to adopt this technology (think of new assembly lines, training of workers, construction of prototypes and samples). These assumptions will deliver a balanced growth path with equal technological progress for all inputs[33] and will avoid scale effects on the equilibrium growth rate, which is in line with the Jones's critique to the first generation of quality-ladder models.[34] Before analyzing the general EMSs of the competition for the market, we briefly present the neoclassical case in which CRS holds in the innovation sector and the market structure is indeterminate.

5.3.2 The Aghion-Howitt model

In their original contribution to the Schumpeterian growth theory, Aghion and Howitt (1992) have assumed that the returns to scale in the innovation process are constant and that there are no fixed costs of the R&D activity. This unrealistic assumption, that has been used by most of the subsequent literature, amounts to $\epsilon \to 1$ with $\eta \to 0$. In such a case, the market structure becomes indeterminate, and we have nothing to say about how many firms invest in R&D and how much each firm invests. As usual, we can assume without loss of generality that there is a single outsider firm investing in each sector, so that the arrival rate of innovation is $\delta_N(\kappa_j) = \phi(\kappa_j)z_i(\kappa_j)$ in sector j. The value of being a monopolist with technology κ_j, (5.25), boils down to:

$$V^M(\kappa_j) = \frac{\Pi(\kappa_j)}{r+\delta_N(\kappa_j)} = \frac{(\mu-1)X(\kappa_j)}{r+\delta_N(\kappa_j)} \tag{5.26}$$

A neoclassical no-arbitrage condition for the patent race κ_j pins down the investment in innovation in the patent race κ_j+1. In particular the no-arbitrage condition $\phi(\kappa_j)z_i(\kappa_j)V^M(\kappa_j+1) = z_i(\kappa_j)$ leads to the same rate of innovation for each sector δ_N, which satisfies:

$$r + \delta_N = \mu - 1$$

This allows one to rewrite the equilibrium growth rate (5.21) as:[35]

[33] Since it is harder to innovate in case of more advanced inputs (that are also inputs whose effective quantity is larger), technological progress is unbiased across sectors. If this was not the case, we would obtain larger investments and a more rapid technological progress for more advanced inputs, something similar to the directed technological change of Acemoglu (2002).

[34] See Jones (1995) and Barro and Sala-i-Martin (2004).

[35] Here we have followed the adaptation of Barro and Sala-i-Martin (2004). The original Aghion-Howitt model assumed a single intermediate good with $\gamma = 0$, which implied $r = \rho$ and $g = [\alpha/(1-\alpha)]\,\delta_N \ln q$.

$$\tilde{g} = \frac{\mu - 1 - \rho}{\gamma + (1 - \alpha)/\alpha \ln q} \tag{5.27}$$

which shows the link between the profit rate and growth. Notice that the growth rate does not exhibit scale effects in the sense that it is independent from the size of the economy. However, this simple characterization excludes any implication for the market structure of the innovation sector and confines the optimal innovation policy to an R&D subsidy/fee.

5.3.3 A Schumpeterian model with EMSs

We now allow for decreasing returns to scale ($\epsilon < 1$) and fixed costs of R&D ($\eta > 0$), and we model competition between firms in the market for innovation in the Nash fashion. As a consequence of the so-called Arrow paradox, under endogenous entry the leader does not invest in R&D, because its best strategy is to stay out from the patent race and enjoy the profits from its current patent until a new innovation will make it obsolete (in the next section we will extend the model to account for innovation by leaders).[36] Competition for innovations is just between outsiders and the scope of this section is to characterize the equilibrium organization of the R&D sector, the number of firms investing and their size together with the usual macroeconomic variables.

The lack of investment by monopolists implies that the value of being a monopolist with technology κ_j, (5.25), boils down to (5.26). Each firm chooses its investment in R&D $z_i(\kappa_j)$ to maximize its gross value (5.24) taking the strategies of the other firms, the value of the next innovation and the interest rate as given. In the Nash equilibrium between N firms, each one of them chooses the same investment $z(\kappa_j)$ satisfying the symmetric first order condition:

$$\frac{\epsilon\phi(\kappa_j)^{\epsilon} z(\kappa_j)^{\epsilon-1} V^M(\kappa_j + 1) - 1}{[\phi(\kappa_j) z(\kappa_j)]^{\epsilon} V^M(\kappa_j + 1) - z(\kappa_j)} = \frac{\epsilon\phi(\kappa_j)^{\epsilon} z(\kappa_j)^{\epsilon-1}}{r + \delta_N(\kappa_j)} \tag{5.28}$$

The gross value of participating to the competition is decreasing in the number of firms, and the endogenous entry condition equates the gross value of each firm (5.24) to the fixed cost of entry:

$$V(\kappa_j) = \frac{[\phi(\kappa_j) z(\kappa_j)]^{\epsilon} V(\kappa_j + 1) - z(\kappa_j)}{r + \delta_N(\kappa_j)} = F(\kappa_j) \tag{5.29}$$

which is the equivalent of (3.26) in a growth context. This condition provides the equilibrium number of entrants N and implies the investment per firm:

$$z(\kappa_j) = \epsilon^{\frac{1}{1-\epsilon}} \phi(\kappa_j)^{\frac{\epsilon}{1-\epsilon}} \left[V^M(\kappa_j + 1) - F(\kappa_j) \right]^{\frac{1}{1-\epsilon}} \tag{5.30}$$

[36] See Acemoglu (2009, Ch. 12) for an extensive discussion of the Arrow paradox.

that is increasing in the value of the innovation net of the fixed cost of entry and independent from the interest rate.

Substituting the endogenous value of innovation (5.26) and using our assumptions in the expression for investment and in the endogenous entry condition, we can fully characterize the EMS in this particular patent race expliciting the investment by each firm and the number of firms.[37] It turns out that the former is increasing in the quality achieved in the single sector, since this implies higher demand and hence higher expected profits for the corresponding intermediate product, while the number of firms and the aggregate probability of innovation in each patent race turn out to be the same in all sectors. To see this, it is easy to use the endogenous entry condition (5.29) to express the effective discount rate $r+\delta_N(\kappa_j)$. Using this to explicit the expected value of innovation (5.26), and substituting in (5.30) we can obtain the equilibrium flow of investment for each firm of sector j:

$$z(\kappa_j)=\frac{\epsilon\eta\left(\mu-1-\eta\right)}{\left[\mu-1-\epsilon\left(\mu-1-\eta\right)\right]}q^{\theta-1}X(\kappa_j) \tag{5.31}$$

This investment level is increasing in the mark up μ, which is exactly at the core of the Schumpeterian idea that monopolistic profits drive innovation of single firms. Moreover, the investment is proportional to the future scale of production and increasing in the degree of returns to scale, ϵ, since this makes investment more productive.[38]

In the equilibrium balanced growth path all firms within any single sector innovate with the same instantaneous probability:

$$\delta_N=\delta_N(\kappa_j)=N[\phi(\kappa_j)z(\kappa_j)]^{\epsilon}$$

for any j, which implies, by (5.20), that growth is a linear function of the number of firms N. Solving for the effective discount rate $r+\delta_N(\kappa_j)$ and using (5.21) we can fully characterize the steady state EMS, the growth rate and the arrival rate of innovations as functions of behavioral, technological and policy variables. In particular, the equilibrium growth rate is:

$$\tilde{g}=\frac{\left[\epsilon\left(\mu-1-\eta\right)\right]^{\epsilon}\left[(\mu-1)(1-\epsilon)/\eta+\epsilon\right]^{1-\epsilon}-\rho}{\gamma+\left(1-\alpha\right)/\alpha\ln q} \tag{5.32}$$

---

[37] Notice that in the Aghion-Howitt model the no-arbitrage condition for the patent race κ_j pins down the investment in innovation in the patent race κ_j+1 without a clear economic intuition. Instead, in our model, the endogenous entry condition for the patent race κ_j pins down the number of firms investing in innovation in the patent race κ_j and, together with their profit maximizing choices, their individual investments in the same patent race.

[38] The effect of higher fixed costs on investment can be shown to be non monotonic, positive for η low, but negative for η high enough: on one side high fixed costs reduce expected profits for a given life of the patent, but on the other, they reduce the innovation rate in the future so as to increase the expected life of the patent.

which is increasing in α, q and μ, while it is decreasing in ρ, η and in γ. As one could expect, the more costly are innovations, the lower is equilibrium growth; increases in the size of the innovations (for instance due to some general purpose technology which enhances technological progress) or in monopolistic mark-ups promote growth, while it can be shown that the relation between growth and ϵ is U-shaped.

The equilibrium arrival rate of innovations is directly proportional to the above growth rate, while the endogenous number of firms in each patent race is:

$$\tilde{N} = \frac{\frac{(\mu-1)(1-\epsilon)}{\eta} + \epsilon - \rho\left[\frac{(\mu-1)(1-\epsilon)+\epsilon\eta}{\epsilon\eta(\mu-1-\eta)}\right]^{\epsilon}}{1+\alpha\gamma\ln q/(1-\alpha)} \tag{5.33}$$

This expression is complex, but emphasizes a relation between the number of firms and a few structural parameters. When the agents are more patient (ρ is lower, that is the discount factor is higher), savings are higher and more firms are created, while a higher size of the fixed costs of entry (a higher η) tends to induce less business creation. A larger expected mark up μ is also leading to a higher number of firms, therefore, as long as a higher degree of substitutability between the intermediate goods (a higher θ) leads to larger mark ups, it also induces more business creation in the long run. These results confirm what we obtained in Chapter 3 for the endogenous number of firms engaged in production activities (here scale effects are absent and exit from the markets is endogenous).

Three other structural parameters affect the number of firms active in the long run. They are all associated with the technological structure: ϵ, α and q. The impact of the degree of returns to scale in the innovative activity (ϵ) on the number of firms is ambiguous.[39] Higher size innovations (with a larger q) are associated with higher growth but fewer firms. Finally, the growth rate is increasing in the factor share of income from the production of intermediate goods (α), but the steady state number of firms is decreasing with it. We can summarize the results as follows:

PROPOSITION 5.1. **The steady state endogenous market structure with competition for the market is characterized by a number of firms increasing in the discount factor, and decreasing in the entry cost parameter, in the degree of substitutability between intermediate goods, in the size of the innovations and in the factor share of income from intermediate goods.**

Contrary to standard models without scale effects, our framework with determinate EMSs for the innovative sectors is consistent with a positive impact of R&D policy not only on the investment in innovation at the firm

[39] Notice that approaching constant returns to scale in our model, the investment by each firm and the number of firms become indeterminate, but the equilibrium growth rate converges to the one of the Aghion-Howitt model.

level and at the aggregate level, but also on the growth rate. Later, we will introduce R&D subsidies and other instruments of fiscal policy and study their impact.

Finally, let us evaluate the impact of other sources of growth on the EMSs and on the rate of innovation.[40] Augmenting the model with an exogenous rate of productivity growth, it turns out that the rate of innovation may increase or decrease (Etro, 2007,c). This result is due to the effect of the endogenous adjustment of the interest rate on the value of innovations. When output growth increases, the first effect is that the profitability of future innovations increases, which tends to promote entry in the R&D activity. However, the second effect is that the interest rate goes up to raise savings and allow consumption growth to increase as well, which has a negative effect on the value of innovations. When the intertemporal elasticity of substitution is high ($\gamma < 1$), a small increase in the interest rate is needed to clear the credit market and the rate of technological progress increases unambiguously, but when the intertemporal elasticity of substitution is low ($\gamma > 1$), the opposite happens. Since realistic values for γ are not too far from unity, we should not be surprised to find out that there is not a strong empirical correlation between growth and R&D investment.[41]

5.4 EMSs with Growth Leaders

As we have suggested in the introduction to this chapter and verified empirically in Section 5.1, an important stylized fact about innovations is that many of them are due to incumbent monopolists and that a lot of the investment in R&D is actually done by these incumbents, at least when they are facing endogenous entry threats.

There are few competing explanations for innovation by incumbents in Schumpeterian growth models. The simplest one, due to Barro and Sala-i-Martin (1995) and Segerstrom (2007)[42] relies on the fact that incumbents may have a technological advantage in the R&D activity. This assumption may be realistic in certain sectors and allows one to study monopoly persistence, but it is basically equivalent to assume the solution of the Arrow paradox rather than solving it. Moreover, taking this view literaly, we should conclude that whenever we observe monopoly persistence it is because the incumbent firm is more efficient than the other firms both at producing and innovating. There are many sectors in which incumbents do not appear to have any cost

[40] The model can be also augmented with a labor input increasing at the constant growth rate of the population to study demographic factors, or with an additional input from exhaustible natural resources (which reduces their stock) to study environmental issues (see Aghion and Howitt, 2009, Ch. 16).

[41] The Schumpeterian framework can be also useful to study the relation between labor and credit market imperfections and growth. See Acemoglu (2009, Ch. 21).

[42] See also Denicolò and Zanchettin (2009).

advantage in the development of innovations compared to the outsiders, and still both the incumbents and the entrants keep investing.

Acemoglu (2008, 2009, Ch. 14) has proposed a different rationale for innovation by leaders. This may be due to the fact that only the incumbents can invest in incremental innovations (because outsiders would infringe their patents through small improvements), while entrants can invest (alone because of the Arrow effect) in more radical innovations. In such a way, both the incumbents and the outsiders invest, and the growth rate depends on their rates of investment weighted by the respective productivity increases. This is a plausible mechanism, but it explains why incumbents may invest in small improvements of their own technologies, which is a trivial activity, and not why they may directly compete against outsiders to obtain radical innovations, which is the key issue.

Aghion and Howitt (2009, Ch. 14) have forcefully advanced an "escape competition" rationale for investment by incumbents under entry, but their models rely on the assumption that a single incumbent faces an exogenous probability of entry (or an endogenous probability that a single rival may replace its leadership). Under endogenous entry of outsiders, the incumbent would not invest as a consequence of the usual Arrow effect (and the escape competition effect would disappear as well).

Here, we propose an alternative explanation for innovation by incumbents, originally introduced by Etro (2004,a), which does not rely on technological advantages or exogenous market structures, but is based on a pure strategic advantage of the incumbents in patent races with endogenous entry of outsiders (and whose implications are consistent with the empirical evidence presented in Section 5.1). Innovation by the incumbents requires two conditions: that they are leaders in the patent races for the next innovations, and that entry in these patent races is endogenous. Under both these conditions the Arrow effect disappears and the optimal strategy of the incumbents requires always a larger investment than the entrants. Such an aggressive strategy is aimed at limiting entry by maximizing the chances to innovate,[43] but in a general equilibrium context with sequential innovations, it leads to further aggregate results that we are about to investigate.

Let us consider the market for innovation described in the previous section, where (5.24) and (5.25) are the objective functions of the entrants and the incumbent patentholder. However, now let us assume that in each patent race the incumbent can commit to the investment in R&D before the outsiders, while the outsiders enter if they foresee non-negative expected profits and choose their investment taking as given the investment of all the other firms. The Stackelberg equilibrium with endogenous entry for each patent race can be derived following the general characterization by Etro (2008,b).

[43] Etro (2007,a) prodives a detailed discussion of innovation by leaders in partial equilibrium. For a critical evaluation of this approach see the book review of the *Journal of Economic Literature* by Zoltan J. Acs (2009, Vol. XLVII, March, pp. 208-210).

When the elasticity of the probability of innovation with respect to the investment ϵ is low, the incumbents invest more than the other firms, and therefore they may remain in the leadership position. When the same elasticity is high (ϵ is close to 1), it is optimal for the incumbents to deter entry investing just enough in R&D to make unprofitable for any follower to engage in R&D activities, and this delivers complete persistence of monopolies. Here we will analyze the first case.

The equilibrium of each patent race is characterized by the first order condition of the outsiders for the maximization of (5.24) given the investment of the leader, and by the endogenous entry condition. The first condition is still given by (5.28), and the second one by (5.29). Together, these conditions imply the same investment of each follower as in (5.30), independently from the investment of the incumbent. The latter affects only the number of entrants, leaving unchanged the aggregate probability of innovation $\delta_N(\kappa_j)$. Substituting the zero profit condition (5.29) in the expression for the value of the incumbent leader (5.25) we have:

$$V^M(\kappa_j) = \max_{z_M \geq 0} \left\{ \frac{[\phi(\kappa_j) z_M]^\epsilon V^M(\kappa_j+1) + \Pi(\kappa_j) - z_M}{[\phi(\kappa_j) z(\kappa_j)]^\epsilon V^M(\kappa_j+1) - z(\kappa_j)} F(\kappa_j) - F(\kappa_j) \right\}$$

whose maximization provides the equilibrium investment of the leader:

$$z_M(\kappa_j) = \frac{\left[\epsilon\phi(\kappa_j) V^M(\kappa_j+1)\right]^{\frac{1}{1-\epsilon}}}{\phi(\kappa_j)} > z(\kappa_j) \tag{5.34}$$

This confirms again the result for which leaders invest more than the other firms when they face endogenous entry threats, the main hypothesis that was successfully tested in Section 5.1. Notice that the investment of each outsider and the number of these outsiders are increasing in the value of being a patentholder and decreasing in the fixed cost of entry, while the investment of the incumbent is independent from the entry cost, therefore the gross rents of the incumbent are increasing in the entry cost.[44]

The full solution for the equilibrium is complicated by the fact that we do not know what is the value of being a patentholder. The problem can be solved using dynamic programming techniques.[45] Etro (2007c) has introduced a way to solve this particular kind of problem, which is likely to emerge

---

44 A consequence of this is that incumbent leaders may have incentives to restrict entry or increase the entry costs to protect their rents, which may reduce aggregate investment and growth. This implication has been used to propose a rationale for the positive relation between democracy and growth: in more democratic countries politicians would be less sensible to the pressure of incumbents to augment entry barriers and protect their rents, which would enhance innovative activities (see Aghion and Howitt, 2009, Ch. 17) For a model of growth and competition between rent seeking groups see Tornell and Velasco (1992). On the political economy of growth see Acemoglu (2009, Ch. 22 and 23).

45 See Stokey and Lucas with Prescott (1989) for an advanced treatment, and Sargent (1987,b).

whenever one is dealing with Schumpeterian models of growth where incumbent monopolists are engaged in R&D activity together with outsiders (see also Segerstrom, 2007).

Let us look for a steady state EMS with constant values for the growth rate, the interest rate, the arrival rate of innovations and the number of firms investing in patent races. To derive the equilibrium values for the value function $V(\kappa_j)$, the functions $z(\kappa_j)$ and $z_M(\kappa_j)$ and the equilibrium values for g, r, δ_N and N, we can adopt the method of undetermined coefficients (see Sargent, 1987,a). Let us guess a functional form for the value function as:

$$V^M(\kappa_j) = V^M(\kappa_j - 1)q^{\theta-1} = \psi \frac{X(\kappa_j)}{r+\delta_N} \tag{5.35}$$

where ψ is a coefficient to be determined, which can be interpreted as the rate of return from leadership. This must be larger than $\mu - 1$, otherwise the value of being a leader investing in R&D would be smaller than the value of being a leader without investing (or, in other words, it would be optimal to stay out of the patent race for the leader). Using our functional form we have:

$$z(\kappa_j) = \left(\frac{\epsilon(\psi-\eta)}{r+\delta_N}\right)^{\frac{1}{1-\epsilon}} q^{\frac{\alpha}{1-\alpha}} X(\kappa_j), \quad z_M(\kappa_j) = z(\kappa_j)\left(\frac{\psi}{\psi-\eta}\right)^{\frac{1}{1-\epsilon}} \tag{5.36}$$

Substituting in the Bellman equation we obtain:

$$\begin{aligned} V^M(\kappa_j) &= \frac{(\phi z_M)^{\epsilon} V^M(\kappa_j+1) + \Pi(\kappa_j) - z_M}{r+\delta_N} - F(\kappa_j) = \\ &= \left[\mu - 1 + \epsilon^{\frac{\epsilon}{1-\epsilon}} \left(\frac{\psi}{r+\delta_N}\right)^{\frac{1}{1-\epsilon}} q^{\theta-1}(1-\epsilon) - \eta q^{\theta-1}\right] \frac{X(\kappa_j)}{r+\delta_N} \end{aligned}$$

whose right hand side contains the mark up from the current innovation and another term which represents the option value of being the leader and having the opportunity to remain the dominant firm in the future innovation: this option has a positive value which directly comes from the leadership. Using our guess (5.35) and solving the value of being a leader for the effective discount rate we have:

$$r + \delta_N = \epsilon^{\epsilon} \left[\frac{(1-\epsilon) q^{\theta-1}}{\psi - \mu + \eta q^{\theta-1}}\right]^{1-\epsilon} \psi \tag{5.37}$$

which provides a negative relation between the effective discount factor and the rate of return from leadership (for ψ small enough): the higher is the effective discount rate, the shorter is the lifetime of an innovation and hence the lower is the value of being an incumbent leader.

Moreover, the zero profit condition for the followers provides another expression for the effective discount rate:

$$r+\delta_N=\left[\epsilon\left(\psi-\eta\right)\right]^{\epsilon}\left[\left(\psi/\eta\right)\left(1-\epsilon\right)+\epsilon\right]^{1-\epsilon} \tag{5.38}$$

This is a positive relation between the effective discount factor and the rate of return from leadership: the higher is the value of being a leader, the larger will be the investment in R&D and hence the probability of innovation and the effective discount rate.

Equating (5.37) and (5.38) we obtain the equilibrium value for ψ which provides all the equilibrium relations. Since $\psi > \mu - 1$, the effective discount rate $r + \delta_N$ must be higher than under Nash competition, and therefore the growth rate - by (5.21) - and the aggregate probability of innovation - by (5.20) - must be higher as well. The incumbency advantage adds a "turbo" to the engine of growth because it endogenously increases the value of innovations associating with them the option to a persistent leadership, which increases aggregate investment and therefore growth:

PROPOSITION 5.2. **The steady state endogenous market structure with competition for the market and leadership by incumbent patentholders is characterized by larger investment of the patentholders and a higher return from leadership than under pure Nash competition.**

Moreover, we can easily verify that the return from leadership ψ, the effective discount rate and the growth rate are all increasing in the mark up μ. An increase in the fixed cost of innovation through η decreases the effective discount rate and hence the growth rate of the economy, but it has ambiguous effects on the value of being a leader. Therefore, under Stackelberg competition in the market for innovations, incumbents invest in R&D more than any outsider. Using (5.21), the equilibrium growth rate becomes:

$$\tilde{g}=\frac{\psi\epsilon^{\epsilon}\left[\left(1-\epsilon\right)/\left(\eta+\left(1+\psi-\mu\right)q^{1-\theta}\right)\right]^{1-\epsilon}-\rho}{\gamma+\left(1-\alpha\right)/\alpha\ln q} \tag{5.39}$$

which is decreasing in η and increasing in μ. More importantly, we have:

PROPOSITION 5.3. **The steady state endogenous market structure with competition for the market and leadership by incumbent patentholders is characterized by a higher growth rate compared to pure Nash competition.**

In conclusion, a model of Schumpeterian growth which incorporates some realistic features of the market for innovation like decreasing marginal productivity of investment, fixed costs of entry and a first mover advantage for the incumbent patentholders, delivers realistic implications for the patterns of innovation. Incumbents do invest in R&D, even more than any other single firm, and their leadership persists with a certain probability, but sooner or later they are replaced by outsiders. The positive effect of entry pressure on the investment of the incumbents enhances growth and the aggregate probability of innovation. Clearly, in this model a policy aimed at increasing IPRs protection for the incumbents is going to increase the aggregate incentives to invest in innovation by both the incumbents and the outsiders.

Acemoglu (2008) obtains an even stronger result in his related model with incremental innovations by incumbents and radical innovations by outsiders. Increasing IPRs protection for the incumbents (or subsidizing them) enhances growth, and also blocking (or taxing) entry of outsiders increase growth, because both policies increase the value of the leadership inducing more innovation: "[e]ntry barriers, by protecting incumbents, increase their value, and greater value of incumbents encourages more R&D investments and faster productivity growth" (Acemoglu, 2009, p. 479).

5.5 Fiscal Policy

The decentralized equilibria derived in the previous sections are typically inefficient because the EMSs can be characterized by a number of firms and an amount of investment per firm different from the socially optimal levels. Fiscal policy can solve these inefficiencies, therefore in this section we first derive the optimal organization of the R&D sector, then we emphasize the differences between this and the equilibrium outcome, and finally we characterize the optimal fiscal policy.

The optimal structure of the R&D activity requires a certain number of R&D laboratories and associated flows of investment that maximize welfare. First of all, it is immediate to derive from the concavity of the arrival rate that it is optimal to allocate equal flows of investment between all the R&D laboratories. Now, let us guess that these flows are linear functions of the future scale of production, let us say $z(\kappa_j) = \varpi X(\kappa_j + 1) = \varpi q^{\theta-1} X(\kappa_j)$ for each firm in sector j, where ϖ is a parameter to be optimally chosen. Let us keep the production of intermediate goods at the level chosen by the patentholders in the decentralized equilibrium.[46]

The resource constraint of the economy must take into account the fixed costs, which are paid only at the beginning of each new patent race. Without loss of generality let us assume that the economy devotes a flow of resources for this purpose in each sector.[47] If the number of sectors N_X is high enough, one can approximate this flows, say $f_j(\kappa_j)$ with those equating their expected present value $f_j(\kappa_j)/[r + \delta_N(\kappa_j)]$ to the fixed cost $F(\kappa_j)$, that is with $f_j(\kappa_j) = \eta X(\kappa_j + 1)$. Using the (5.17) and (5.18), we can rewrite the resource constraint as:

[46] We are basically solving for a second best allocation. As well known, the first best allocation would be obtained by subsidizing sales in such a way that their price equates the marginal cost.

[47] We may think of a perfectly competitive banking sector specialized in venture-capital financing. Banks finance the fixed cost for the investment in a new technology and investors commit to a flow of payment until the new technology is obtained.

$$Y = C+\sum_{j=1}^{N_X} X(\kappa_j)+\sum_{j=1}^{N_X}\sum_{i=1}^{N} z_i(\kappa_j)+\sum_{j=1}^{N_X}\sum_{i=1}^{N} f_j(\kappa_j) = C+X+N(\varpi+\eta)q^{\frac{\theta}{1-\theta}}X$$

Using the fact that $X = \alpha Y$, we can derive an expression for consumption holding at each point in time:

$$C_t = X_t\left[\frac{1-\alpha}{\alpha} - N(\varpi+\eta)q^{\theta-1}\right]$$

According to (5.20), growth is determined by the rate of innovation as $g = N\left[\phi(\kappa_j)z(\kappa_j)\right]^{\epsilon}(\theta-1)\ln q$, or:

$$g(N) = (\theta-1)\,\varpi^{\epsilon}N\ln q \tag{5.40}$$

which emphasizes the direct relation between the number of firms researching in each sector and the growth rate. From the expression for consumption in time $t=0$ we have:

$$C_0(N) = X_0\left[\frac{1-\alpha}{\alpha} - N(\varpi+\eta)q^{\theta-1}\right] \tag{5.41}$$

Putting together these two relations, the intertemporal utility, which is finite as long as $\rho > (1-\gamma)g(N)$, can be written as:

$$U = \int_0^{\infty}\frac{C_t^{1-\gamma}}{1-\gamma}e^{-\rho t}dt = \frac{C_0(N)^{1-\gamma}}{(1-\gamma)\left[\rho-(1-\gamma)g(N)\right]} \tag{5.42}$$

Finally, substituting initial consumption and the expression for growth in (5.42), we can summarize the social planner problem as:

$$\max_{\{N,\varpi\}}\frac{{X_0}^{1-\gamma}\left[\frac{1-\alpha}{\alpha}-N(\varpi+\eta)q^{\theta-1}\right]^{1-\gamma}}{(1-\gamma)\left[\rho-(1-\gamma)\varpi^{\epsilon}(\theta-1)N\ln q\right]} \tag{5.43}$$

which emphasizes the basic trade-offs. A higher number of firms or a higher flow of investment per firm imply a higher growth rate of consumption but with a lower initial consumption level (and the time preference rate and the elasticity of substitution govern this trade-off in a standard fashion), but the weights on benefits and costs are different for the two choice variables. If an interior solution exists, the first order conditions for the social planner problem (5.43) with respect to ϖ and N are:

$$\varpi: \quad \frac{Nq^{\theta-1}}{\frac{1-\alpha}{\alpha}-N(\varpi+\eta)q^{\theta-1}} = \frac{\epsilon\varpi^{\epsilon-1}(\theta-1)N\ln q}{\rho-(1-\gamma)\varpi^{\epsilon}(\theta-1)N\ln q}$$

$$N: \quad \frac{(\varpi+\eta)q^{\theta-1}}{\frac{1-\alpha}{\alpha}-N(\varpi+\eta)q^{\theta-1}} = \frac{\varpi^{\epsilon}(\theta-1)\ln q}{\rho-(1-\gamma)\varpi^{\epsilon}(\theta-1)N\ln q}$$

Dividing one by the other we obtain $\varpi^{GR} = \epsilon\eta/(1-\epsilon)$, which implies the optimal flow of investment in R&D per firm of sector j:

$$\tilde{z}^{GR}(\kappa_j) = \frac{\epsilon\eta}{1-\epsilon} q^{\theta-1} X(\kappa_j) \tag{5.44}$$

Let us compare (5.44) with the equilibrium flow of investment (5.31) for a given level of expected production. While the equilibrium investment of each firm increases in the profit margin μ, the optimal investment is independent from that and it is positively correlated with the elasticity of expected revenue with respect to the flow of investment. Moreover, it is immediate to verify that the equilibrium investment is always below the optimal level:

PROPOSITION 5.4. **The decentralized equilibrium with Nash competition in the market for innovations always implies a sub-optimal flow of investment in R&D per firm.**

In other words, the EMS of the R&D activity is biased toward too small firms. Clearly this does not mean that there is not enough R&D activity, because there may be too many small firms, but just that the division of total investment is inefficient. It would be efficient to increase the investment of each single firm in R&D. When growth is led by technological progress, a country with an industrial structure characterized by small firms achieves inefficient results, and in particular it could grow more (or enjoy a larger welfare) if its firms were increasing in size. This general conclusion may shed new light on the problems of countries that do not grow much, are characterized by many small firms investing too little in R&D and lack large and innovative corporations.

This form of dynamic inefficiency is absent in traditional models of endogenous growth, where the economy may grow above or below an optimal benchmark, but cannot increase the growth rate without giving up to some of the current consumption: when marginal productivity in the R&D sector is decreasing, the EMS of this sector creates this inefficiency. Only a proper R&D policy can solve this problem through an R&D subsidy s^z on the investment in innovation by all the firms. Introducing this, the investment rate per firm becomes:

$$z(\kappa_j, s^z) = \frac{z(\kappa_j, 0)}{1 - s^z}$$

which is of course increasing in s^z. Using this with (5.31) and (5.44), one can easily derive the subsidy which induces the optimal investment per firm in the decentralized equilibrium:

$$s^z = \frac{\eta(\theta-1)}{1-\epsilon+\eta\epsilon(\theta-1)} \in (0,1) \tag{5.45}$$

The optimal subsidy is increasing in θ and η because more substitutability between goods and higher fixed costs reduce the effective markups and require larger subsidies to obtain the optimal investment. Moreover, the optimal subsidy is increasing in ϵ: the more elastic is the probability of innovation to investment, the larger should be the subsidy. In other words, high levels of ϵ suggest that in the decentralized equilibrium innovations are undercompensated (compared to the social optimum). For this reason, the estimate of this parameter is crucial to evaluate the optimal protection of innovations.

Let us now look at the number of firms investing in R&D, which allows us to obtain a complete characterization of the optimal EMS. From the first order conditions we obtain the optimal number of R&D laboratories as (the maximum between 1) and:

$$\tilde{N}^{GR} = \frac{1}{\gamma}\left[\frac{(1-\alpha)(1-\epsilon)}{\alpha\eta q^{\theta-1}} - \frac{\rho}{(\theta-1)\ln q}\left(\frac{1-\epsilon}{\epsilon\eta}\right)^{\epsilon}\right] \tag{5.46}$$

which is decreasing in ϵ at least for ϵ high enough: this leads to the fact that when the marginal productivity of the investment does not decrease much with R&D investment, it is optimal to have just one laboratory investing a lot. However, here we focus on the case of decreasing marginal productivity in which the optimal number of laboratories is larger than one. Comparing (5.46) with its equilibrium counterpart (5.33), it can be verified that the decentralized EMS implies too few (many) firms for any γ smaller (larger) than a cut-off. This result has a simple intuition: when γ is low, it is optimal to choose a high growth rate of consumption, therefore the social value of innovations, which is what drives growth, is high. On the other side, the private value of innovations depends on market features which are independent from consumers preferences (except for an indirect channel going through the interest rate). Consequently, for low enough γ, the social value of innovations is larger enough than the private value and the optimal number of firms becomes larger than the equilibrium number.

Finally, substituting in our expression for growth, we obtain that the optimal growth rate is:

$$\tilde{g}^{GR} = \frac{1}{\gamma}\left[\epsilon^{\epsilon}\left(\frac{1-\epsilon}{\eta}\right)^{1-\epsilon}\left(\frac{1-\alpha}{\alpha}\right)\left(1-q^{1-\theta}\right) - \rho\right] \tag{5.47}$$

which decreases with η and is higher than the equilibrium growth rate for any γ small enough.[48]

The optimal steady state EMSs can be achieved with two policy tools (derived in Etro, 2008,a), a positive R&D subsidy s^z, which targets the optimal

[48] Approaching constant returns to scale in our model (that is when $\epsilon \to 1$ and $\eta \to 0$), this converges to the optimal growth rate in the Aghion-Howitt model:

$$\tilde{g}^{GR} \to \frac{1}{\gamma}\left[\left(\frac{1-\alpha}{\alpha}\right)\left(1-q^{1-\theta}\right) - \rho\right]$$

which corresponds to the one of Barro and Sala-i-Martin (2004) without scale effects.

allocation of resources between investors, and a capital income tax τ which targets the optimal number of firms so as to replicate the optimal EMSs. If the capital income tax reduces the net interest rate of the representative agent to $r - \tau$, the new equilibrium growth rate becomes:

$$\tilde{g} = \frac{\left[\frac{\epsilon(\mu-1-\eta)}{1-s^z}\right]^{\epsilon}\left[\frac{(\mu-1)(1-\epsilon)}{\eta}+\epsilon\right]^{1-\epsilon} - \tau - \rho}{\gamma + (1-\alpha)/\alpha \ln q}$$

The optimal fiscal policy requires the same R&D subsidy derived above in (5.45) and a capital income tax equating this growth rate to the optimal one derived in (5.47). Of course the same outcome could be achieved with an entry fee or with a profit tax. Summarizing, we have:

PROPOSITION 5.5. **The optimal fiscal policy requires a positive R&D subsidy to investment and a capital income tax to achieve the optimal endogenous market structure of the market for innovations and the optimal growth rate.**

The necessity of two fiscal instruments to achieve optimality is deeply depending on the introduction of EMSs. In the limiting neoclassical case of CRS in the innovation technology, which is the traditional focus of the literature, the size and the number of firms do not matter and an R&D subsidy alone can achieve optimality.

Finally, notice that in the presence of a leadership of the incumbent, the optimal EMSs can be obtained with a similar policy, but with a lower (and still positive) R&D subsidy for the market leader.

The message of this section is quite in contrast with the usual models of Schumpeterian growth, where the optimal R&D policy may imply taxation or subsidization of the R&D investment (see Grossman and Helpman, 1991, or Barro and Sala-i-Martin, 1995). When we take into account the organization of the market for innovations, we obtain a more intuitive result, for which innovating firms should always be subsidized (to increase their size), even when growth is too high. Moreover, the need for a R&D subsidy in steady state is in line with the spirit of the results we obtained in Chapter 3 on the necessity of positive sale subsidies to achieve long run optimality.

Finally, the message of this section is in radical contrast with one of the main policy prescriptions of the neoclassical model, for which steady state capital income taxation should be zero (because in that framework the taxation of the return on capital affects the marginal productivity of capital and distorts capital accumulation). In the presence of EMSs, capital income taxation (or equivalently an entry fee or a profit tax) reduces the net return of the investment in business creation, which is typically beneficial when decentralized entry is inefficient (see Peretto, 2008a, for a similar point and further investigations).

5.6 Monetary Policy

Endogenous growth models are well equipped to face monetary issues in growing economies, but they have been rarely used for this purpose. In this section, following Etro (2007,c), we show that an inverted-U relation between inflation and growth can emerge when monetary frictions affect the EMS of the market for innovations, and we characterize the long run optimal monetary policy.

Consider a closed economy with the same technological features of the previous sections. For simplicity, in the rest of the chapter we confine the analysis to the particular case with $\alpha = 1 - 1/\theta$. Let us introduce real money balances $m_t = M_t/P_t$, where M_t is nominal money issued at a growth rate $\sigma = \dot{M}_t/M_t$, which is the policy tool in the long run, and P_t is the price level for the final good, which is perfectly flexible and changes at the rate of inflation $\pi_t = \dot{P}_t/P_t$. Let us adopt the Sidrausky (1967) approach and extend our utility function to:

$$U = \int_0^\infty \left(\frac{C_t^{1-\gamma}}{1-\gamma} + \frac{\chi m_t^{1-\xi}}{1-\xi} \right) e^{-\rho t} dt \quad \text{with } \gamma,\, \chi,\, \xi > 0 \tag{5.48}$$

Separability implies that the optimal consumption growth is still given by (5.9), which in steady state becomes:

$$g = \frac{r - \rho}{\gamma}$$

while money demand is:

$$m_t = \left(\frac{\chi C_t^\gamma}{r + \pi_t} \right)^{\frac{1}{\xi}} \tag{5.49}$$

Equating the latter and the money supply delivers the equilibrium price level at each point in time. This implies that on a balanced growth path the endogenous level of inflation is constant if and only if:

$$\pi = \sigma - \frac{\gamma g}{\xi} \tag{5.50}$$

In general equilibrium, monetary policy and inflation may affect the return rate and the growth rate. This is not the case when the prices of the intermediate goods perfectly adjust to changes in the price level of final goods: in that case inflation is superneutral. Under flexible prices, the optimal monetary policy is still given by the Friedman Rule (derived in the neoclassical approach of Chapter 1), which sets the nominal interest rate $i = r + \pi$ as close as possible to zero. Since the equilibrium growth rate implies $r = \gamma g + \rho$, the optimal nominal interest rate $i^{FR} = 0$ implies an optimal rate of money growth given by the following generalized Friedman Rule:

$$\sigma^{FR} = \frac{g\gamma(1-\xi)}{\xi} - \rho \tag{5.51}$$

Summarizing:

PROPOSITION 5.6. **With perfect price flexibility, monetary policy does not affect the real economy and the growth rate, but the optimal monetary policy requires a positive growth rate of money balances when $\xi < 1$ and the growth rate of output is larger than $\xi\rho/\gamma(1-\xi)$.**

Accordingly, even if the optimal growth rate of money balances is negative for small output growth, under reasonable conditions the theory is consistent with a realistic optimal policy of increasing real money balances. For future reference, notice that under perfectly flexible prices, the optimal policy is independent from χ.

Now, let us introduce nominal frictions for the intermediate goods, while leaving perfect price flexibility for the final goods. When competition in the market for intermediate goods takes place in the choice of prices and these prices are sticky in the New-Keynesian tradition (Rotemberg, 1982, a,b), new consequences emerge. First of all, when prices are sticky, surprise inflation increases demand of intermediate goods and hence production, a point clearly made by Barro and Tenreyro (2006) in a related but static environment:[49] this would create a case for monetary activism at least as a countercyclical policy. Unfortunately, there is another effect of inflation, and this is a permanent one. In the presence of sticky prices, inflation affects monopolistic profits in real terms and therefore the incentives to invest. The effects of inflation and price variability on investment are at the forefront of policy debates, and this is a framework to think of them in a more rigorous way.

We now introduce price stickiness in the spirit of the Calvo (1983) model, with only a fraction of the firms adjusting their prices, and the remaining firms keeping their prices constant. Contrary to the usual Calvo pricing, however, let us endogenize the probability of adjustment assuming that firms adopt new optimal prices only for the new goods, while old goods keep a constant nominal price until they are replaced by new ones.[50] Under this realistic assumption (at least when the replacement of goods is rapid enough), a monopolist innovating at time $t = 0$ sets the price of its good at a level $P_0\mu$

[49] For instance, if inflation is expected to be zero, a positive inflationary surprise will decrease the real price of intermediate goods (until firms change them), temporary boosting production. Here, the impact of the shock vanishes gradually since new vintages of intermediate goods with new prices gradually arrive in the market. Clearly this may induce problems of time inconsistency on the expectations of zero inflation, as in the Barro and Gordon (1983,a,b) framework. See Barro and Tenreyro (2006).

[50] This state-dependent strategy is the optimal one if the fixed menu cost of price adjustment is high enough and if the average length of a leadership is short, that is if we are dealing with simple innovations.

and keeps it at this level until a new vintage is on the market with a new price. Then, the initial value of this intermediate good can be derived as:

$$V^M(\kappa_j) = \int_0^\infty \left(\frac{(\theta-1)P_t}{\theta\mu P_0}\right)^\theta q^{\kappa_j(\theta-1)} \frac{AL(\mu P_0 - P_t)\, e^{-[r+\delta_N(k_j+1)]t}}{P_t} dt$$

where $P_t = P_0 e^{\pi t}$. Developing the integral we have:

$$V^M(\kappa_j) = \frac{AL\left[(\mu-1)(r+\delta_N(k_j+1)-\pi\theta)-\pi\right] q^{\kappa_j(\theta-1)}}{[r+\delta_N(k_j+1)-\pi(\theta-1)]\,[r+\delta_N(k_j+1)-\pi\theta]}\left(\frac{\theta-1}{\theta\mu}\right)^\theta \tag{5.52}$$

which is an inverted-U curve in the inflation rate and it is maximized at a level $\hat{\pi}$, which is positive if μ is large enough. The incentives to innovate are driven by (5.52) in the same way as before (even if firms decide their real investments while changing the nominal one according to the inflation level). Since the equilibrium effective discount rate, the aggregate investment and the growth rate are directly related to the value of intermediate goods (5.52) by the endogenous entry condition in the market for each innovation, they inherit the same non-monotonic relation with inflation. Therefore price stickiness generates an inverted-U relation between the inflation rate and the endogenous growth rate:[51]

$$g = g(\pi) \quad \text{with } g(\pi) \gtrless 0 \text{ for } \pi \lessgtr \hat{\pi} \tag{5.53}$$

Summarizing, we have:

PROPOSITION 5.7. **In the presence of price stickiness, there is an inverted-U relation between inflation and growth due to the effects of inflation on the incentives to invest in business creation.**

Notice that this result is in contrast with the Tobin effect (Tobin, 1965) for which inflation stimulates investment and growth: this happens only for low levels of inflation, while in our model higher levels of inflation erode profits and reduce investments. The last outcome may provide a channel for the negative relation between inflation and growth emphasized in the empirical literature on the determinants of growth. The balance growth path is characterized by (5.50) and (5.53) for a given policy of constant money growth σ.[52]

[51] This effect may be quite relevant. Imagine that prices are constant for a year, the mark up is at 20% and the inflation rate at 5%. Then, assuming $\alpha = 0.5$, after one year the flow of profits is reduced by more than 20%. If the average lenght of competition for the market is one year, the value of innovation is reduced by about 10% because of inflation. As in the New-Keynesian literature on business cycles, small price frictions imply that demand shocks can have large macroeconomic consequences. However, here the consequences are permanent.

[52] Notice that for intermediate levels of money growth there are two equilibrium growth paths. The inefficient one is characterized by high inflation and low

In this context, the utility maximizing policy can be quite complex, but it simplifies in case the two tools of fiscal policy (R&D subsidies and capital income taxes) can be used to solve the inefficiencies in the market for innovation (so that, absent price stickiness the model achieves an efficient growth rate). In such a case, the optimal inflation rate must be between the Friedman rule level $\boldsymbol{\pi} = -r$ and the growth maximizing level $\boldsymbol{\pi} = \hat{\boldsymbol{\pi}}$, depending on the weight given to money in the utility function χ. Interestingly, the growth maximizing inflation rate boils down to zero when pricing strategies are optimally chosen by monopolists. Indeed, the optimal nominal price that maximizes (5.52) generates the optimal monopolistic mark up:

$$\mu^M(\boldsymbol{\pi}) = \frac{\theta}{\theta-1}\left[1 + \frac{\boldsymbol{\pi}}{r + \delta_N(\kappa_j+1) - \theta\boldsymbol{\pi}}\right] \tag{5.54}$$

which exhibits a simple positive relation with the inflation rate, contrary to the mark up (3.49) obtained by Bilbiie, Ghironi and Melitz (2008,b) in a related model (with endogenous entry in the competition in the market and continuous but costly adjustment). The optimal mark up is also decreasing in the interest rate and in the expected rate of innovation (and of endogenous price adjustment). In this case, the value of innovation becomes:

$$V^M(\kappa_j) = \frac{q^{\kappa_j(\theta-1)}AL}{\theta-1}\frac{\left[r + \delta_N(\kappa_j+1) - \theta\boldsymbol{\pi}\right]^{\theta-1}}{\left[r + \delta_N(\kappa_j+1) - \boldsymbol{\pi}(\theta-1)\right]^{\theta}} \tag{5.55}$$

which is maximized exactly by $\hat{\boldsymbol{\pi}} = 0$. Therefore, also the growth rate is maximized under price stability. Assuming that $\chi = 0$, we can conclude that: under price stickiness, when firms choose optimally their nominal prices taking inflation into account, optimal monetary policy requires price stabilization, $\pi^{Opt} = 0$. This is equivalent to the optimal rate of monetary growth:

$$\sigma^{Opt} = \frac{\gamma g(0)}{\xi} > 0$$

or to the optimal nominal interest rate $i^{Opt} = \gamma g(0) - \rho$. Summarizing, we have:

PROPOSITION 5.8. **In the presence of price stickiness, adopting the optimal fiscal policy, the optimal monetary policy requires a positive growth rate of money balances, consistent with zero inflation and linearly increasing in the growth rate.**

An explicit solution for the growth rate can be obtained when $\theta = 2$ and $\epsilon \to 1$ with $\eta \to 0$:[53]

growth. This self-fulfilling stagflation has a simple intuition: if high inflation is expected, firms reduce investment decreasing growth, which generates high inflation in turn. However, such a path is unstable.

[53] For consistency, here we assume $\phi(\kappa_j) = 1/\left(\frac{\theta}{\theta-1}\right)^{2\theta} q^{(\kappa_j+1)(\theta-1)}ALP_t$ at time t.

$$g(\pi) = \frac{\pi + 1/2 + \sqrt{1/4 - \pi} - \rho}{\gamma + 1/\ln q} \tag{5.56}$$

which is clearly a parable with a maximum in correspondence of zero inflation.

Notice that this result is quite robust: introducing money through a cash-in-advance constraint or real resource costs (or endogenizing the timing of price changes introducing small menu costs) would not affect the optimality of zero inflation (as long as monopolists can optimally choose their prices and the timing of price changes).

However, the optimality of zero inflation is not robust when the tools of fiscal policy are not available. For instance, assume that the flexible price growth rate is above the optimal level because the EMS is characterized by too many firms. Imagine first that R&D subsidies and capital income taxes are not available: in such a case a non-zero inflation rate could reduce the incentives to invest in R&D pushing down the growth rate toward its optimal level. In the example above with $\theta = 2$, assume that γ is large enough that the optimal growth rate $\tilde{g}^{GR} = \left[(q-1)/q - \rho\right]/\gamma$ is lower than $g(0)$. Then, the optimal growth rate can be obtained simply by adopting the following positive inflation rate $\pi^{opt} = \varrho(1-\varrho)$ with $\varrho = \sqrt{\rho/\gamma(q-1) + (\gamma-1)/\gamma q}$.

More interesting phenomena can be detected introducing strategic interactions in the competition in the market and taking into account persistent leadership. We leave these for future research. The bottom line of this discussion is that whenever policy affects the value of innovations even marginally, it affects the (stock market) value of firms and the associated process of business creation and growth and, therefore, it has permanent consequences: monetary policy should take this in consideration.

5.7 Growth and International Trade and Finance

A large part of the literature on endogenous growth has been focused on the understanding of the determinants of growth in a open economy context with different kinds of spillovers between countries (see Grossman and Helpman, 1991).[54] In this section we extend our growth models to an open economy framework to characterize the functioning of the market for innovations in a global framework. The extension of the endogenous growth model *à la* Romer (1990) to a multicountry scenario with trade of goods shows that growth increases in the degree of openness and exhibits scale effects, and it leads to a characterization of the endogenous size of the national markets driving growth. The extension of the Schumpeterian model to a multicountry scenario with trade in both goods and capital shows that global growth tends to be driven by innovations of firms in the largest economy and it is

[54] See also Krugman (1987), Young (1991), Rivera-Batiz and Romer (1991) Baldwin and Forslid (2000) and Aghion and Howitt (2009, Ch. 15).

enhanced by both the relative size of this economy and the degree of openness. Finally, we evaluate R&D policy showing that governments may want to subsidize domestic firms for profit shifting reasons (as in the theory of strategic export promotion analyzed in Chapter 4), while global efficiency requires international coordination of multiple instruments.

Let us consider $\mathbf{N} \geq 2$ countries endowed with the same technology and preferences as before, but different TFP and population levels. Final goods freely move across borders, but there can be frictions in the trade of intermediate goods: in particular, assume that for 1 unit of intermediate goods sent to a foreign country, $d \leq 1$ units arrive to destination because of iceberg transport costs or protectionism (but one may also think of losses due to incomplete protection of foreign IPRs): the parameter d can be interpreted as a measure of the degree of openness. These trade frictions imply that foreign demand is smaller than domestic demand for each intermediate good produced at home. As a consequence, the expected discounted value of the profits with innovation κ_j by a firm from country i becomes:

$$V_i^M(\kappa_j) = \frac{(\mu-1)\left(\frac{\theta-1}{\theta\mu}\right)^\theta q^{\kappa_j(\theta-1)}\left(A_iL_i + d^\theta \sum_{f=1,f\neq i}^{\mathbf{N}} A_fL_f\right)}{r_i+\delta_N(\kappa_j)} \tag{5.57}$$

which is a generalization of (5.26) with r_i interest rate in country i. Let us derive the implications for both the endogenous growth model *à la* Romer (1990) and for the Schumpeterian growth model.

5.7.1 Endogenous size of national markets and global growth

Consider first the model with innovation in new varieties of intermediate goods in the absence of technological differences between sectors ($\kappa_j = 0$ for all j). In the international trade tradition, we assume that both labor and capital do not move across countries, while final and intermediate goods are traded. The varieties of intermediate goods are sold in all the countries forever ($\delta_N(0) = 0$), therefore firms from every country have incentives to invest and innovate for the global market. The only difference compared to the closed economy version is that investment is going to be higher in larger countries. More precisely, given the fixed cost of innovation F, the endogenous entry condition $V_i^M(\kappa_j) = F$ determines the equilibrium interest rate in each country and its growth rate:

$$\tilde{g}_i = \frac{(\mu-1)\left(\frac{\theta-1}{\theta\mu}\right)^\theta\left(\frac{A_iL_i}{F} + d^\theta \sum_{f=1,f\neq i}^{\mathbf{N}} \frac{A_fL_f}{F}\right)-\rho}{\gamma}$$

which is increasing in the size of the country A_iL_i. Let us assume a common global productivity A and define the scale factor $h = [(\theta-1)/\theta\mu]^\theta AL_W/F$ and the relative population $b_i = L_i/L_W$ for country i, where $L_W = \sum_{f=1}^{\mathbf{N}} L_f$ is world population. Then, we can rewrite the growth rate of country i as:

$$\tilde{g}(d, b_i) = \frac{(\mu - 1)h\left[d^\theta + b_i(1 - d^\theta)\right] - \rho}{\gamma} \tag{5.58}$$

which boils down to the growth rate of the closed economy in case of perfect market integration ($d = 1$) or in case of a single country ($b_i = 1$). Notice that the absolute scale effects disappear when the fixed costs of production are proportional to the size of the world economy, but the differences in growth rates between countries of different sizes remain as long as there are trade frictions: "relative scale effects" emerge.

The growth rate is increasing in the degree of openness ($\partial g/\partial d > 0$), but the benefits of openness diminish in the relative size of the country ($\partial^2 g/\partial d \partial b_i < 0$). Both these implications find a strong empirical support and have a simple rationale (Alesina and Spolaore, 2003): smaller countries benefit more from opening up to trade because they can exploit intra-industry trade with larger foreign markets, while larger countries benefit less because they already enjoy a large market. This result suggests that there may be important politico-economic motivations behind the determinants of the size of national markets b_i. These have been investigated by the economic theory of political geography. The crucial trade-off is between the growth benefits of country size and the costs of heterogeneity in preferences over the provision of a national public good, which are positively correlated with country size. Globalization strengthens political forces leading to separatism and smaller national markets, that is to high $\mathbf{N}$ (notice that in a symmetric equilibrium $b = 1/\mathbf{N}$ for any i).

Until now we have considered the degree of openness as exogenous. However, in the long run (which is our focus in this chapter), this is endogenous too, because it depends on trade policy and investments in transport infrastructures, and it can affect the endogenous size of the markets. Etro (2003,a, 2006) has investigated the trade-off between the growth benefits of openness and its social costs (in terms of losses for workers and firms in selected sectors, uncertainty costs, and lost revenues from tariffs). The immediate result is that smaller countries tend to prefer larger degrees of openness and larger countries prefer protectionist policies that reduce openness, something similar to what we found in earlier analysis of trade policy. However, this leads to the possibility of multiple long run equilibria, some with small national markets and high openness (high $\mathbf{N}$ and high d) and some with large national markets and low openness (low $\mathbf{N}$ and low d).

Finally, structural policies are endogenous as well in the long run, and they can affect the endogenous size of the markets. Etro (2006) has endogenized also the size of national public spending in function of the size of countries and of preference heterogeneities.[55] The analysis confirms the pos-

[55] For related investigations see Wrede (2004), Chiang and Mahmud (2008) and Gregorini (2009). When augmented with monetary frictions, the multi-country model in the text can be used to analyze monetary policy as well as the role of monetary unions (see Alesina and Barro, 2002).

sibility of multiple equilibria which display a negative correlation between size of national markets and both openness and size of the public sector: there can be equilibria with globalization backlash associated with large national markets, small governments, high protectionism and low growth, and other equilibria with smaller national markets, bigger governments, high openness and sustained growth.[56]

In conclusion, our analysis of global growth with endogenous size of the national markets suggests that the coordination of trade policy aimed at enhancing openness can be fundamental to promote global growth and prosperity.

5.7.2 World technological leadership

Consider now the Schumpeterian framework in which international firms invest to improve the quality of the existing intermediate goods produced by any firm around the world. Alternatively think of a model of trade in goods and technologies between the North with higher productivity levels and the South with lower productivity levels. In the international finance tradition now let us open countries to trade not only in goods but also in capital, so that a unique interest rate emerges in the global equilibrium. In this way, all the possible differences in the growth rates are eliminated by means of capital movements toward better return rates, just like in the neoclassical approach, but as we will see with completely different implications for the direction of capital flows.

Since the value of innovations (5.57) is increasing in the size of the country where a firm is active, it is easy to verify that the endogenous allocation of R&D investment is always biased toward the largest country. More precisely, under Nash competition in the market for innovations, only its firms will invest in R&D, while under Stackelberg competition incumbent firms from other countries may keep investing and retain the leadership for a while, but they will lose it sooner or later in favor of firms from the largest country. Therefore, for any initial allocation of the technological frontier, the engine of world growth ends up in the largest economy, which gradually conquers the technological leadership in all sectors through its innovative firms.

The world economy must be characterized by a constant growth rate for all countries which increases in the relative size of the leading country, say $\bar{b} = \max(L_f A_f)/(\sum_{f=1}^{\mathbf{N}} L_f A_f)$. For instance, consider Nash competition with $\epsilon \to 1$ and $\eta \to 0$. The growth rate, as a function of the relative size of the leading country and of the degree of openness, becomes:[57]

[56] Despite the latter kind of equilibrium tends to be Pareto superior, stable equilibria may be characterized by excessive globalization, too small national markets and excessive public spending.

[57] For consistency, now we assume: $\phi(\kappa_j) = \left[\left(\frac{\theta-1}{\theta}\right)^{2\theta} q^{(\kappa_j+1)(\theta-1)} \left(\sum A_f L_f\right)\right]^{-1}$.

$$g(d,\bar{b})=\frac{(\mu-1)\left[d^{\theta}+\bar{b}(1-d^{\theta})\right]-\rho}{\gamma+1/(\theta-1)\ln q} \tag{5.59}$$

which boils down to the growth rate of the closed economy in case of perfect market integration. Even if this model does not exhibit absolute scale effects (exactly as its closed economy version), relative scale effects emerge again, in the sense that the larger is the leading economy compared to the rest of the world, the higher is the growth rate: $\partial g/\partial \bar{b}>0$. A consequence of this form of relative scale effects is that the positive relation between openness and growth survives: $\partial g/\partial d>0$.

In equilibrium, the largest country develops a comparative advantage in the intermediate goods sector. The clear consequence is that in the long run this country must export these intermediate goods and import final goods. Notice that, at each point in time, the sum of the trade surplus in intermediate goods, national savings and net capital inflows has to be matched by a deficit in trade of final goods and by investments. This bring us toward completely different implications from those of the neoclassical model of international finance.

Since the world interest rate has to equate global savings and global investments, it follows that the R&D investments of the leading country must be entirely financed by the savings of the rest of the world. This may help explaining the Lucas paradox (Lucas, 1990) concerning why capital does not fly to poor countries and why often there is a flow of resources moving in the opposite direction. Summarizing, we have:

PROPOSITION 5.9. **In a open economy context with trade frictions, the largest country has a comparative advantage in the innovation sector and, in the long run, it leads alone the technological frontier, exports intermediate goods, imports final goods and attracts foreign capital to finance investment. Growth increases in the degree of openness and in the relative size of the largest country.**

Such an scenario may appear too extreme to be realistic, nevertheless, in a stylized way, it appears in line with the growth experience of the last decades. This was characterized by large R&D investments and high rate of technological progress in the U.S., by high U.S. imports of final goods which allowed other countries to grow as well, exporting final goods and importing American technology, and by impressive capital flows toward the U.S. financing the large American current account deficits. Of course, such a process can end when another country is creating a market of a larger size: a gradual migration of the technological frontier toward the country that is catching up (with leapfrogging sector by sector) takes place when the population differential compensates for the productivity differential.[58]

[58] However, notice that introducing higher productivity growth for the country that is catching up would increase the expected value of its market and would anticipate the migration of the technological frontier before the two markets have the same size.

5.7.3 Optimal R&D policy in the open economy

As in the closed economy model, also the decentralized international equilibrium derived above is dynamically inefficient and a system of R&D subsidies and capital income taxes (or profit taxes) would be welfare improving. However, countries have conflicting interests and the optimal R&D policy would not emerge in a decentralized way. Our model can be useful in assessing what would be the optimal unilateral R&D policy for a single country and what kind of coordination would be needed to achieve global efficiency.

For simplicity, consider the case of perfect market integration ($d = 1$), in which firms from any country are engaged in R&D activity. In such a case, the optimal R&D policy can be characterized exactly as in the closed economy. In particular, to restore efficiency in the market for innovations one should always subsidize firms, and to reach the optimal growth level one would also need appropriate capital income taxes or entry fees/subsidies. However, it is difficult for a country to commit to permanent policies of this kind.[59] When a country decides a subsidy for a firm engaged in R&D activity in a particular market, the perceived impact on global growth is negligible, and the only rationale behind subsidization is profit shifting in favor of the same domestic firm (as in Brander and Spencer, 1985): unfortunately, this leads to inefficient unilateral policies.

In our framework, the problem of the optimal unilateral R&D policy is a particular case of a more general problem of optimal strategic export promotion investigated in Section 4.9 on the basis of Etro (2010). Applying Prop. 4.10, we notice that R&D subsidies are optimal because they provide the domestic firm with the incentive to invest aggressively in R&D, exactly as if this firm was a leader in the patent race. Since this model is nested in the general case of Section 4.9, we can substitute in (4.60) for our general equilibrium values of the interest rate, the value of innovations and the fixed cost of entry so as to obtain the optimal R&D subsidy:

$$s_H^z = \frac{\eta(1-\epsilon)(\theta-1)}{[1-\eta(\theta-1)]^{\frac{\epsilon}{1-\epsilon}}[1-\epsilon+\epsilon\eta(\theta-1)]-\epsilon\eta(\theta-1)} \in (0,1) \qquad (5.60)$$

As already noticed, this policy just shifts profits from one country to another, crowding out investments by foreign firms in favor of the domestic subsidized firm, while aggregate growth is unaffected. Summarizing, we have:

[59] Imagine a situation where countries could commit to a permanent R&D subsidy. In our world without frictions, all countries would agree on what is the optimal policy. Nevertheless, each country would have an incentive not to subsidize R&D so as to "import" growth from foreign technological progress and enjoy more consumption. Moreover, the strategic interdependence of the policy tools is now more complex. In an interesting investigation, Impullitti (2007) has examined a related model with two countries showing that R&D subsidies are strategic complements. He has also provided some numerical simulations for the Nash equilibrium subsidies and the optimal ones: his results suggests that the gains from R&D policy coordination can be quite large.

PROPOSITION 5.10. **In an open economy context without trade frictions, the optimal unilateral policy for a single country is a positive R&D subsidy for the domestic firms active in the international market for innovations, but the optimal coordination of R&D policies requires both a positive R&D subsidy and a capital income tax.**

Once again, we are left with the need for international policy coordination, in this case of the innovation policy. While some countries have been trying to coordinate the support of R&D activities at a supra-national level, the heterogeneity of views and the lack of binding commitments have limited these efforts (think of the Lisbon Agenda of the European Union). Further coordination for the protection of IPRs at the global level could enhance the global incentives to invest in R&D in a substantial way.

5.8 Conclusions

In this chapter we have developed a model of creative destruction where the engine of growth was in the microeconomic structure of the patent races leading to innovations. The analysis of EMSs in the competition for the market has emphasized a number of new factors that characterize the growth process, as the inefficient organization of R&D investments in a decentralized equilibrium, the fundamental role that market leaders play in the innovation process, the distortionary effect of inflation on the endogenous growth process and the tendency of countries to implement inefficient innovation policies. We have analyzed the role of fiscal, monetary and innovation policy at improving the allocation or resources in the endogenous structures of the innovative markets and we have also suggested the need for a better coordination of global policies for innovation.

At the end of this path, researchers are left with an obvious but difficult task, that of developing a model that includes all the main factors that we studied in the last three chapters: a model of growth in a multicountry framework where shocks lead to realistic business cycles. The development of such a model where EMSs are fully characterized could shed light on many questions and contribute to a better understanding of all the three separate issues. However, such a difficult task is left for future research.

6. Economic Principles and Policy for the New Economy

Now you listen to me. I want details and I want them right now. I don't have a job, I have no place to go. You're not in the mood? Well, you get in the mood!" **George Costanza**

In this chapter we review the main results of the EMSs approach to macroeconomics. For simplicity, we summarize them in ten principles, half of which have a positive nature, in the sense that they describe crucial aspects of the way the economy works, and half of which have a normative nature, in the sense that they provide policy implications. The discussion is going to be largely informal and try to link our theoretical results to the current economic debates on macroeconomic issues and on relevant market structures, in particular of the New Economy.

The development of the EMSs approach relies on a general critique of the neoclassical approach: the main neoclassical assumption, perfect competition, is not only a bad approximation of realistic market conditions, but also a misleading one if we want to understand the aggregate behavior of the economy and derive correct policy implications. The first point is almost self-evident: large size of the markets and entry do not lead to market structures in anyway close to the perfectly competitive ideal. If we think about any major market of the global economy (automobiles, telecommunications, pharmaceuticals, computers, semiconductors, software, online advertising) we find high levels of concentration and sometimes only a few big players. Strategic interactions, variable mark ups and entry are crucial aspects of real world markets and they play no role in the neoclassical paradigm (or in the extensions to monopolistic behavior with constant mark ups and exogenous entry).

In real world markets, profitability is what attracts entry and investment. In turn, entry strengthens competition and reduces the mark ups through strategic interactions. The EMSs approach suggests that these factors lead to new mechanisms of propagation of the shocks over the business cycle, to additional consequences on the long run performance of the economy, to novel sources of gains (and losses) from trade associated with the globalization process, to a different role of market leaders in driving technological progress

F. Etro., *Endogenous Market Structures and the Macroeconomy*,
DOI: 10.1007/978-3-540-87427-0_6, © Springer-Verlag Berlin Heidelberg 2009

and growth, and to new forms of dynamic inefficiencies in the process of business creation.

Our hope is that the introduction of EMSs in the macroeconomic analysis can also shed new light on a number of policy issues. While our approach confirms the optimality of a countercyclical fiscal policy and a price stabilizing monetary policy, it provides alternative motivations for these policies. The former should optimize the process of business creation over the business cycle and requires a supply-based fiscal policy with countercyclical tax rates on sales and labor income. The latter should avoid the negative effects of nominal frictions on the process of business creation, and especially on the process of innovation driving growth. Finally, the EMSs approach provides new predictions for the optimal trade policy, for the role of exchange rate policy and for R&D policy, predictions that are often in radical contradiction with the traditional results. For instance, our approach shows the general optimality of unilateral export promoting policies as export subsidies, against traditional results which are typically in favor of export taxes.

In the rest of the chapter we discuss these principles one by one, and associate them with digressions on related applied issues, with a special attention to the endogenous structure of global markets that are relevant for the macroeconomic analysis. We recommend the reader to focus directly on the digressions of his or her interest.

6.1 Short-run EMSs and the Competition Effect

The novelty of the EMSs approach to macroeconomics relies on its analysis of the structure of markets that populate modern economies. Perfect competition, which is the standard way to model competition in the neoclassical theory, requires that firms can be created at no cost, act as price-takers, in the sense that they do not perceive themselves as affecting market prices with their choices, and in equilibrium they sell goods at the marginal cost of production, obtaining always zero profits. In such a framework, the market structure is indeterminate, in the sense that we have nothing to say about how many firms should be in the market and how much each one should produce in equilibrium. Even the concept of (stock market) value of a firm as the discounted sum of its future profits has no sense, since expected profits are zero.

Contrary to the neoclassical approach, the EMSs approach departs from the perfectly competitive paradigm, and introduces more realistic forms of competition between firms choosing their prices or their production levels and interacting in a strategic way. Moreover, this approach takes in consideration that firms decide rationally whether to enter or not in a market according to the profitability conditions: in particular the technological conditions are generalized to include positive fixed costs of entry, so that only a few firms enter in each market and only if they foresee enough gross profits to cover the fixed costs. The combination of these ingredients leads to markets where the strategies of the firms, their number and even their (stock market) value can be endogenously characterized in the short and long run as functions of the structural parameters of the economy.

Our approach can also depart from the neoclassical assumption that investment builds physical capital which is then rented by the firms and used as a factor of production together with labor. That was a good starting point to describe production in the industrialization phase, characterized by the dominance of the secondary sector based on the accumulation of physical capital (manufacturing) and by the social conflict between capital and labor. However, that assumption is not such a good one to describe production in the modern age, dominated by the tertiary sector (services) and by the New Economy. In this age creativity, know-how and innovations are the main inputs needed to bring new products to the market, and a well developed financial sector and protected IPRs allow smart ideas to become business activities and start up companies to reach easily a high (stock market) value without having built any stock of physical capital (yet). This is even more evident when we think of growth, a process driven mostly by capital accumulation during the industrialization phase, but largely driven by business creation and innovations in high-tech sectors in the current industrialized world.

On the basis of these considerations, the EMSs approach takes a different route from the neoclassical one in assuming that investment mainly creates new firms or new products which complement or replace the old ones. The mechanism of business creation works through a simple channel: the expectation of future profits induces entrepreneurs to invest in the creation of new products. However, this mechanism has already *in nuce* a counterbalancing effect. Entry of new firms increases the number of competitors, and these become more aggressive. They may actually compete in different ways, for instance by choosing their prices (Bertrand competition) or the quantity of production (Cournot competition) or in more complex ways in the presence of leaders, asymmetries, heterogeneity between firms and so on. In all these cases, entry strengthens competition, which reduces the mark ups and the profits that other firms can expect from entering in the same market. Ultimately, in equilibrium, each firm must choose the profit maximizing strategy given those of the other firms, and the number of firms must be such that there are no other firms with incentives to enter.

The characterization of the aggregate equilibrium with EMSs emphasizes a new mechanism of propagation of the shocks in the short run that can be relevant to explain the business cycle beyond the standard mechanisms emphasized by the neoclassical approach of Kydland and Prescott (1982). Consider a positive and temporary shock to the economy. The positive impact on consumption leads to an increase in profits for the active firms, which attracts subsequent entry of new firms. Entry is only gradual and temporary, but it strengthens competition between all the firms, which leads to a gradual and temporary reduction of the mark ups and, through a general equilibrium mechanism, to an increase of the real wages. The temporary reduction of the mark ups induces agents to substitute future consumption with current consumption, i.e. to temporarily increase consumption. At the same time, the temporary increase in the real wage induces agents to increase labor supply. The consumption boom and the work boom, in turn, have a feedback effect on profits, entry, competition and mark ups, which magnifies the impact of the shock. Of course, this mechanism could not take place in situations where mark ups were zero (as in the perfectly competitive framework) or in situations where mark ups were positive but constant and entry was independent from the profitability conditions (as in the New Keynesian literature). We can summarize this novel mechanism with the following principle:

PRINCIPLE 1. IN THE SHORT RUN, THE EMSS LINK DEMAND & SUPPLY CONDITIONS WITH ENDOGENOUS ENTRY AND MARK UPS. A POSITIVE SHOCK TO THE ECONOMY ATTRACTS ENTRY, STRENGTHENS COMPETITION REDUCES MARK UPS AND INCREASES REAL WAGES, WHICH BOOSTS CONSUMPTION AND LABOR SUPPLY AND PROPAGATES THE SHOCK.

An important consequence of this principle, is that the economy is characterized by procyclical entry of firms and countercylical mark ups, two patterns that are well documented in the data (see Section 3.1) and that remain largely unexplained in the neoclassical theory or in its extensions to monopolistic behavior. The simulation of the simplest possible model calibrated on the U.S. economy (in Chapter 3) has shown that the EMSs approach allows us to mimic the variability of the main aggregate variables as output, consumption, investment, labor supply and profits at least as well as a more complex neoclassical model in the Real Business Cycle tradition, and to do even better at replicating second moments of the U.S. data.

6.1.1 The boom of the 90s

If the mechanism of propagation of the shocks suggested here represents an important component of the business cycle, it may allow us to reinterpret the reaction of economies to shocks, and in particular to clarify what is going on in the current recession. With this purpose in mind, let us first look at what happened during the 90s, when a positive and permanent technology shock hit the global economy: it was the beginning of the so-called New Economy, founded on new general purpose technologies associated with PCs and the Internet. When the impact of the diffusion of these technologies on the global productivity became evident, growth opportunities opened up and profit expectations gradually improved in many sectors, and not just in the sectors of the New Economy. In particular, most service sectors, which represent the large majority of business in the developed world, benefited from the cost-reducing impact of new software and hardware, and a heavy process of business creation and innovation took place in these sectors.

The expectations on profitability and growth, mirrored by the stock market have been positive for such a long period (during the years of the Clinton Administration) that they drove one of the longest period of sustained growth. Observers and economists started welcoming a "new era" in which business cycle fluctuations were limited and could be smoothed with the standard tools of fiscal and monetary policy, and the growth rate became the main interest of policymaking and economic theory. Growth was high not only in the U.S., but also in all those countries where the market forces (of business creation) were working freely. As a matter of fact, U.S. consumption and imports started driving foreign growth, and U.S. investments (in business creation and innovation) attracted capital from the rest of the world.

The American consumption boom went as far as to reach a rate of savings out of disposable income close to zero in 2000 and a rate of indebtedness out of disposable income at 140 %.[1] While a similar extreme behavior could be rationalized on the basis of high growth expectations, it persisted when these expectations changed at the turn of the century (with the stock market crash first, and then with the terroristic attacks of September 11, 2001), and turned into a pathological incapacity to save, leading to serious imbalances. These can be summarized in three main critical consequences of the consumption boom: 1) the excessive imports of foreign goods maintained a large deficit in the foreign accounts, 2) the excessive borrowing in terms of easy mortgages put upward pressure on real estate prices (see Figures 6.1-2), and 3) the excessive leverage of equity capital[2] in the financial sector induced excessive risk taking and drugged stock market prices (see Figures 6.1-2). Some commentators, led by Shiller (2005), have argued that the dot-com boom (peaking in 2000) and the real estate boom (peaking in 2006) could be explained only in part by structural factors,[3] but also by cultural and psychological factors associated with a sort of "irrational exuberance". Whether a large part of these booms were "bubbles" or not, the American imbalances could only be corrected with a drastic depreciation of the Dollar, with the house price crash and with the financial crisis.

---

1 Total indebtedness of the U.S. economy surpassed 350 % of GDP in 2006, most of which owed by the private sector, while public debt was around 60 % of GDP (Di Noia and Micossi, 2009).

2 The ratio assets/equity reached an average of 30 for U.S. investment banks (up to 33 for Morgan Stanley, 32 for Merrill Lynch, 31 fo Lehman Brothers, 22 for Goldman Sachs) and of 33 for large European cross-border banks (up to 53 for Deutsche Bank and 52 for UBS).

3 According to Shiller (2005), structural factors leading to the stock market boom included the cultural and political (fiscal) changes favoring business creation, the new information technologies, a supportive monetary policy, the expansion of defined contribution pension plans in the stock market, the growth of mutual funds and even the expansion in media reporting of business news and analysts' over-optimistic forecasts.

6.1.2 The current global recession

During 2008, the global economy has entered in its deepest economic recession since the Great Depression of 1929.[4] Multiple factors have caused this crisis, including the rapid but temporary increase in the oil price during the first half of the year (a negative supply shock similar to those of the 70s, though shorter; see Figure 6.3). However, there is a large consensus that the main source of the global recession was in the losses emerged from the subprime crisis associated with the bursting of the U.S. housing boom in 2006, and with the consequent stock market crash and the depression of consumer confidence (a negative demand shock similar to that of the Great Depression). What the debate is mainly about, however, is the mechanism that has propagated and deepened the crisis: is it revised expectations on future prosperity? or irrational depression? a collapse of consumer demand due to the wealth losses? a pure credit crunch? or bad or insufficient policy reactions? Before advancing our hypothesis, let us establish the facts.

For a decade before the real estate downturn in the U.S., loan incentives and a long-run trend of rising housing prices encouraged Americans to assume mortgages with the hope that they could refinance at more favorable terms later. However, once housing prices started to drop, refinancing became more difficult, and in front of a fall in prices by 25 % or more (especially in towns like Boston, Los Angeles or Miami), many borrowers ended up with negative equity, that is with a mortgage worth more than the house, and became insolvent. The number of borrowers in default kept increasing with the housing bust (and now with the crisis of the real economy), but the worst consequences were going to happen in the financial sector.

The so-called "subprime crisis" was exactly due to the high default rates on subprime and other adjustable rate mortgages made to higher-risk borrowers with lower income or worse credit history than prime borrowers. The market for subprime lending reached a fifth of total U.S. mortgage market. The potential losses in the event of a real estate downturn would have not been such a big problem if risks were properly taken into account by the mortgage brokers: this was not the case since they were able to repackage the returns on these mortgages, bundle them together and sell them in different slices to financial institutions, even under high ratings that had no relation with the underlying reality of the actual subprime mortgages. In a recent book on the role of animal spirits in driving the economy Akerlof and Shiller (2009) notice that, as long as housing prices kept increasing this was "an economic equilibrium that econmpassed the whole chain, from the buyers of the properties, to the originators of the mortgages, to the securitizers of

[4] The Great Depression started with the stock market crash on October 29, 1929, and was characterized by a sequence of bank failures, a tight monetary policy, and a long deflation, with U.S. GDP falling from $ 103 bn in 1929 to $ 56 bn in 1933, and unemployment increasing from 3 % to 25 %. Only with the New Deal of public spending by President Roosevelt, the recovery started.

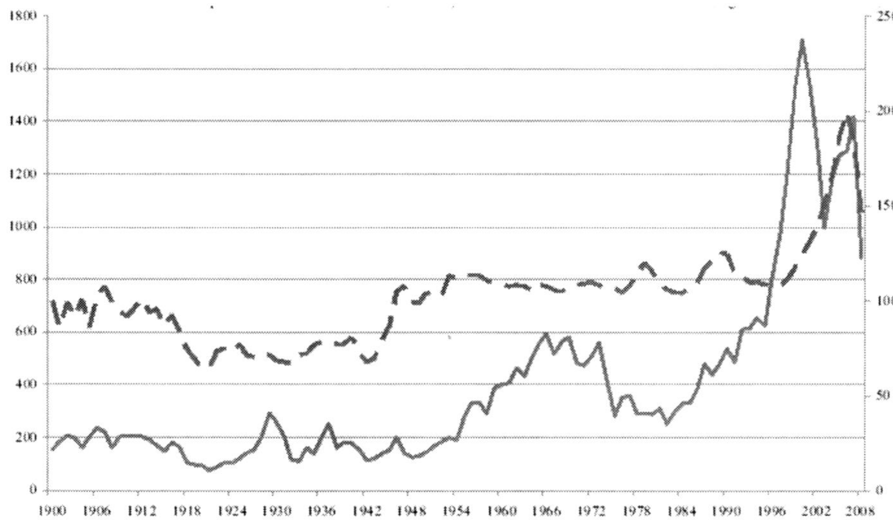

Fig. 6.1. Real S&P 500 Composite Price Index (solid, left scale) and Home Price Index (dashed). Source: Di Noia and Micossi (2009).

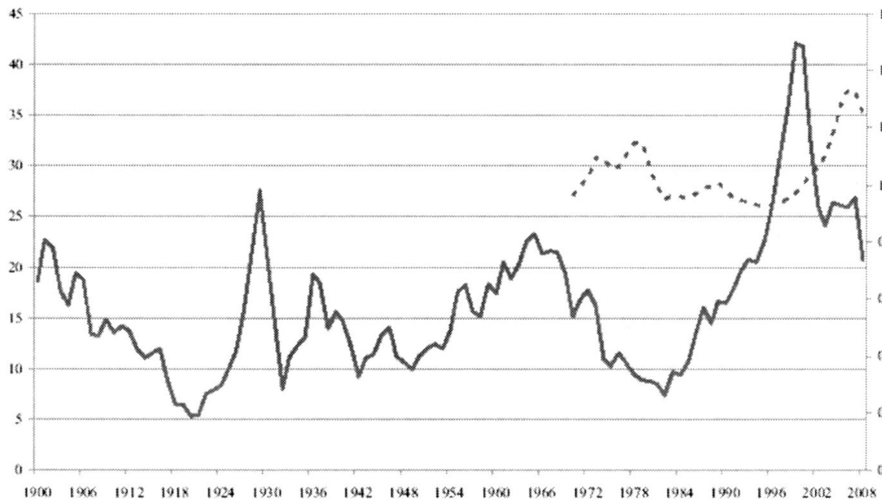

Fig. 6.2. Real S&P 500 Price/Earnings, 10 yrs Moving Average (solid, left scale) and Home Price/Rent (dashed). Source: Di Noia and Micossi (2009).

the mortgages, to the rating agencies, and finally to the purchasers of the mortgage-backed securities. They each had their own motives. But those at the beginning of the chain - those who took on the mortgages and the houses they could not afford, and those who were the ultimate holders of the debt - were buying a modern form of snake oil."

The wide (and unregulated) diffusion of derivatives based on these risky assets spread the related losses throughout the American and international financial markets, with effects that were largely ignored by rating agencies and that emerged in their magnitude only later, gradually and everywhere.[5] In February 2008 a highly leveraged British bank, Northern Rock, had to be nationalized because of its heavy liquidity problems which triggered a bank run. Since then, a number of American and European financial institutions that were widely engaged in the securitization of mortgages started facing similar problems. In March 2008, Bear Stearns had to be acquired by JP Morgan Chase with the assistance of the Fed. In July, one of the largest mortgage lenders in the U.S., Indy Mac Bank, collapsed. In September 2008, the U.S. Government placed the huge mortgage lenders Fannie Mae and Freddie Mac into federal conservatorship and bought 80% of the major insurance company AIG. On September 15, 2008, the investment bank Lehman Brothers filed for Chapter 11 (the largest bankruptcy in U.S. history) after the Bush Administration refused to bail it out (probably to avoid further moral hazard problems). In the meantime, Merrill Lynch was acquired by Bank of America (and Wachovia by Wells Fargo) and the two remaining large investment banks, Goldman Sachs and Morgan Stanley, were converted to traditional banks, concluding the era of investment banking,[6] and maybe also the Anglo-Saxon dominance of the global banking sector.[7]

---

[5] In February 2007 HSBC was the first bank to write down its holdings of subprime-related mortgage backed securities, by $ 10.5 billion. The current estimates suggest that American financial institutions will have to write off at least $ 1.5 trillion of their holdings of similar "toxic assets" (but a similar amount may be hidden in European institutions).

[6] Large bonuses and stock options compensations for investment banking executives are fundamental incentive schemes, but they have been also responsible for promoting excessive risk taking and debt leverage in the last decade (though this is not a good reason to limit them *ex post*, as proposed by many observers ignoring the adverse selection of workers that would follow up). Contrary to common beliefs, the same critique may not hold for the hedge funds, whose managers typically own a large percentage of their funds so as to take large risks, but without perverse moral hazard problems.

[7] In 1999 only two American financial institutions had a market capitalization above $ 100 billion, Citigroup (with $ 151 bn) and Bank of America (with $ 113 bn), followed by the British HSBC and the Lloyds. In March 2009, three Chinese banks lead the ranking: the Industrial & Commercial Bank of China ($ 175 bn), China Construction Bank ($ 129 bn) and Bank of China (113 $ bn). At the time of writing, the market capitalizations of Citigroup and Bank of America are respectively $ 14 bn and $ 40 bn.

In the Fall 2008 the crisis entered an acute phase characterized by a stock market crash, the failure of prominent banks, efforts by American and European authorities to bailout distressed financial institutions, lack of confidence and further defaults. As Krugman (2008) has noticed, "the result of this self-reinforcing process was, in effect, a massive bank run that caused the shadow banking system to shrivel up, much as the conventional banking system did in the early 1930s. Auction-rate securities, in effect a banking sector providing $ 330 billion worth of credit, disappeared. Asset-backed commercial paper, another de facto banking sector, dropped from providing $ 1.2 trillion in credit to providing only $ 700 billion." The lack of confidence froze interbank lending worldwide and induced a substantial reduction of lending to firms. Subsequent announced and implemented nationalizations[8] spread additional fear and lack of confidence in the markets and led to further stock market crashes at the beginning of 2009.[9] Larry Summers, Director of the White House's National Economic Council, has calculated that worldwide wealth lost about $ 50 trillion in the last year and half, more or less two thirds of global GDP.

The Obama Administration is reacting to the crisis with an unprecedented expansionary fiscal policy (see Section 6.6) and with a plan by the Treasury Secretary Tim Geithner to deal with the "toxic assets" that are clogging up bank balance sheets,[10] while the Fed, led by the main academic

[8] Temporary and partial nationalizations of European banks include in chronological order Roskilde Bank (Denmark), Fortis (Belgium, Netherlands and Luxembourg), Glitnir, Landsbanki and Kaupthing (Iceland), Carnegie (Sweden), Parex Banka (Latvia), Commerzbank and Hypo (Germany), Anglo Irish Bank (Ireland), RBS and Lloyds (U.K.).

[9] A clever proposal to reverse this process and revitalize the stock market and investment came from Ricardo Caballero (How to Lift a Falling Economy, *The Washington Post*, p. 19). His proposal was that governments should pledge to buy "up to twice the number of shares currently available, at twice some recent average price, five years from now. (Obviously the specific numbers are only an example). While the policy is about future (and unlikely) interventions, the immediate impact would be enormous. In particular, it would turn around the negative stock markets dynamics, and it would allow banks to raise private capital." As all smart proposals, it has not received much attention.

[10] The plan provides mechanisms to price these troubled assets (whose real value is hardly known), and huge public subsidies (by the Federal Deposit Insurance Corporation and the Fed) to encourage on one side investors to buy them and on the other side banks to sell them. There are two problems with the plan. The first is an accounting problem: these transactions force banks to record losses on their portfolios, while they may prefer to carry loans at the original face value on their balance sheets and set aside only a percentage of the future losses. The public subsidy is aimed a solving this problem by overpricing the assets (with public money and risk), but it may not be enough. The second problem is economic: banks may try to sell only the truly worthless toxic assets, and this adverse selection may lead buyers not to bid for them. The cooperation of public authorities and banks in the choice of the assets to sell may solve this problem. Only the success of the Geithner plan and further bank re-capitalization will

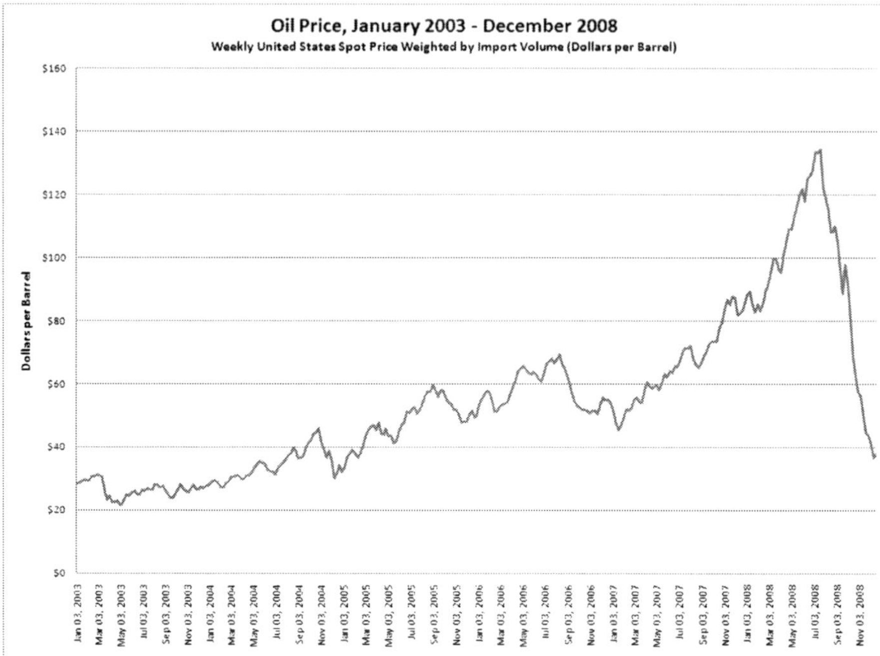

Fig. 6.3. The 2008 Oil Shock

expert of the Great Depression, Ben Bernanke, has implemented an equally unprecedented expansionary monetary policy (see Section 6.7). The rest of the Western world has followed a similar road adopting deficit spending policies and reducing drastically the interest rates. Meanwhile, the financial crisis has extended its disruptive impact from financial institutions to countries (in particular small European countries outside the E.U. or inside it but not in the Euroarea) whose financial accounts and currencies were (and will keep being) under heavy pressure, in particular Iceland, the Baltic Republics (Latvia in particular), Hungary, Bulgaria, Romania and Ukraine.

In this scenario, the supply side is playing a key role. The melt down of stock market capitalizations reduced even more the incentives to invest and create new business activities. In the U.S., venture capital investment, a key source (and indicator) of innovation and business creation (amounting to 0.22% of American GDP in 2007), went down by a third in the last quarter of 2008 (compared to the previous year). Other traditional forms of investments in business creation collapsed as well, possibly limited by a credit crunch. This weakness of the investment process transmitted the financial

lessen the need for the U.S. government to intervene directly in the banking sector.

crisis to the real economy. First, firms reduced their production levels, exhausted their inventories and stopped hiring new workers, then they started to close factories and fire workers. At the end of 2008, consumer demand in the Western world started decreasing as well, quickly for durable goods (as it always happens during recessions), and slowly for the other goods. In the last quarter of the year most Western countries were officially in recession. During 2009 we expect an impressive number of firms to go out of business: only in Europe two hundred thousand SMEs are expected to go bankrupt in 2009,[11] a third more than usual, and many others will drastically restrict their business or stop exporting. At the time of writing, the estimates for GDP growth are negative for the entire globe, with our guesses at - 3 % for the U.S., - 4 % for the Euro area (worse for Germany, better for France), - 6 % for Japan, and + 6.5 % for China (everywhere with inflation rates very close to zero). We expect unemployment rates to increase (at a lower speed in the rest of the year) toward 10 % in U.S., and to overcome 10 % in the Euro area, which may imply a slow return toward steady state levels in the next years.

Market structures have experienced two main phenomena. On one side investment has decreased, business creation and R&D spending have been often limited or postponed, many multi-brand firms have reduced the number of brands supplied on the market, other firms simply stopped exporting to selected countries, some others have merged or are trying to merge with direct competitors,[12] and other firms have gone bankrupt. The consequent drop in net firms entry has led to a reduction in the number of firms or products within many sectors. On the other side, surviving firms have undergone a process of rationalization and job cuts. This process has been quite spectacular in certain global and highly concentrated sectors, starting with the automotive and electronic markets, where most of the leading manufacturers announced job cuts almost at the same time.[13] But of course, it is in smaller and local markets that business destruction leads to substantial reductions in the number of active firms. Likewise, many suppliers of large companies (for instance in the automotive and electronic markets) and downstream firms go

[11] See the article by Guido Tabellini (Le Scelte dei Governi, *Sole 24 Ore*, February 1, 2009).

[12] Beyond the banking sector, think of the pharmaceutical market, with the huge deals between Roche and Genentech, between Pfizer and Wyeth and between Merck and Schering-Plough.

[13] The CEO of FIAT, Sergio Marchionne envisions a global automotive market with a market structure characterized by only six major players, each one able to produce at least 6 million vehicles (one each from U.S., Japan, Germany, France and China plus another European-American). At the time of writing, General Motors is at the border of bankruptcy and could only survive thanks to government aid (and reducing the number of brands and selling Opel), Chrysler could survive only if bought by FIAT and only Ford is in a better position in U.S., Toyota is suffering heavy losses in Japan, and all the European producers are asking for strategic state aids (already arrived for Peugeot and Renault).

bankrupt as a consequence of the problems of the leading companies. This is going to increase the levels of concentration in many markets, allowing the remaining firms to exploit the only ways to cover the fixed costs of production in the presence of a smaller aggregate demand: first, by reducing the remuneration of labor and real wages when it is feasible, or by reducing employment otherwise, and, second, by gradually increasing their mark ups.

The recession is generating a reduction of labor income and profits, but a relative increase of mark ups, and this has a crucial consequence: a further reduction in aggregate consumption and total employment, generating an additional depression of the aggregate demand and a collapse of trade across countries. Surviving firms are reducing production and cutting jobs, contributing to the raise of the unemployment rate. This is the competition effect of the EMSs approach in action, though of course it is working in reverse: in a recessionary context net business destruction and contraction of sales deteriorate the division of surplus between input remuneration and mark ups (positive profits with lower sales must be associated with higher margins) and contribute to depress demand. In turn, this reduction in demand is going to exert a negative feedback effect on profits and stock market evaluations, which has a major negative impact on business creation, competition and trade, so as to propagate the recession over time and space.

6.2 Steady-state EMSs and Entry

Even if the number of goods and firms, the mark ups and all the aggregate variables as output, consumption, investment, labor force and profits are changing over the business cycle, in the long run their pattern must be determined by the structural parameters of the economy.

In the neoclassical approach (Solow, 1956), the wealth and growth of nations depend on the productivity of the workers (and their machines) and on the evolution of this productivity, on their propensity to work which determines the size of the labor force in the economy, and on their propensity to save which determines the sizes of investment and ultimately the stock of physical capital of the economy. Of course, there is not a unique path toward economic progress. Some countries have based their prosperity on the achievement of high levels of productivity, others have obtained prosperity by working more and others have reached the same prosperity through high savings rates (because their citizens were patient and available to give up to current consumption for the well being of the following generations). Moreover, as suggested by Acemoglu (2009), there are other factors that affect the long run performance of economies, including luck, geography, culture, and formal and informal institutions.[14]

[14] See the Intertic Lecture by Dixit (2008) for a wide discussion of this topic.

Taking as given the long run impact of these fundamental factors on the production possibilities, we claim that there is a second order impact that these and a few other technological and behavioral factors exert on the market structures and consequently on the long run performance of the economies. The EMSs approach has characterized the average structure of productive markets in steady state (number of firms, individual production and mark ups) depending on a few structural parameters and on the form of competition.[15] Our baseline investigation (in Chapter 3) has emphasized five main determinants of the long run market structures and of the other aggregate variables.

The first determinant is given by scale factors as the level of productivity (or the size of the population): higher productivity (or larger population) leads to a larger size of the demand inducing more business creation, which in turn increases the steady state number of firms and enhances competition while reducing the mark ups. We have noticed that the relation between the size of the markets and the number of firms should be less than proportional, reflecting the strengthening of competition associated with more competitors: this hypothesis is strongly supported by preliminary empirical evidence (see Section 4.1).

The second determinant of the steady state EMSs is the size of the barriers to entry: when these are high, the profitability of entry is low and the long run equilibrium is characterized by high concentration and high mark ups. Of course, markets characterized by high sunk costs of entry due to technological conditions naturally lead to equilibria with few active firms, and this does not represent a problem in itself. However, the introduction of a general purpose technology which reduces the fixed costs of entry is going to positively affect business creation and therefore competition and output production. Finally, to the extent that the entry barriers are artificial, in the sense that they are due to product market regulations, we can conclude that reforms leading to deregulation reduce concentration and mark ups in the long run, with a positive impact on the performance of the economy as a whole.

The third structural factor determining the nature of the EMSs in the long run is a behavioral factor, the way people discount the future. This degree of patience is what determines the propensity to save of the households, which in turn affects the equilibrium in the credit market. When agents are more patient, their large supply of savings reduces the interest rate, which means that the discounted sum of future profits is higher: this attracts more entry,

[15] For a given number of firms, competition tends to be stronger when the firms choose their prices against each other rather than when they choose their production levels. Nevertheless, stronger competition reduces the expected profitability and therefore attracts a lower number of firms, that in turn works toward less intense competition between the active firms. Which one of the two effects dominates in the long run is not obvious, but our investigation shows that, *ceteris paribus*, competition in prices leads to steady states with fewer firms and lower mark ups.

strengthens competition and ultimately reduces the mark ups. Therefore, more patient agents lead to a higher number of firms in the steady state.

The fourth element is the rate of business destruction due to exogenous reasons: when the risk of bankruptcy is high, the expected value of business creation is lower and business creation is limited. Therefore, in case of a high rate of default there are only few firms in the long run (but with a high rate of turnover), and they apply a high mark up to their goods.

The fifth determinant of the long run equilibrium emphasized in our framework is the degree of substitutability between goods. Higher homogeneity of the goods induces stronger competition between the firms and leads to lower mark ups, which in turn attracts a limited number of firms in the markets. On the contrary, when goods are highly differentiated, competition is relaxed and mark ups are higher, but this attracts more firms. Of course, markets with different levels of substitutability between goods coexist in the real world, and we should think of these relations only as general tendencies characterizing different markets.[16]

Summarizing the results obtained until now with a focus on the impact on the mark ups, we have:

PRINCIPLE 2. IN THE LONG RUN, THE STEADY STATE EMSS ARE CHARACTERIZED BY MARK UPS INCREASING IN THE COST OF ENTRY AND IN THE RATE OF BUSINESS DESTRUCTION AND DECREASING IN LABOR PRODUCTIVITY, IN THE DISCOUNT FACTOR AND IN THE DEGREE OF SUBSTITUTABILITY BETWEEN GOODS.

Notice that the steady state structure of the markets determines not only what is produced and at which price it is sold, but also how much of it is consumed, which is what matters for our understanding of the behavior of the economy and for its reaction to structural changes. The impact of the main structural parameters on long run consumption under competition in prices in the markets is the following: higher productivity, more substitutability between the goods and more patience ultimately lead to larger consumption bundles, while higher costs of entry and higher rates of business destruction lead to smaller consumption bundles in the long run.[17]

The following two case studies discuss the role of business creation over the business cycle (in the U.S. manufacturing sector), and in reaction to a technology shock (the introduction of a general purpose technology as cloud

[16] A last structural parameter that can affect the steady state is the elasticity of labor supply. When this is larger, agents tend to work more, which tends to allow entry of a larger number of firms and to strengthen competition.

[17] Things are not that simple under different modes of competition because these may lead to persistent inefficiencies on which we will return below. Notice also that we are loosely referring to the consumption bundle of the agents, which is enhanced by the number of varieties available (to appreciate the difference, notice that more patience leads to richer consumption bundles because higher savings allow the economy to produce more goods, but the agents consume less of each good).

Product Entry and Exit in the U.S.

Period	9-year 1994 - 2003	4-year 1999 - 2003	1-year Median
Entry Rate	0.78	0.50	0.25
Creation	0.64	0.37	0.09
Entrant Relative Size	0.49	0.56	0.30
Exit Rate	0.72	0.46	0.24
Destruction	0.37	0.18	0.03
Exiter Relative Size	0.23	0.23	0.09
Ratio Share Common (t/t-1)	0.57	0.77	0.93

Notes: Entry Rate = number of new products (*t*) / total number of products (*t*)
Exit Rate = number of disappearing products (t-1) / total number of products (*t*-1)
Creation = Value(new products, *t*) / Total value (*t*)
Destruction = Value(disapp. products, *t*-1) / Total value (*t*-1)

Fig. 6.4. Business creation. Source: Broda and Weinstein (2009)

computing). The aim is to exemplify how EMSs can play a crucial role for macroeconomic analysis.

6.2.1 Entry and business creation in the U.S.

Our first two principles of the EMSs approach have characterized the role of entry of firms in the short and long run. Since the phenomena of entry, business creation and business destruction have been largely neglected by the neoclassical approach and by the related empirical research at the macroeconomic level, it is important to collect new evidence on their nature. In a recent fundamental work, Broda and Weinstein (2009) provide the first large-scale examination of product creation and destruction, documenting the nature, extent and cyclicality of product entry and exit in the U.S. manufacturing sector (and evaluating the elasticity of substitution within and between sectors). They use a unique dataset that contains the universe of products with barcodes in sectors that cover around 40 % of all expenditures on goods in the U.S. Consumer Price Index (therefore they do not refer to general business creation, which includes any kind of business or service, but to genuine product creation or innovation). Figures 6.4 to 6.6 reproduce their main results. The first one reveals that almost 80 % of the products that existed in 2003 did not exist in 1994, and they comprised 64 % of expenditures in 2003. The value of the products that disappeared in the same period was much smaller, about 37 % of expenditure in 1994. This suggests that new

Net and Gross Rates by Module, Size-Weighted Averages Ranked by SUM (over 1-year)

MODULE DESCRIPTION	RANKING (1)	CREATION (2)	DESTRUCTION (3)	NET ENTRY (2) - (3)	TURNOVER (2) + (3)
VIDEO PRODUCTS PRERECORDED	1	0.56	0.09	0.47	0.65
CAMERAS	2	0.44	0.18	0.26	0.62
COMPUTER SOFTWARE	3	0.32	0.19	0.13	0.51
TELEPHONE AND ACCESSORY	4	0.25	0.11	0.14	0.36
VACUUM AND CARPET CLEANER APPL	5	0.25	0.14	0.11	0.39
CANDLE AND CANDLE IN HOLDER	6	0.30	0.15	0.14	0.45
DISPOSABLE DIAPERS	7	0.17	0.06	0.12	0.23
STORAGE AND SPACE MANAGEMENT	8	0.19	0.10	0.09	0.29
KITCHEN UTENSIL AND GADGET	9	0.18	0.10	0.08	0.28
NUTRITIONAL SUPPLEMENTS	10	0.14	0.06	0.08	0.20
SUGAR-GRANULATED	100	0.01	0.01	0.01	0.02
FRANKFURTERS-REFRIGERATED	99	0.01	0.01	0.00	0.02
BUTTER AND SPREADS	98	0.02	0.00	0.01	0.02
CHEESE - COTTAGE	97	0.01	0.01	0.00	0.02
SAUSAGE-BREAKFAST	96	0.02	0.01	0.01	0.03
SEAFOOD-TUNA-SHELF STABLE	95	0.02	0.01	0.02	0.03
EGGS-FRESH	94	0.03	0.01	0.02	0.04
DINNERS-FROZEN	93	0.04	0.00	0.03	0.04
MARINARA SAUCE	92	0.03	0.01	0.02	0.04
PIZZA-FROZEN	91	0.04	0.01	0.03	0.04

Note: Select top 100 MODs by average value and then rank top 10/bottom 10 by SUM

Fig. 6.5. Business creation by sectors. Source: Broda and Weinstein (2009)

products, mostly produced by previously existing firms from the same or another sector (innovation by leaders), and sometimes produced by fully new entrants, systematically displaced market share from the available products. This is an important indication that new products are of a higher quality and contain a component of innovation. Broda and Weinstein (2009) show that, in a typical year, 40 % of household's expenditures are in goods that were created in the previous four years, and 20 % of expenditures are in goods that will disappear in the following four years.

Moreover, market turnover is higher in innovative sectors. Figure 6.5 reports the ranks of the top and bottom ten sectors (between a hundred main sectors) in terms of turnover, where Broda and Weinstein (2009) define turnover as the sum of creation and destruction in the sector. Pre-recorded videos, cameras, and computer software are characterized by high entry/exit process and this is not surprising given the high degree of technological innovations in these sectors.

More important for our purposes is the analysis by Broda and Weinstein (2009) on the cyclical behavior of entry. Their Figure 6.6 shows that net business creation is strongly procyclical, with more products being introduced in expansions and in product categories that are booming: an additional 1 % growth in consumption of a particular sector is associated with 0.3 % increase in the share of new goods. Gross business creation is strongly procyclical and covaries more with demand in booms than contractions, while the opposite is true for business destruction, which is countercyclical, but responds more strongly in recessions than in booms. However, most of the procyclicality of net business creation comes from the procyclicality of creation.

Other recent studies have analyzed empirically the cyclical nature of the timing of new product introductions in U.S. manufacturing,[18] finding that business creation varies more in nonseasonal frequencies than in seasonal frequencies, and responds more to fluctuations in aggregate demand than in market demand, and in business cycles than in seasonal cycles. All these results support the thesis that entry is an important procyclical phenomenon that brings new and better products in the market and that can have a crucial impact on the behavior of the aggregate economy.

The American economy is a friendly environment to business creation as few other economies around the world. In the last decade an average of 550,000 new small businesses were created in the U.S. every month. Since the beginning of the current crisis this number has been in free fall, and without doubts this is having a strong impact on market structures and on the aggregate production.

6.2.2 Cloud computing: the impact of a General Purpose Technology

The introduction of a general purpose technology (GPT) can provide a fundamental contribution to promote growth, competition and business creation. This was the case of the Internet in the 90s, but a new interesting example is now given by the introduction of "cloud computing",[19] an Internet-based technology through which information is stored in servers and provided as a service (Software as a Service) and on-demand to clients (from the "clouds" indeed). Its impact will be spectacular on both consumers and firms. On one side, consumers will be able to access all of their documents and data from any device (the personal laptop, the mobile phone, an Internet Point..), as they already do for email services. On the other side, firms will be able to rent computing power (both hardware and software) and storage from a service provider, and to pay on demand, as they already do for other inputs as energy and electricity. The former application will affect our lifestyles, but the latter will have a profound impact on the cost structure of all the industries using hardware and software, and therefore it will have an indirect but crucial impact on business creation and on the macroeconomy.

The EMSs approach provides the tools to evaluate the economic impact of the introduction of cloud computing. Before showing this, however, we need to describe further the nature of this new GPT. Many hardware and software companies are currently investing to create new platforms able to attract customers "on the clouds". These "cloud platforms" provide services to create

[18] See Axarloglou (2003)

[19] This section is based on Etro (2009,a).

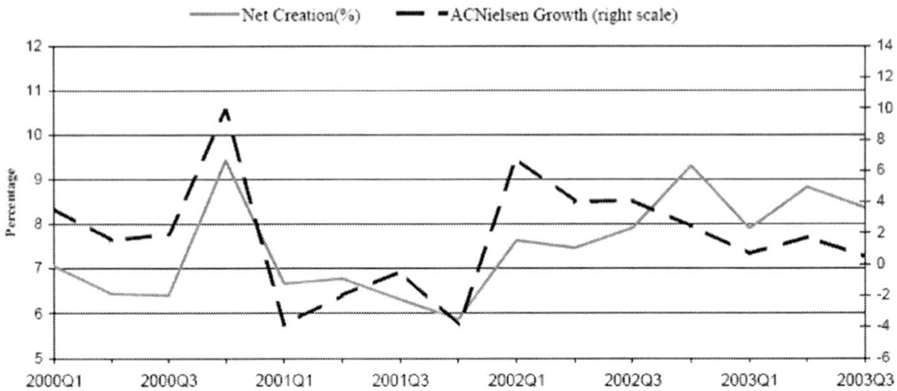

Fig. 6.6. Sales growth and net creation (Q4/Q4growth rates). Source: Broda and Weinstein (2009)

applications in competition or in alternative to "on-premises platforms", the traditional platforms based on an operating system as a foundation, on a group of infrastructure services and on a set of packaged and custom applications. The crucial difference between the two platforms is that, while on-premises platforms are designed to support consumer-scale or enterprise-scale applications, cloud platforms can potentially support multiple users at a wider scale, namely at the Internet scale.

Currently we are only in a phase of preparation with a few pioneers offering services that can be regarded as belonging to cloud computing. A recent study of the International Data Corporation has examined the role of IT cloud services across five major product segments representing almost two-thirds of total enterprise IT spending (excluding PCs): business applications, infrastructure software, application development & deployment software, servers and storage. Out of the $ 383 billion that firms have spent in 2008 worldwide for these IT services only $ 16.2 billion (4%) could be classified as cloud services. In 2012 the total figure was expected to be at $ 494 billion and the cloud part at $ 42 billion, which would correspond to 9% of customer spending, but also to a large part of the growth in IT spending. The majority of cloud spending is and will remain allocated to business applications, with a relative increase of investment in data storage.

Many large high-tech companies are building huge data centres loaded with hundreds of thousands servers to be made available for customer needs in the near future. The first mover in the field has been Amazon, that provides access to half a million developers by way of Amazon Web Services. Through this cloud computing service, any small firm can start a web-based business on its computer system, add extra virtual machines when needed and shut them down when there is no demand: for this reason the utility is called Elastic Cloud Computing.[20]

Google is also investing huge funds in data centres. Already nowadays Google Apps provides word processing and spreadsheet applications online, while the software and data are stored on the servers. Even the search engine of Google or its mapping service can offer cloud application services. For instance, when Google Maps was launched, programmers easily found out how to use their maps with other information to provide new services.

Other software and hardware companies have been actively investing in cloud computing.[21] Social networks have moved in the same direction turning into social platforms for consumer based applications, with Facebook in the front road (with its 200 million or more subscribers and an impressive amount of information available). Yahoo! is developing server farms as well. Oracle has introduced a cloud based version of its database program. Microsoft has started later but with huge investments in the creation of new data centres. In the fall of 2008, the leading software company has introduced a cloud platform called Windows Azure, currently available only in a preview version. Azure is able to provide a number of new technologies: a Windows-based environment in the cloud to store data and to run applications, an infrastructure for both on-premises and cloud applications, a cloud based database, and an application tool which allows to synchronize and constantly update data across systems joined into a "mesh". Moreover, Windows Azure provides a browser-accessible portal for customers: these can create a hosting account to run applications or a storage account to store data in the cloud, and they can be charged through subscriptions, per-use fees or other methods.[22]

The battle for the clouds between these companies is going to reshape the ICT market structure as PC distribution did in the 80s. But according to the Economist[23] "cloud computing is unlikely to bring about quite such a dramatic shift. In essence, what it does is take the idea of distributed computing a step farther. Still, it will add a couple of layers to the IT stack.

---

[20] For instance, Animoto, an application that produces videos from user-selected photos and music, has been a successfull business of this kind. When Animoto was launched on the leading social network Facebook, it was forced by exponentially increasing demand to bring the number of machines used on the Amazon Web Services from 50 to 3500 within three days, something that would have been impossible without relying on a cloud platform.

[21] Notice that cloud computing implies outsorcing of both software and hardware, therefore it should not be surprising that hardware producers like Dell, Hewlett-Packard, Cisco and IBM are investing in the field as well. Even producers of consoles for videogames may switch to games in the clouds in the near future.

[22] While many applications and services can perform well either on-premises or in the cloud, Microsoft is working on a wider range of combinations, which enables developers and customers to manage applications and data in the clouds, or on-premises, or via some combination of both (Software plus Services).

[23] 2008, *Economist*, "Where the Cloud Meets the Ground", October 25th, p. 387.

One is made up of the cloud providers, such as Amazon and Google. The other is software that helps firms to turn their IT infrastructure into their own cloud, known as a 'virtual operating system for data centres' ... Will this prospective platform war produce a dominant company in the mould of IBM or Microsoft that is able to extract more than its fair share of the profits? Probably not, because it will be relatively easy to switch between vendors... Nor is it likely that one firm will manage to build a global cloud monopoly. Although there are important economies of scale in building a network of data centres, the computing needs of companies and consumers vary too widely for one size to fit all." Moreover, the need of creating network effects in the development of a cloud platform will keep low the margins for a while and will maximize the speed of diffusion of cloud computing between firms at the global level.

In front of this rapid evolution, it is crucial to understand the economic impact of the introduction of this GPT. Its main economic benefit is associated with a generalized reduction of the fixed costs of entry and production, in terms of shifting fixed capital expenditure in ICT into operative costs that depend on the size of demand. This contributes to reduce the barriers to entry especially for SMEs and intensifies the business creation process. The consequences on the endogenous structure of markets with large needs of hardware and software is going to be substantial, with entry of new firms, strengthening of competition, reduction of the mark ups, and with an increase in average and total production.[24]

We have employed an adapted version of the model of Chapter 3 with accumulation of ICT capital (hardware and software) and fixed costs of entry in terms of the final good, to estimate the impact of a gradual diffusion of cloud computing. This is translated into a slow reduction of the fixed cost of entry, which endogenously generates further investments in business creation. The calculations based on a model calibrated on E.U. countries show a significative impact. Starting from conservative assumptions on the gradual reduction of the fixed entry costs due to the adoption of cloud computing, the

[24] Another important benefit is associated with the creation of multidimensional network effects due to the new possibilities of product creation in the clouds, that is between companies exploiting in different ways the potentialities of cloud computing through the same platform or different ones. This is related to another new possibility, the rapid adoption of changes: it is not uncommon, that applications in the clouds are modified on a daily base (to accommodate new requirements, or enable new economic venues), which is impossible with on-premise solutions. It is important to notice that the aggregate role of these network effects can be relevant but it is extremely difficult to measure. Finally, cloud computing is going to introduce the possibility of a) sharing resources (and costs) among a large pool of users, b) allowing for centralization of infrastructures in areas with lower costs, and c) allowing for peak-load capacity increases (generating efficiency improvements for systems that are often only 10-20% utilised). These features will lead to additional savings in energy and to greater environmental sustainability, whose measure, however, is again subject to large uncertainty.

exercise suggests that the diffusion of this innovation may induce the creation of 100-400 thousand new SMEs in the whole Euroarea, adding a few decimal points to the growth rate and about a million new jobs in the medium run.

Part of the positive effects of cloud computing are going to be positively related to the speed of adoption of the new technology. For this reason, our investigation suggests that policymakers should promote as much as possible a rapid adoption of cloud computing. Concrete possibilities include fiscal incentives and a specific promotion of cloud computing in particular dynamic sectors. For instance, governments could finance, up to a limit, the variable costs of computing for all the (domestic and foreign) firms that decide to adopt a cloud computing solution. Moreover, they could introduce business-friendly rules for the treatment and movement of data between their country and foreign countries. These policies may be studied in such a way to optimize the process of adoption of the new technology and to strengthen the propagation of its benefits within the country.[25] International policy competition for the subsidization of cloud computing solutions would generate positive spillovers across countries, and some coordination at the E.U. level would be welcome.

As the Economist (*ibidem*) claims, "[I]nternet disrupted the music business; Google disrupted the media; cloud-based companies could become disrupters in other inefficient industries." One of the few positive aspects of recessions is that the market selects efficient and innovative firms able to invest in new technologies to be exploited in better periods. This may turn out to be at the basis of the recovery from the current recession.

6.3 Gains from Trade and the Effects of Globalization

Globalization, defined as the increase in trade in goods and factors of production and associated with the reduction of natural and artificial trade barriers, is one of the main phenomena of the last decades, with implications and motivations that go far beyond economic factors. Understanding the impact of increasing (and decreasing) openness is one of the major aims of macroeconomic theory.

It is often claimed that globalization leads to lower prices for the consumers but also to business destruction at the local level. However, the new

[25] Moreover, in a context as the European one, smaller countries would be able to obtain larger gains from similar policies at least in the initial phase, because they would easily attract foreign investments from larger countries. In a period of increasing limits to other forms of fiscal competition, a policy of subsidization of cloud computing (without discrimination across firms of different member countries) could generate substantial capital flows toward countries with good general infrastructures. For instance, early adoption of these policies by small E.U. countries as Luxembourg or Malta could attract large investments and create wide effects in terms of output growth and job creation in these countries.

trade theory summarized by Helpman and Krugman (1985) has emphasized that opening up to trade leads mainly to other forms of gains and other forms of impact on business: it keeps prices at the same level and it does not affect the number of active firms, while it increases the total number of goods available for consumption at the local level. This generates what are usually called the "gains from variety" due to openness.

The EMSs approach emphasizes a related but more complex mechanism (see Chapter 4). When a country opens up to trade, say with a bordering country, the domestic firms start competing with the foreign ones for both the domestic and the foreign market (that become an integrated market in the absence of trade frictions). This strengthening of competition leads to a reduction of the mark ups and therefore of the prices of all the goods. Profitability and entry are affected in three ways: first, each firm serves two markets rather than one, which enhances profitability; second, each firm is sharing each market with a larger number of firms, which reduces profitability; third, stronger competition reduces mark ups and profitability. The net effect determines the impact on the number of firms active in each market. In the absence of asymmetries between the firms, the first two effects balance each other and the overall impact of opening up to trade is a reduction of the total number of firms, which implies business destruction at the local level. Of course, the number of firms active in the integrated market increases, but less than proportionally. The associated competition effect generates price reductions and what we called the "gains from competition".

The extent of this mechanism crucially depends on the type of traded goods. At one extreme we have perfectly differentiated goods with competition in prices: for these goods, all the gains from opening up to trade derive from an increase in the number of consumed varieties and not from price changes, while business destruction is absent. This is the typical situation that Krugman had in mind when talking about the gains from intra-industry trade: globalization makes different brands of cars available for all consumers, which enlarges their options.

At the other extreme, however, we have homogenous goods with competition in quantities: for these goods, all the gains from trade derive from lower prices, but business destruction is heavy.[26] This is probably what happened in most markets during the phase of intense globalization of the last years, and it is not surprising that many supporters of the globalization process have constantly emphasized the price-reducing impact of openness while associating the business destruction effect with a healthy reallocation of labor across firms and sectors. Nevertheless, when the labor market works imperfectly, the social costs associated with this process of job reallocation can be quite relevant. Summing up the main insights, we have:

[26] Of course, in intermediate situations the gains from trade derive from both lower prices and more varieties, and business destruction is partial.

PRINCIPLE 3. GLOBALIZATION BRINGS GAINS FROM TRADE BY STRENGTHENING COMPETITION, REDUCING MARK UPS AND PRICES, AND INCREASING THE NUMBER OF AVAILABLE GOODS, BUT IT INDUCES BUSINESS DESTRUCTION AT THE LOCAL LEVEL.

Finally, globalization strengthens the interdependence between economies, and leads to faster propagation of the shocks across countries. In a global world, we can re-examine the impact of shocks and their international propagation through the EMSs approach. Imagine that the domestic economy is hit by a temporary positive productivity shock. Such a shock increases domestic consumption and profits, which attracts entry of new firms in the domestic economy and strengthens competition. Domestic entry increases labor demand at home compared to abroad, and exerts upward pressure on the domestic wages, which causes the relative price of non-traded goods at home to increase relative to foreign. As a consequence, the increase in domestic productivity generates a real exchange rate appreciation due to higher prices of domestic non-tradables and imports and to a larger demand for domestic goods, which leads to persistent deviations from the purchasing power parity (in line with the evidence). Meanwhile, the positive impact on the profits of the foreign firms and the temporary price reductions do exert an indirect expansionary effect abroad, which contributes to propagate the boom across countries. The opposite mechanism works in case of a negative shock, and the rapid diffusion of the 2008 crisis from the U.S. to the rest of the world has been one of the most dramatic examples of the deep interdependence of the modern economies.

Looking at these phenomena in a global perspective, it emerges quite clearly that macroeconomic policy at the national level is always less effective when it is not coordinated at a supra-national level, but also that there are large gains from coordinating policies within international unions or global agreements. The recent financial crisis has urged the coordination and the tightening of financial regulation at a global level to reduce the systemic risks associated with the tendency of unregulated financial intermediaries to overleverage their capital and to undertake excessive risk). Since only few countries would give away their sovereignty on this matter, it is at least important that most countries could agree on a number of basic principles (maybe through the Financial Stability Forum currently directed by Mario Draghi) and allow stricter regulation on a unilateral or bilateral basis. More or less as the W.T.O. has done to coordinate the process of tariff reductions toward free trade.[27] The future of globalization is deeply related

[27] More generally, Alesina, Angeloni and Etro (2005) have shown that such a two-level system is politically viable and can improve the outcome of international coordination. For a similar proposal on financial regulation see Dani Rodrik (2009, *Economist*, "A Plan B for Global Finance", *Economic Focus*, March 14th, p. 72).

with the willingness and the ability of developed and developing countries to coordinate properly their policies.

Globalization, income distribution and inequality. Our investigation of the impact of openness has emphasized a competition effect, for which globalization strengthens competition and reduces the prices, and a business destruction effect, for which an endogenous increase in the equilibrium size of international firms leads to the foreclosure of smaller local business. In the absence of labor market imperfections, this business destruction effect is inconsequential for the workers: they switch jobs reallocating their work between a smaller number of larger firms that have a larger share of the global markets.

Nevertheless, job destruction due to globalization can have dramatic effects in the presence of labor market rigidities, both in terms of unemployment and income inequalities. When labor market frictions are relevant, some of the workers hit by globalization may find it hard to switch to different jobs in different sectors and when social insurance and unemployment benefits are limited they would bear substantial losses while looking for a new job. Meanwhile, the economy may suffer heavily because of the reduction and misallocation of the labor force.

Our distinction between different cases above can help us to understand better the consequences of globalization. One could think of sectors producing highly differentiated goods with competition in prices as sectors where innovation and design are the fruit of skilled labor (characterized by higher human capital): in these sectors business destruction due to globalization is limited. On the other side, sectors with homogenous goods and competition in quantities can be seen as sectors characterized by standard production processes employing low-skilled workers: for these sectors, we have seen that business destruction due to globalization is radical.

Therefore, in the presence of labor market imperfections, the consequences of globalization can be worse for the low-skilled workers and can potentially lead to their unemployment and to larger income disparities between them and the skilled workers. For this reason, the gains from trade due to globalization and associated with lower prices are often criticized for their consequences on income distribution and for the increase in the inequality between low productivity workers and high productivity workers.

The EMSs approach provides further insights on issues of income distribution that we should emphasize in this context. Contrary to the neoclassical approach, our approach is able to generate an interesting evolution for the distribution of income between extra profits (in the form of dividends and capital gains) and labor income (or remuneration of the inputs in general). In the short run, a boom is associated with an increase of both forms of income but with a larger increase for the latter due to the stronger competition between firms. *Vice versa*, in a downturn, the mark ups tend to increase and the fraction of income accruing to the factors of production shrinks (the procyclicality of the labor share would require further empirical investigations).

In a long term perspective, the process of globalization may generate a positive effect on labor income, since the associated increase in competition may shift part of the surplus from profits toward labor income. Again, in the presence of labor market frictions, it may take time before globalization exerts its benefits on the workers, but the ultimate impact is going to be positive. For these reasons, the protectionist tendencies that are emerging during the current crisis may have negative consequences in the medium-long run for the global economy.

6.4 Growth Driven by Leaders

Understanding the determinants of growth and the reasons for which differentials in growth rates are so large is fundamental to foster economic progress around the world. While growth in developing countries is largely associated with the process of industrialization through established technologies and with the accumulation of physical and human capital in the neoclassical sense, growth in the Western developed countries is largely driven by the continuous process of expansion of the technological frontier.

Technology-driven growth is mainly due to the innovations of firms investing in R&D to create new or better products and replace the existing ones (Romer, 1990, and Aghion and Howitt, 1992). Profits, associated with innovations and temporarily protected through IPRs, provide the incentives to invest in R&D for firms in high-tech sectors, and the structure of the competition between these firms is a crucial element of the engine of growth. The EMSs approach has widely examined this structure, endogenizing strategic interactions in the investment choices and the entry process, and it has been also focused on the peculiar role of market leaders.

Entry of firms in the competition for the market is primarily driven by the size of the expected profits from innovations, and therefore it is directly related to the strength of IPRs protection. This protection (as the legal enforcement of contracts and the protection of physical property rights) is one of the founding elements of the free market economies,[28] and possibly the most important, since dynamic growth (rather than static wealth) relies on it. Of course, entry changes the nature of the strategic interactions between investors, increasing aggregate investment and reducing the expected profitability (per firm) at rates that depend on the substitutability or complementarity between the investments of the firms. The entry process is exhausted when the expected return on R&D investment is the same as the return on alternative investments, and this equilibrium determines the aggregate rate of technological progress.

[28] See the Intertic Lecture by Dixit (2008) on this topic.

The endogeneity of entry plays a crucial role in the decision of technological leaders to invest in R&D to perpetuate their status: leaders tend to invest less than their rivals and to reduce the aggregate investment when they face an exogenous number of competitors, but they tend to invest more and to increase the aggregate investment when they face an endogenous number of investors trying to replace the leading technology. For this reason, it can be highly misleading to evaluate the competitiveness of a dynamic market on the basis of its concentration level or of the market share of its leader without concern for the entry conditions (this is a common mistake done by antitrust authorities).

The hypothesis for which leaders invest more in R&D if and only if they face a strong entry pressure has been tested in Chapter 5 on a unique dataset of the German manufacturing sector, the Mannheim Innovation Panel, which includes a wide number of firm level data and answers to a survey with a special focus on innovation.[29] A novel aspect of this research is given by the fact that the same firms provide a subjective view on the key determinants of R&D intensity, the entry pressure and the leadership. Rather than determining arbitrarily the size and composition of a market, assigning a degree of entry intensity in a discretionary way, and assigning a status of leadership on the basis of predetermined variables, the questionnaire of the Mannheim Innovation Panel allows one to ask the same firms to identify the size of their main market, the existence of an endogenous threat of entry in the market and the identity of the leader in the market. Using standard control variables as employment, capital intensity, stock of IPRs and sector dummies, the econometric investigation provides interesting results. Apparently, incumbent leaders invest slightly less than the average followers, with a difference that is not significant. However, this apparently uniform behavior hides a very different reality, which emerges when we control for endogenous entry pressure. The presence of such an entry pressure leads the average firm to invest less, but when the firm is an incumbent leader in its main market, this leader increases its investment in R&D above the investment of the average firm.[30] This confirms the main predictions of the EMSs approach, for which market leaders are more innovative if and only if they are pressured by endogenous entry.

In the presence of sequential innovations, the consequence of the impact of the entry pressure on the investment of the leaders is somewhat paradoxical: the likelihood of persistence of leadership is high exactly when there is endogenous entry in the competition for the market (and not when there is high market power of the leader). Therefore, growth is mainly driven by the investments of the market leaders to perpetuate their positions under the pressure of rival firms. We can summarize this result with:

---

[29] See Czarnitzki, Etro and Kraft (2008).

[30] For a fascinating introduction to the discriminating power of econometric analysis see Levitt and Dubner (2005).

PRINCIPLE 4. GROWTH IS DRIVEN BY THE PROCESS OF BUSINESS CREATION AND ENHANCED BY A STRATEGIC LEADERSHIP OF PATENTHOLDERS IN THE COMPETITION FOR THE MARKET, WHICH CREATES ENDOGENOUS PERSISTENCE OF THEIR TECHNOLOGICAL LEADERSHIP.

As we have seen, the New Economy is key to the engine of world growth because it is often associated with GPT innovations that spread their benefits in other sectors. For instance, this is the case of innovations in hardware, software, online business and advertising. Moreover, we have suggested that the leaders of these sectors contribute to drive global technological progress. Etro (2007,a, Chapter 6) provides a wide discussion of the software market in the perspective of the EMSs approach, with particular attention to the role of Microsoft as a leader forced by endogenous entry pressures to invest a lot in R&D. The software market is not only characterized by endogenous entry at the global level in the competition *for* the market, but also in the competition *in* the market, which forces the leader to implement aggressive strategies (low pricing, bundling) with competitive purposes: for this reason, the large market share of Microsoft should not be necessarily seen as a consequence of market power.[31]

Leaving that reference to the interested reader, here we want to present a microeconomic digression on a related market of the New Economy, whose structure is characterized by endogenous entry in the competition *for* the market (through investments in innovation), but not in the competition *in* the market (not through prices). This market will probably play an important role in future economic scenarios.

6.4.1 Online advertising: market structure and innovation

Advertising is at the basis of any process of business creation and diffusion, and its efficiency is crucial for all sectors. It is calculated that worldwide spending on total advertising is currently around \$ 600 billions, of which at least 40 billions are spent in the online field, half in the U.S., a third in the E.U. Since 1994, when HotWire sold the first banner for advertising,

[31] In a comment for the *Financial Times* (Microsoft's New Openness is Welcome Change, February 28, 2008, p. 11), the author of this book has expressed concern toward the negative attitude of EU policymakers toward the innovative role of leaders facing entry pressure as Microsoft: "Neelie Kroes, the competition commissioner, has explicitly cited as one of her objectives a radical reduction of the alleged monopolist's market share... Such an objective may reflect a dangerous misunderstanding of this dynamic market. As a wide academic literature has shown, a large market share in the high-tech sector can be the consequence of low prices charged by a leader in a market with network effects and endogenous entry, and not of unconstrained market power. Therefore, antitrust enforcement to promote consumer welfare should not be driven by the aim of reducing a company's market share, but should instead preserve competitive entry in the market."

and 1995, when Infoseek introduced search-based advertising, online ads have been constantly growing in all of their different forms: search advertising associated with search engines (40 % of online advertising), display advertising (30 %), classified listings on web sites (20 %), email advertising (2 %) and others. The current crisis has stopped only temporary the growth of the online market. This is destined to increase its share in total advertising for the following reasons: the Internet is rapidly growing and the large majority of websites generates revenues from advertising; other devices as mobiles and televisions will be always more often connected to the Internet; software innovation allows more efficient mechanisms to reach targeted consumers on the basis of the characteristics of search through the keyword bidding system and of the websites visited (contextual advertising), and in perspective on the basis of the characteristics of the Internet users (behavioral advertising).

Advertising in general is a two-sided market, in the sense that there are platforms that attract viewers and sell their access to the advertisers. These typically multi-home and pay for access based on the amount of space they obtain and who they reach. Online advertising provides new ways to attract viewers, and new and better ways to link advertisers to their targets. This happens in the two main sub-markets of online advertising, search and display advertising, whose structures will be the focus of our attention.

Let us start with *search advertising*, which accounts for about 40 % of online advertising. Many search queries on the Internet are a potential source of business transaction, therefore advertisers place their ads next to search engine results through second price auctions on the keywords that match the content of their ads and lead Internet users to click on their advertisement: charges are typically for each click on the ad (PPC, price per click), and the highest bid for each keyword association wins, but the price is given by the second highest bid. Therefore, the revenue per search, RPS, on a particular platform is given by $RPS = APS \cdot CPA \cdot PPC$, where APS is the Ads per search and CPA is the clicks per Ad.[32]

The market is characterized by strong network effects in the sense that the search engine that reaches most Internet users is more valuable to advertisers. Consequently there is a strong competition for the market to develop the leading search engine, which leads the dominant firm to invest a lot in R&D and maintain its leadership, but a limited competition in the market between the main search engines, which allows these, and the dominant one in particular, to apply substantial mark ups. As well known, Google is currently the leading search engine, with about 60 % search traffic in U.S., against almost 20 % of Yahoo! and less than 10 % of Microsoft's Live, but with even higher market shares, often above 90 %, in other parts of the world.[33]

[32] Google adopts a sophisticated quality scoring technology that displays and orders ads based on expected revenue ($CPA \cdot PPC$) to maximize RPS; Yahoo! and Microsoft have introduced sophisticated quality scoring algorithms more recently.

[33] The exceptions are Japan where Yahoo! is the leader, China with Baidu, South Korea with NHK, and Russia with Yandex. Notice that Yahoo! has been the

Competition *for* the market is strong and crucial since access to search engines is free and simple, and most users employ the search engine that is regarded as the most valuable (for the quality of the search results and ad relevance), even if they often use more than one. In the absence of substantial product differentiation (limited to search tools and page layouts), network effects and lock in effects naturally lead to a single dominant player in the market for search advertising. Today Google dominates this lucrative business and the lack of genuine entry pressure in the competition *in* the market allows it to apply higher mark ups to advertisers than its main competitors. Google AdWords (launched in 2000) charges double prices than its two main rivals (Yahoo! and Microsoft) and accounts today for about 70 % of search advertising revenue worldwide.

Let us move to the second main field of online advertising, *display advertising*. Part of this market concerns the direct placement of banner ads on third-party publishers, which represents the direct (or reserved) channel, the valuable ad inventory that large web publishers directly negotiate with the advertisers (through their direct sale forces). However, most of the advertising space available on large websites and typically all of the space available on small and medium size websites cannot be sold in direct negotiations. Therefore, most of it is sold through indirect intermediaries that buy the so-called "remnant" ad inventory from publishers and sell it to advertisers. This can be seen as a separate market from the direct channel. Google and its recent acquisition DoubleClick play a major role in the market for intermediation services for remnant ad inventory. Google provides vertically integrated intermediation platform between online web publishers and advertisers: Google's AdSense, which publishers use to dedicate ad space to Google contextual ads, reaches more than 80 % of the ad revenue in the indirect channel with integrated ad networks. The Google platform targets advertising to the relevant websites, and pays the web publishers with a large percentage of its revenues; meanwhile advertisers buy inventories from the platform. DoubleClick offers an ad serving and ad management product, DART, for publishers and advertisers. Such a publisher tool manages the inventory of a website, receives the ads from ad networks and delivers them in the relevant inventory (according to the behavioral history of Internet users), usually at a fixed cost per thousand impressions which is a small percentage of the price that the web publisher charges on the advertisers. The market share of ad revenue served by DoubleClick in the indirect channel with non-integrated ad networks is around 75 %. Since almost 60% of online advertising taking place through the indirect channel adopts integrated intermediation, after the 2007 merger, Google with DoubleClick controls about 80 % of the worldwide market.

---

leader in the U.S. until 2002. Before 2000 Yahoo! was followed by Altavista, in 2001 by Microsoft and in 2002 by Google, which subsequently took the lead.

6.4.2 Google economics

As the market for search advertising, also the market for display advertising is characterized by endogenous entry in the competition *for* the market, but not in the competition *in* the market, with Yahoo! and Microsoft providing the only alternatives to Google in the short-medium run.[34] In this scenario, the merger between Google and DoubleClick substantially relaxed price competition and allowed them to increase their mark ups on publishers and advertisers. As a reaction, in February 2008 Microsoft made an offer to buy Yahoo! and create a stronger search platform with the size and the network effects needed to compete with Google. Yahoo!'s board has stopped this attempt by exploring strategic alternatives, including a commercial deal with the market leader (to deliver relevant Google ads alongside Yahoo!'s own search results), which would have reduced competition even further.[35] After the withdrawal of that deal in November 2008 under the pressure of the U.S. Department of Justice, the position of Yahoo! in the market has

[34] Switching to a different tool involves high sunk costs in terms of substantial investments in software, in training the staff, coding all of the publisher's web pages, creating novel datasets, transferring ad campaigns to the system and so on, with all the associated business risk. For the same reason, multi-homing (with multiple non-integrated ad networks) is highly inefficient in this case. These high switching costs and the difficulty of building alternative high quality intermediation services represent a substantial barrier to entry of new firms in the short and medium run, which is the relevant time frame in such a rapidly evolving market.

[35] In an oped for the british newspaper *The Scotsman* (April 30, 2008), the author of this book has expressed concern for the antitrust implications of Google moves, starting from the merger with DoubleClick: "It would be odd if this did not lead to price increases: before the merger, competitive forces kept online advertising rates under control (DoubleClick could not increase prices because many consumers would have quickly switched to AdSense, and Google was similarly disciplined by the prospect of customers switching to DoubleClick's products); following the merger, these competitive constraints no longer apply. For consumers and advertisers, the consequences of the current consolidations are uncertain. A merger between Yahoo! and Microsoft would create synergies in research and development without affecting the prices of either company's main products, which are complements rather than substitutes. It would allow the two to join forces and develop search engine capabilities and online services that could constitute a genuine, competitive alternative to Google ... By contrast, a Google/Yahoo! tie-up, or even limited outsourcing of advert placement by Yahoo! to Google, such as that announced by Yahoo! two weeks ago in the US, would radically reduce competition, while generating no significant 'efficiencies' to benefit advertisers and consumers. Any combination of the two would likely violate EU anti-monopoly rules, with an outsourcing deal that sidelines Yahoo! as a competitor allocating approximate 90 per cent of search advertising to Google, virtually ending prospects for competition. Even more importantly from an anti-monopoly point of view, locking Yahoo!'s search query share and online traffic into Google's ad platform would ensure that no-one could reach the scale necessary to mount a credible competitive alternative." See also Etro (2009,b).

become even weaker, and Google has remained dominant and unconstrained by endogenous entry threats.

Future alliances and technological advances will crucially shape the structure of this crucial market for the global economy. Competition and innovation (by the leader and the followers as well) in a sector like this will exert a positive impact on the other sectors and on the aggregate economy, especially in a time of crisis as the current one. For this reason it is important that a strong innovative pressure on Google remains in the competition *for* this market, and that a stronger competitive pressure emerges in the competition *in* the market of online advertising.

6.5 Dynamic Inefficiency

How much a country should save? By saving a lot a country can invest a lot and substantially improve the conditions of the future generations, but meanwhile consumption of the current generations is penalized. By saving less, a country privileges current consumption over growth. Apparently the right choice is just a matter of preferences, but this is not always true. There is a limit beyond which a country should never go: excessive savings may create such a large potential production that the country will not be even able to absorb it in the long run (Phelps, 1961). In these situations, the country could make current and future generations better off by simply reducing the rate of savings (this would expand current consumption without penalizing future consumption). Dynamic inefficiencies of this kind are only a theoretical possibility in the neoclassical framework, but they become a concrete chance in a world where markets are characterized by more realistic EMSs.

When we depart from a perfectly competitive world to introduce imperfect competition, we can expect that some forms of inefficiencies emerge. Nevertheless, it is not obvious in which directions they should go in general equilibrium: in the competition in the market, should we expect excessive concentration and mark ups? or too many firms and too small mark ups? Likewise, in the competition for the market, should we expect a number of firms and an aggregate investment in R&D above or below the efficient levels? The general answers to these questions are much more clear-cut than what one could expect on the basis of casual intuitions. It turns out that under fairly general conditions (see Chapter 3 and 5), the number of firms is excessive both under competition in the market and for the market and, under more restrictive conditions, this form of excessive entry may lead to persistent negative consequences, namely dynamic inefficiency.

The excessive entry result is related to the fact that, whenever there are few firms, entry of a new one exerts a strong negative externality on the profitability of the incumbents, but this externality is irrelevant for the actual choice of entry. This phenomenon is well known in the theory of industrial organization as the business stealing effect, but it makes its appearance in a general equilibrium context with implications for long run consumption levels exactly when we introduce EMSs, that is strategic interactions and endogenous entry.

The dynamic inefficiency result emerges in the presence of competition in quantities in the market when the agents are extremely patient. In such a case, agents save a lot, which leads to excessive investments in business creation and to the proliferation of too many small firms. Consider the particular case of firms producing homogenous goods: the creation of new firms is a poor way to increase production (relative to an increase in the individual production of the existing firms), since it requires a pure waste of resources in fixed costs of entry. Because of this waste, consumers could be better off reducing their savings to increase current consumption without reducing long run consumption. A similar phenomenon takes place in the competition for the market: savings are wasted in excessive creation of firms duplicating investments in R&D to replace each other, and a reallocation of resources could increase current consumption without penalizing long run growth. Of course, when investment creates steady growth, the damages of the inefficiency of the market structure are quite dramatic. Countries characterized by excessively small firms could grow at higher rates if they experienced a process of concentration leading to larger firms. Summarizing, we have:

PRINCIPLE 5. IN THE LONG RUN, THE EMSS CAN BE CHARACTERIZED BY DYNAMIC INEFFICIENCY IN TERMS OF EXCESSIVE SAVINGS AND TOO MANY SMALL FIRMS ACTIVE IN THE COMPETITION IN THE MARKET AND FOR THE MARKET.

Italian SMEs: too much of a good thing?. While the problem of the size of firms may appear a theoretical curiosity rather than a concrete threat, a possible example of this pathology in the mechanism of business creation has emerged in a country traditionally characterized by a high savings rate and whose growth performance has been quite poor in the last two decades: Italy. The industrial structure of this country has been traditionally characterized by a large number of SMEs whose innovative capacity is quite limited (compared to larger corporations) and whose focus, in the last two decades, has gradually moved away from the top high-tech sectors (at least relative to other developed countries and with relevant exception of the automotive sector). The reasons for the lack of growth in the size of Italian firms have been usually associated with the family-based structure of Italian capitalism or with credit rationing problems, but the endogenous tendency toward small innovative firms suggested here may be part of the story (also because one of the possible solutions, that is R&D subsidization, has been always limited in Italy compared to other Western countries).

It is probably not by chance that the Italian endogenous response (without proper equivalents around the world) to this problem has been the coordination of some innovative activities through "*industrial districts*", that is organizations of multiple SMEs from the same sector, typically located in the same geographical area, to invest in R&D and other productivity enhancing activities on a larger scale.[36]

However, we cannot avoid to stress that the poor growth performance of Italy has other fundamental sources as well. The social and cultural environment has scarcely rewarded risk taking, mobility and meritocracy, while it has protected rents and gerontocracy (and sometimes even nepotism). The political class has been quite slow in the introduction of reforms aimed at enhancing competition, reducing barriers to entry, increasing labor market flexibility and reacting to the challenges of globalization especially in the less developed southern regions. Strong ideological labor unions at the national level and a parliamentary system unable until recently to express coherent governments with clear agendas have been additional burdens in the reforms' path. Also the large size of the public sector and the inefficiency of its bureaucracy have heavily handicapped the business activity. Last but not least, investment in human capital and in R&D have been well below the European average and scarcely promoted.

Italy faced the current crisis with a high private savings rate (corresponding to a high net wealth of households), a stable real estate market,[37] a low unemployment level, a dynamic export sector focused on luxury goods, and a solid banking system which avoided substantial consequences of the financial turmoil, but the structural weakness of its economy led it into negative growth ahead of the others. The Berlusconi government has not followed the expansionary fiscal policies of other countries to limit the increase of the debt-GDP ratio, which is already large, but this may end up being a wise choice for the future as long as the promised path of structural reforms will be pursued. These reforms should be aimed in the medium-long run at reducing the size of the public sector on one side and labor taxation on the other side, at eliminating residual rents associated with labor protection and pensions' treatments while introducing general unemployment benefits, and at strengthening market liberalization and subsidizing firms that invest in R&D and export.

6.6 Fiscal Policy

The departure from a perfectly competitive environment and the emergence of inefficiencies lead naturally to a new role for fiscal policy in the EMSs

[36] Notice, however, that the lack of large corporations in high-tech sectors is also a problem in other European countries compared to the United States.

[37] During 2008, only Italy and France recorded a small increase in housing prices, against -18.2% in the U.S. (Case-Shiller national index), -17.6% in the U.K., -2.5% in Germany, -1.8% in Japan (*Economist*, March 21st, p. 79). We expect a downward correction for Italian and French real estate prices.

approach (see Chapters 3 and 5). For instance, as a consequence of the dynamic inefficiency emphasized in the previous section and due to excessive investment in business creation, our theory suggests that it would be optimal to correct this steady state distortion with a production subsidy that leads firms to increase their individual production, and an R&D subsidy that leads firms to invest more in innovative activities. The former instrument should be associated with a wage subsidy to implement the optimal allocation of resources (investment and labor) in the competition in the market. The latter instrument should be associated with an appropriate capital income or profit taxation, or with an appropriate entry fee/subsidy to target the right number of firms active in the innovative activity and reach the optimal growth rate.

However, fiscal policy plays a more interesting role outside of the steady state. The reason is that, contrary to a New-Keynesian environment where firms behave as independent monopolists, strategic interactions and endogenous entry lead to a complex impact of the fiscal tools along the transition path to the steady state. In particular, fiscal policy needs to change along the gradual process of entry because the inefficiencies in the allocation of resources are deeper when the number of firms is low and they are smaller when the number of firms is closer to the efficient level. The same efficient number of firms trades off the advantages of increased product variety and quality with the disadvantages of the costs of business creation. When competition in the market is still characterized by a small number of firms it is optimal to subsidize production and labor supply, or to reduce sales taxes and wage taxes. Since the number of firms is positively correlated with output along the transition path and over the business cycle, this implies that fiscal policy and also the optimal tax rates should be countercyclical.

Apparently, the result on the countercyclicality of the optimal fiscal policy is in line with a wide consensus in both the neoclassical and Keynesian approaches. However, here it derives from different reasons and it has more radical implications.

In the Keynesian approach it is the stabilizing role of government intervention on the demand side that leads to a countercyclical fiscal policy implemented with deficit spending in recessions and budget surpluses in boom (Keynes, 1936). In the neoclassical approach, a countercyclical fiscal policy is the result of the principle of tax smoothing, for which constant tax rates are optimal to minimize the tax distortions on the supply side, so that the public deficit increases in recessions and decreases in booms (Barro, 1979).

In the EMSs approach the distortions are endogenously induced by the strategic interactions and by entry choices in decentralized markets, they get worse in recession when competition is weaker, and they are reduced in boom, when competition is stronger. Therefore, the optimal fiscal policy minimizes these market distortions with an expansive role that must be stronger in recession and weaker in boom (under both lump sum taxes and distorsive taxes). In other words, the optimal fiscal policy has to minimize the market distortions rather than the tax distortions, and it has also to stabilize the economy through the supply side rather than the demand side. Finally, the optimal tax rates on sales and labor income should be reduced during recessions and increased during booms: this result on countercyclical tax rates is in contrast with the traditional tax smoothing principle and in favor of an active policy on the supply side. We can summarize these results as follows:

PRINCIPLE 6. THE OPTIMAL FISCAL POLICY REQUIRES COUNTERCYLICAL TAXES ON PRODUCTION AND LABOR, R&D SUBSIDIES AND CAPITAL INCOME TAXATION.

The general necessity of capital income taxation or profit taxation emerges in an endogenous growth context, where fiscal policy has to target both the efficient EMSs and the efficient long run growth rate. This result is in radical contrast with one of the main policy prescriptions of the neoclassical model, for which steady state capital income taxation should be zero. In the neoclassical framework the taxation of the return on capital affects the marginal productivity of capital and distorts capital accumulation, therefore it is optimal to avoid it in the long run. In the presence of EMSs, the taxation of capital income affects the net return of investment in business creation, and a reduction of this return obtained through this form of taxation is beneficial when decentralized entry is excessive and needs to be limited.

6.6.1 Fiscal policy in the U.S. and the E.U.

In the last decades U.S. fiscal policy has been often used in a countercyclical way, leading to surpluses during booms and deficits during recessionary phases. The most recent examples were the tax cut adopted by Bush at the beginning of 2008 and the recovery plan adopted by Obama at the beginning of 2009, the largest American fiscal stimulus package of all times.

At the beginning of 2008 the widespread dispersion of credit risk and the unclear effects of the subprime crisis on the financial institutions reduced the incentives to invest and to demand (and obtain) credit. Commercial papers for the finance of corporate business (mainly firms' working capital) collapsed,[38] leading to negative implications for business and job creation. The risks of further impact on the real economy and the real estate downturn were primary determinants of the Economic Stimulus Act signed by the Bush Administration. This $152 billion package introduced tax rebates to low- and middle-income U.S. taxpayers and tax incentives to stimulate investment, and it relaxed prudential controls over government-sponsored mortgage lenders with the purpose of keeping real estate finance flowing. Unfortunately, the margins to stimulate aggregate demand were quite limited. Given the high current account deficit of the U.S. and the weak Dollar of

[38] Between the third quarters of 2007 and 2008 U.S. nominal GDP was still growing by 3.4 %, but commercial paper outstanding was decreasing by 25 %.

the time, the only chance for such a fiscal package to stimulate the economy was by gradually restoring the incentives to create new business or to export. This did not happen because, as we have seen, the financial crisis got worse in the second half of 2008.

In front of such a dramatic scenario, the new President Barack Obama has launched a massive Recovery Plan of $ 787 million spread over 2009 to 2013, including mainly new public spending in education, health care, energy and infrastructures, a tax cut, an expansion of unemployment benefits and other social welfare provisions. The American Recovery and Reinvestment Act is clearly inspired by a Keynesian philosophy and aimed at raising aggregate domestic demand at the cost of increasing the public deficit (and adopting some protectionist measures, as "Buy American" policies).[39] The problem is that given the already high propensity to consume of the American citizens, an additional stimulus to private consumption may be partially ineffective: rather than spending, most Americans will simply repay their debts or start saving. It is true that a plan of additional public spending may increase aggregate demand, but on one side the associated additional burden of public debt may crowd out part of this increase, and on the other side the marginal return of additional public spending may be quite low in terms of business creation (relative to the return of private spending). Finally, it takes time before a fiscal stimulus exerts any impact on the economy, especially when largely based on long run public investments.

The IMF estimates that the size of the U.S. stimulus directly active in 2009 should be at least 2 % of GDP: together with the rescue packages financed by the government, this expansionary fiscal policy is expected to increase the U.S. public deficit beyond 13 % of GDP in 2009, and to substantially increase American public debt toward 100 % of GDP for the years to come. In spite of such an impressive and unprecedented policy of deficit spending, we doubt that this will be the key factor to trigger a quick recovery. Even if the financial crisis has not eroded the physical capital or affected productivity, we believe that an expansion of the aggregate demand alone is not going to bring production back to its full-employment level in a short time (contrary to what both the Keynesian approach and the neoclassical approach may suggest). The reason is that the structure of the aggregate supply changes during a recession: when net business creation falls, when firms restrict pro-

[39] In January 2009, a petition of economists including Kenneth Arrow, Guillermo Calvo, Lawrence Klein, Paul Samuelson and Robert Solow has endorsed the stimulus package, while Joseph Stiglitz and Paul Krugman criticized it for being too small. On the other side, a group of economists including James Buchanan, Edward Prescott and Michele Boldrin has signed an open letter (published by the Cato Institute) to President Obama to criticize the package: "Notwithstanding reports that all economists are now Keynesians and that we all support a big increase in the burden of government, we the undersigned do not believe that more government spending is a way to improve economic performance. More government spending by Hoover and Roosevelt did not pull the United States economy out of the Great Depression in the 1930s."

duction, when market structures become more concentrated and mark ups tend to increase, then the supply side is not going to satisfy anymore the same full-employment level of production as before. It takes time and further investments to recover the earlier supply levels, and a simple demand-based fiscal policy cannot succeed alone in the short term: there is the need for a supply-based fiscal policy.

Under the current conditions, we believe that it would have been better to focus the stimulus more on private investment, in particular enhancing the incentives to create new business. The U.S. have been consuming more than they were producing for more than a decade thanks to substantial imports: this imbalance needs to be cured by increasing production and exports (and not by decreasing imports with passive protectionist measures). Fiscal incentives for new enterprises, lower sales taxes and corporate taxes, export subsidies and R&D subsidies would have a better chance to promote the recovery in this moment.

As well known, continental Europe is characterized by a larger role of the government compared to the U.S., a richer (but sometimes less efficient) welfare state, stronger automatic stabilizers and more rigid labor markets with stronger unions. For this and other reasons, including differences in the political systems, the role of discretionary fiscal policy for stabilization purposes has been sometimes more limited. Moreover, the creation of the E.U. has set additional limits to the discretionality of national governments in adopting countercyclical fiscal policies (even if their rigid implementation has been avoided in the last years). The same E.U. has a limited spending capability and a limited role in coordinating fiscal policies, a role that would have been precious during the current crisis, as many observers have noticed.[40] The reaction of European governments to the recession has been weaker than in the U.S., with small stimulus packages in France and Italy (respectively 0.7 % and 0.3 % of GDP) and larger ones in Spain (1.1 %), United Kingdom (1.4 %) and Germany (1.5 %), but this does not take into account the role of automatic stabilizers:[41] therefore, also the European fiscal expansion has been in line with traditional Keynesian prescriptions of a substantial deficit spending.[42]

---

[40] On October 1, 2008 a group of economists including Alberto Alesina, Richard Baldwin, Willem Buiter, Francesco Giavazzi, Guido Tabellini and also the author of this book signed an open letter (published by Vox) to ask for European-level actions to supplement and coordinate national actions in facing the crisis.

[41] Data are from the IMF and refer to planned interventions directly aimed at stimulating the recovery in 2009. Large stimulus packages have been implemented also in Saudi Arabia (3.3 % of GDP), China (2.0 %), Canada (1.5 %), South Korea (1.5 %) and Japan (1.4 % but with more to come, in spite of the world largest debt-GDP ratio, above 170 %). Of course, additional (coordinated) efforts may emerge during the rest of the year.

[42] Overall, public finances are going to largely deteriorate since 2009, with public deficits expected well beyond the Maastricht limit of 3 % of GDP: around 10 % in U.K., Ireland and Spain (where, however, initial debt levels are low compared

Most European packages are mainly focused on supporting aggregate demand through new investments in domestic public infrastructures and support to domestic firms in bad conditions. Also in this case, limited attention has been given to the support of business creation and we are not aware of any attempt to reduce tax rates on the supply side even temporarily.

The IMF has calculated that more than a trillion Dollars will be concretely invested in stimulus packages worldwide during 2009 - much more has been (and probably will be) promised for the following years also in a coordinated manner. To have an idea of the size of these efforts, notice that 2008 worldwide nominal GDP is estimated in \$ 78 trillion, of which 18.9 produced in the European Union, 14.3 in the United States, 4.8 in Japan, 4.2 in China, 1.7 in Russia and Brasil, 1.6 in Canada, 1.2 in India, 1.1 in Mexico and one trillion Dollars in Australia.[43] This means that from a global perspective we are in front of a stimulus package of more than 1 % of world GDP, a percentage that may increase in the next months, possibly through international fiscal coordination. However, we believe that even the huge size of this unprecedented global stimulus is not a sufficient condition to trigger the recovery. What matters is the way this money will be spent, whether to push aggregate demand and public investments only (as now seems to be the case), or also to boost aggregate supply, business creation and trade.

6.6.2 A supply-based fiscal policy

The EMSs approach suggests exactly the necessity of an intervention on the supply side rather than on the demand side during recessions. According to the conventional wisdom, when the market activity declines one can stimulate the aggregate demand by artificially augmenting private and public spending, so as to force firms to produce more. This is the typical recipe given by (Keynesian) economists, but there is an alternative way to look at the problem and find solutions. Loosely speaking, market demand increases not only when available income is higher, but also when prices are lower (relative to the wages), that is when the mark ups are lower. In a world of constant mark ups (as the neoclassical one with perfect competition or the New-Keynesian one with monopolistic firms) a mark up reduction can never occur, but in a world with endogenous mark ups and entry this can happen whenever policy stimulates business creation. Any policy aimed at promoting entry and innovation is going to strengthen competition, reduce the mark ups and increase

to the Maastricht limit of 60 % of GDP), and on average above 5 % in the Euroarea.

[43] Data are from CIA World Factbook updated on March 19 at the official exchange rate. Within Europe, Germany has an estimated GDP of \$ 3.8 trillion, France 3, U.K. 2.8, Italy 2.4 and Spain 1.7. The differences are smaller after adjusting for PPP: the E.U. remains the largest world market with \$ 14.8 trillion, followed by U.S. with 14.3, China with 7.8, Japan with 4.3 and India with 3.3 (as a consequence of the current crisis, we expect that the catch up of China on U.S. will occurr at the beginning of the twenties, at least in PPP terms).

the real wages. This increases the aggregate supply and attracts demand. In front of the limited success of policies aimed at supporting aggregate demand to promote the recovery, we may start thinking seriously about policies aimed at supporting aggregate supply.

Fiscal policy can promote business creation acting either on expected profitability or on the fixed costs of entry. On the first element, one can act on corporate taxes, capital income taxation and other taxes whose incidence is born by the production side, including production and labor taxation. In particular, a heavy but temporary reduction of the indirect taxation on sales and direct taxation on profits could generate substantial positive effects on consumption and profits, and therefore on entry and production, and could induce significant reductions of the mark ups. Moreover, the impact on the economy of these tax cuts could be quite rapid, as opposed to the slow impact of increases in public spending (especially for infrastructures) and of cuts on general income taxes. On the second element of the supply side, the fixed costs of entry, one can act on subsidies for business creation, on temporary reductions of regulatory constraints to new business activities, especially in key sectors (as the construction sector), on temporary tax exemptions for new SMEs, on R&D subsidies aimed at promoting new and innovative products and on export subsidies aimed at extending business activities abroad.

The need for a supply-based fiscal policy is by no means new. However, it has been largely neglected in macroeconomic theory, which has been often biased toward a demand-based fiscal policy in the Keynesian tradition (which assumes a flat aggregate supply equation) or toward a neutral tax smoothing policy in the neoclassical tradition (which assumes a vertical aggregate supply and limits the scope of fiscal policy to the minimization of tax distortions). In front of the policy ineffectiveness of repeated plans of fiscal stimulus based on support to the aggregate demand, it could be useful to point out the opportunity of a larger, and maybe complementary, support to the aggregate supply.

6.7 Monetary Policy

The neoclassical approach to monetary policy in a frictionless economy generates the neutrality of inflation for the real economy and suggests the optimality of a zero nominal interest rate associated with a negative inflation rate and a negative growth rate of money supply (Friedman, 1968). The New-Keynesian literature has shown that in the presence of monopolistic behavior by an exogenous number of firms and nominal rigidities in price-setting, the real allocation of resources is affected by inflation (see Mankiw, 1985, and Blanchard and Kiyotaki, 1987). In this environment, the optimal policy requires monetary authorities to actively manage the nominal interest rate committing to a rule that links the latter to inflationary expectations

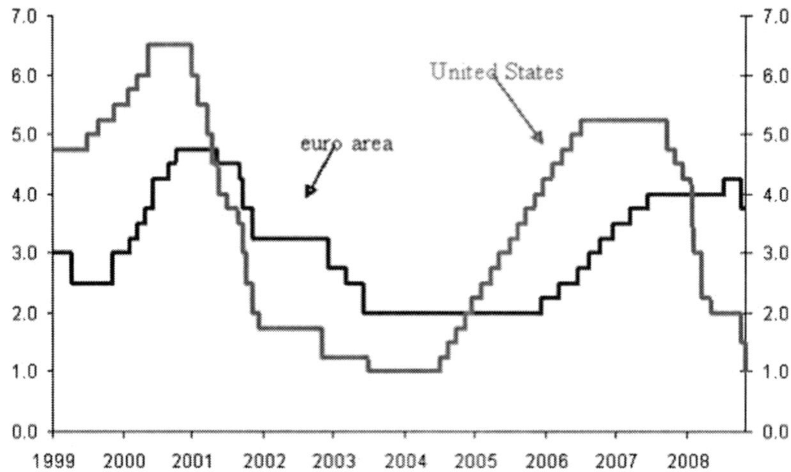

Fig. 6.7. Official interest rates of Fed and ECB (January 1999 - November 2008).

and the output gap, as in the Taylor rule (see Taylor, 1993, and Rotemberg and Woodford, 1997).

When the structure of the markets is endogenous, inflation and monetary policy have new roles compared to those emerging under monopolistic behavior by an exogenous number of firms. In particular, nominal rigidities affect the real profits of each firm with consequences both on the strategic interactions between them and, most of all, on the process of entry of new firms. The EMSs approach has emphasized static and dynamic consequences of this, studying the role of monetary shocks on the economy and revisiting the characterization of the optimal monetary policy.

The basic principle derived in Chapters 3 and 5 can be summarized as follows:

PRINCIPLE 7. THE OPTIMAL MONETARY POLICY HAS TO IMPLEMENT THE INFLATION RATE WHICH MINIMIZES THE DISTORTIONS ON THE PROCESS OF BUSINESS CREATION DUE THE NEGATIVE EFFECT OF NOMINAL RIGIDITIES ON THE EXPECTED PROFITS.

The new role of inflation and, consequently, of monetary policy emerges in a basic general equilibrium framework where inflation acts as a distortionary tax on firms profits and biases the allocation of resources between production of goods and business creation against the latter. Under these conditions, as long as fiscal policy can take care of the inefficiencies in the market structures, the optimal monetary policy should simply avoid mark up non-synchronization by implementing producer price stability (while leaving the mismeasured consumer price index to fluctuate because of the endogenous entry process).

Moreover, the EMSs approach could be useful also to study optimal monetary policy along the business cycle in the absence of optimal fiscal policy. Then, during recessions monetary policy should stimulate entry (which is below the optimal level) through reductions of the nominal and real interest rates aimed at increasing the (stock market) value of firms and promoting investments in business creation, and during booms it should limit excessive investments in business entry with a tight monetary policy. In practice, since output stabilization requires the stabilization of the entry process and, therefore, of the (stock market) value of the firms, we suggest that including equity price stabilization in the targets of the monetary authority could be useful (indeed many central banks have been occasionally active in the stock market with this purpose in the last years).

When growth is endogenous and depends on technological progress, inflation plays a similar role in distorting the business creation process, but it now leads to more radical consequences. In a Schumpeterian world, what drives the aggregate investment of the firms in R&D is the profitability of the future innovations, which depends on the level of protection of the intellectual property rights, but also on the appropriability of the rents from innovation through profit maximizing strategies. When nominal price changes are costly, profit maximizing strategies cannot be implemented in a systematic way. Firms choose mark ups that are increasing in the expected rate of inflation between the current price adjustment and the next one, but their profits are gradually eroded until that adjustment occurs. Even if the negative impact on the real profits is small because price adjustments are frequent, this impact negatively affects the incentives to invest for all the firms, reducing long run growth. Of course, this induces a non-negligible effect on the aggregate economy over time: accordingly, non-zero inflation decreases the rate of technological progress with permanent consequences on welfare. Therefore, when the objective is simply to maximize growth, price stabilization remains optimal.

More generally, the distorsive role of inflation on the incentives to invest in R&D induces an inverse-U relation between inflation and growth: in particular, and in line with the evidence, for moderate and high levels of inflation, the growth rate is decreasing in the inflation rate because the incentives to invest in innovation are reduced. Notice that, when the decentralized rate of growth under price stabilization is not optimal (as typically is the case), and there are distortions that cannot be fully eliminated through the fiscal incentives, a non-zero rate of inflation can be welfare maximizing. This outcome is more realistic since Central Banks around the world, including the Fed and the ECB, target low but positive inflation rates, and not zero inflation.

6.7.1 Monetary policy in the U.S. and the E.U.

In the last two decades most monetary authorities have formally committed to anti-inflationary policies, often adopting explicit inflation targets. Nevertheless, some of them have been also engaged in policies that were clearly aimed at output stabilization. These policies have been typically implemented through increases in the nominal interest rates in front of inflationary expectations and reductions of the same rates in front of reductions of the inflation. According to the leading view, when the reactions of the nominal interest rates are strong enough (according to the Taylor rule, a 1.5 % change of the interest rate for a 1 % change of the inflation rate), they affect the real economy through the impact on the real interest rate (which is the difference between nominal rates and expected inflation). For instance, in front of increased inflationary expectations, a temporary increase of the nominal interest rate, which increases the real interest rate as well, is expected to reduce current consumption (and investment in business creation), which slows down the economy and tends to reduce the inflation. On the other side, in front of a slowdown of the economy, a reduction of the interest rates is expected to promote consumption (and business creation) so as to trigger the recovery.

With the Chairman Alan Greenspan and his follower Ben Bernanke, the U.S. Federal Reserve has been quite active in the stabilization of the American economy,[44] implementing a tight monetary policy in booms and an expansionary one in recessions. For instance, a drastic reduction of the interest rates has been implemented in the aftermath of September 11, and another one during the last two years to contrast the current recession, arriving to nominal rates close to zero (see Figure 6.7). After reaching this lower bound of the interest rate policy (nominal interest rates cannot be negative), a further expansionary policy requires a direct increase of money supply. Therefore, the Fed has implemented a form of "credit easing" by pumping new liquidity into markets. In the last months, it has expanded its discount operations in particular with the creation of special loan facilities (as the Term Asset-Backed Loan Facility), has issued direct injections of capital into the main banks,[45] some of which are still in troubles, has promoted direct lending from government-sponsored enterprises and has bought corporate debt. The hope is that this aggressive intervention will manage to re-start the process of investment and business creation, and with it the recovery.

Following the Fed, all the main central banks have reduced their nominal interest rates. The Bank of England, led by Mervin King, has been the first monetary authority to announce a policy of "quantitative easing", that is to buy long-term government bonds (for planned £ 75 billion) and, at the beginning of March, it launched a reverse auction with investors as sellers, rather than buyers, of U.K. "gilts" to the central bank. In mid March, also the Fed announced it would buy long term U.S. Treasury bonds (starting

---

[44] See Greenspan (2007) for a fascinating account of his reaction to the stock market crashes of 1987, 1998 and 2001.

[45] Bank of America, Citigroup, JP Morgan Chase, Goldman Sachs, Morgan Stanley, Bank of New York Melllon and Wells Fargo.

with $ 300 billion). The European Central Bank, led by Jean-Claude Trichet, has reduced the interest rates in a less aggressive way (see Fig. 6.7), and has not adopted forms of unconventional monetary policy until now. In particular, the European monetary authority has not been engaged in outright purchases of private securities or unsecured lending to the private sector[46] or in purchases of public debt, but has adopted a different approach: it makes available unlimited credit to banks at the official rate (at 1.5 % at the time of writing, to be reduced in the next months) with short term maturities (up to six months at the time of writing, instead of one week). This credit is provided against eligible euro-denominated collateral defined in a very liberal way, so that there is a large short-term liquidity in the Euroarea and the unsecured overnight interbank rate is quite close to the American one. As a result, money supply has been growing at substantially high rates, especially for the Dollar and the Pound. Nevertheless, at the time of writing monetary policy appears to have no traction at all on the real economy.

In front of the apparent resistance of the real economy to react to these forms of monetary stimulus, two are the possible motivations. The simplest one is that after all monetary policy is not as effective as New-Keynesian theory would claim, for instance because the reactions of the monetary authorities are by now fully expected and already discounted, and therefore cannot be effective (Lucas, 1972; Sargent and Wallace, 1975): if this is the case and the real economy follows its path in a way that is largely independent from monetary actions, central banks should focus on controlling the inflation rate around a favorite level, rather than taking the risk of creating undesired inflation in the medium run. The alternative hypothesis is that a suboptimal policy has been implemented. According to many economists,[47] during the last decade the Fed set excessively low interest rates compared to the optimal Taylor rule, and did not contrast the equity boom at the end of the 90s with a properly tight policy. This would have been at the roots of the current crisis inducing excessive debt and risk taking within the economy and generating the well known problems of the financial sector. Moreover, the low interest rates kept penalizing savings and postponing a solution to the current account deficit problem, which remains a critical aspect of the American situation. On the top of this, the Fed is now increasing money supply at a very high rate, which may create substantial inflation and depreciation of the Dollar in the medium term (see Section 6.9), and may also change the role of the central bank in harmful ways in the absence of a quick recovery. According to John Taylor, "the success of monetary policy during the great moderation period of long expansions and mild recessions was not

[46] The rationale behind this has probably to do with the different structure of the European financial system compared to the Anglosaxon one: European firms and households rely more on banks than on capital markets, so it is better to lend freely to banks so as to help cap lending rates rather than to lower the cost of capital in securities markets.

[47] See Taylor (2009).

due to large discretionary interventions, but to following predictable policies and guidelines that worked."[48]

Contrary to the Fed, the European Central Bank has traditionally followed a less aggressive management of the interest rates to stabilize the economy, paying much more attention to the control of inflation in the Euro area. Even at the beginning of the financial crisis, European rates remained above the American rates for a while. Only during 2008, when residual inflationary pressures were over, the European monetary authority has started reducing the interest rates to contrast the recession, and a policy of quantitative easing has not been adopted yet. This is not surprising for a more heterogenous area where opposite shocks often occur in different member countries and where the monetary authority is largely independent (and not even backed) from fiscal authorities. In the second part of the year the European Central Bank may contemplate following the Fed and the Bank of England to buy private securities or government bonds (and it would be politically hard to decide which member countries' bonds to purchase), but only in case of a further deterioration of the recession and concrete possibilities of deflation.

Many economists have claimed that the current recession undermines the relevance of the neoclassical approach to macroeconomics and can only be explained within the (New-)Keynesian approach. However, we believe that the real test of the Keynesian approach will emerge from the success or the failure of the demand-based policies that are currently implemented to trigger the recovery, both on the fiscal and monetary front. As of now, the results are poor, exactly as they have been during the "lost decade" of ineffective expansionary fiscal and monetary policies in Japan. Time will give its verdict.

6.8 Trade Policy

The implications of the EMSs approach for trade policy are against many forms of passive protectionism, including import tariffs and quotas, which are aimed at restricting trade volumes. However, under certain conditions, they are in favor of what could be called an active protectionism, based on strategic export promotion, which may distort trade, but does not restrict trade volumes.

Our analysis of trade policy confirms that positive import tariffs represent the optimal unilateral policy for the domestic market both in the traditional case of an exogenous number of firms and in the case of an endogenous number of firms competing in the domestic market. In both cases, the optimal tariff tends to zero under perfect competition, that is when the (exogenous or endogenous) number of firms increases. In this sense, the qualitative predictions of the neoclassical trade policy are not changed by the endogeneity of

[48] *Financial Times*, "The Threat posed by Ballooning Federals Reserves", March 24th, 2009, p. 9.

the market structures. Neither are the standard critiques to these predictions: import tariffs often lead to retaliation, and equilibria with multiple countries adopting tariffs and quotas end up reducing the gains from trade. Passive protectionism ultimately hurts global welfare.

In spite of this, under certain conditions, the EMSs approach supports active forms of promotion of domestic firms in the international competition, as export promotion, export subsidies, R&D subsidies and protection of intellectual property rights for the exporting firms. It is interesting to focus on this novel aspect starting from the rationale for export subsidies.

What is the optimal trade policy with respect to exporting firms? How much should we invest to promote international demand of domestic products? There is a lot of debate about these questions between policymakers. This is not surprising since also at a theoretical level there are not clear or unambiguous answers. Common wisdom on the benefits of export subsidization largely departs from the implications of trade theory. While export promotion is often seen as welfare enhancing at least in the short run and often supported by governments, theory is hardly in favor of its direct or indirect implementation. In the standard neoclassical approach with perfect competition, the scope of trade policy is to improve the terms of trade, that is the price of exports relative to the price of imports, and, as long as a country is large enough to affect the terms of trade, it is optimal to tax exports (since this is equivalent to set a tariff on imports). In case of imperfect competition, a second aim of strategic trade policy is to shift profits toward the domestic firms. Therefore, a large body of literature, started by Brander and Spencer (1985) and Eaton and Grossman (1986), has studied international markets with a fixed number of firms to determine the optimal profit-shifting policy. Also in this case the optimal unilateral policy is an export tax under price competition, because the tax induces the domestic firm to increase its price and relax competition, which augments its profits more than enough to compensate for the social cost of the subsidy. Under quantity competition, an export subsidy could be optimal, but only under restrictive conditions (on demand and on the number of domestic firms).

Even if the World Trade Organization forbids export subsidies (except for agriculture), different forms of direct or indirect export promotion are widespread. Governments strongly support exporting firms, they often hide forms of export promotion behind nationalistic pride, and consider the conquer of larger market shares abroad as a positive achievement in itself. The E.U. coordinates trade between its members and the rest of the world in a similar spirit, and subsidizes exports of agricultural products and the aircraft industry (Airbus is probably one of the main examples of strategic trade policy). France supports its "national champions" with public funding. Italy has a long tradition of public support of the "Made in Italy". Japan has adopted a policy of targeted export promotion through its Ministry of Economy, Industry and Trade. Korea and other East-Asian countries have implemented export promoting policies for decades. Heavily protected South-American countries have tried to subsidize manufactured products in which they could develop a comparative advantage (and not only those). Even U.S. has implemented strong forms of export subsidization through tax exemptions for a fraction of export profits, foreign tax credit and export credit subsidies.

As we have seen in Chapter 4, the EMSs approach provides a new theoretical argument for the general optimality of export subsidies whenever the domestic firms compete in international markets where entry is endogenous (notice that free entry is a realistic assumption since a foreign country without a domestic firm in the market can only gain from allowing entry of international firms). Under EMSs, export subsidization becomes the best unilateral policy under quantity and price competition. The intuition is simple. While firms are playing some kind of competition in the foreign market, the government can always give a strategic advantage to its domestic firms with an appropriate policy. When entry is endogenous, an export tax would lead the domestic firms to increase their prices or to reduce their production levels. On impact, this would induce the other firms to increase their prices or production levels respectively, so as to increase their profits. However, the increase of these profits would attract entry *ex ante* until any extraprofits were eliminated. As a result of this, the domestic firms would end up worse off and the policy would fail. To the contrary, it is optimal to adopt a policy that induces the domestic firms to be aggressive, that is to expand production or (equivalently) reduce prices. This behavior limits entry of international competitors and increases the market shares of the domestic firms. Such an outcome can only be induced by subsidizing exports. In Chapter 4 we have derived the optimal unilateral subsidy under different market conditions. It turns out that the optimal subsidy is inversely related to the elasticity of foreign demand: ironically, when goods are homogenous the optimal export subsidy is identical (but with an opposite sign) to the optimal export tax emerging in the neoclassical trade theory. In both cases, the distortions due to the policy increase with demand elasticity, therefore high elasticity recommends lower intervention. However, the neoclassical policy is aimed at increasing the price of exports, while the policy recommended by the EMSs approach is aimed at decreasing the price of exports.

The general lesson about trade policy can be summarized as follows:

PRINCIPLE 8. THE OPTIMAL UNILATERAL TRADE POLICY HAS TO PROMOTE DOMESTIC FIRMS COMPETING WITH FOREIGN ONES THROUGH EXPORT SUBSIDIES UNDER ANY FORM OF COMPETITION (WITH THE OPTIMAL EXPORT SUBSIDY WHICH IS INVERSELY PROPORTIONAL TO THE ELASTICITY OF FOREIGN DEMAND).

Against passive protectionism, in favor of export subsidies. As we have seen, the EMSs approach does not provide support for passive protectionism in the form of import tariffs or other trade restrictions, but it is in favor of export subsidies (and not generic state aid) because they do not restrict trade volumes, but they actually tend to increase them. The same argument can be applied to other forms of indirect export promotion, as policies which boost demand or decrease transport costs for the exporting firms: as long as these policies increase the marginal profitability of the domestic firms, there is a strategic incentive to use them unilaterally. Ultimately, the scope of export policy is just to conquer market shares abroad and shift profits from firms of other countries towards domestic firms. If we interpret globalization as the opening up of new markets to international competition we can restate the main principle as follows: in a globalized world, there are strong strategic incentives to conquer market shares abroad by promoting exports and adopting forms of active protectionism. The positive aspect of the latter is that it does not restrict trade, but it actually promotes it.

According to the World Trade Organization, the volume of world trade in merchandising is expected to fall by 9 % during 2009, the largest decline since World War II. Meanwhile, passive protectionism is rising and the fear is that it may aggravate the crisis, as it happened in the Great Depression, when the American Smoot-Hawley Act of 1930 increased U.S. import duties and provoked widespread retaliation, which reduced global trade by a quarter. Notice that at that time trade was less developed and was mainly inter-industry trade, while today it represents a larger fraction of GDP and it is mainly intra-industry trade: this leads to a deeper interdependence of economies and worse consequences of import tariffs and other forms of passive protectionism. We believe that in a period of crisis and globalization backlash as the current one there is an additional reason why certain forms of export promoting policies for high-tech and differentiated products could be welcome.[49] They would redirect the protectionist tendencies away from the adoption of import tariffs and quotas, toward a form of intervention (export subsidies) that can revitalize intra-industry trade and help the recovery.

The typical argument against foreign export subsidies is that subsidized foreign firms exert unfair competition against unsubsidized domestic firms. It really sounds like the typical argument in favor of passive protectionism: since more cost-efficient foreign firms exert unfair competition toward less cost-efficient domestic firms, we should adopt import tariffs. We believe that both arguments are flawed. In both cases, subsidized or more efficient foreign firms end up selling goods at lower prices with clear gains for the domestic consumers. The only difference is that in the former case foreign governments are paying for those gains, and in the latter case foreign workers are receiving lower wages to provide those gains: ultimately the costs are abroad and the gains are at home. Therefore, adopting import tariffs and forbidding export subsidies simply reduces consumer welfare to protect domestic profits. It is

[49] Notice that the WTO allows export subsidies for agricultural goods, that are usually homogeneous goods traded under perfect competition. We recommend the opposite policy: forbid (or at least reduce gradually) export subsidies in inter-industry trade and allow those in intra-industry trade.

Fig. 6.8. Euro-Dollar nominal exchange rate (January 2000-March 2009).

quite surprising that many economists and the same World Trade Organization keep condemning both import tariffs and export subsidies as if they were equivalent policies.

6.9 Exchange Rate Policy

In a fixed exchange rate regime, governments undertake occasional devaluations with the specific aim of supporting the competitiveness of domestic firms in international markets: for this reason we talk about competitive devaluations. In a flexible exchange rate regime this is not possible, but for the same reasons a depreciation of the nominal exchange rate, due for instance to an expansionary monetary policy, is expected to support the competitiveness of domestic firms exporting abroad.

In spite of this common wisdom, economic theory has been rather ambiguous on the role of the nominal exchange rate. In the frictionless neoclassical macroeconomic approach, exchange rate policy is neutral just like monetary policy, and it cannot affect real variables. In the Keynesian approach, as long as the devaluation improves the trade balance in the medium-long run, it boosts aggregate demand with an expansionary effect on the domestic country and possibly a negative effect on the trading partners. The result is confirmed within the modern approach of the New Open Economy Macroeconomics (Obstfeld and Rogoff, 1996), but not even under all circumstances (a devaluation could be even welfare decreasing when the reduction in the purchasing power of the domestic agents is strong enough). Finally, partial equilibrium models of competition in international markets with firms engaged in pricing to market (Dornbusch, 1987) show that devaluations can start price wars between firms active in different countries, which ultimately leads to lower profits for all of them. The intuition for the last mechanism is simple: after a devaluation, exporters reduce their prices in foreign currency thanks to the better exchange rate, but this induces their rivals to do the same in order to defend their market shares and profitability, with the consequence that all the firms end up with reduced prices and profits.

In front of these contrasting theoretical positions it is difficult to make sense of the common wisdom according to which devaluations provide a positive strategic advantage on the international markets with net benefits for the devaluating country. However, the EMSs approach provides a consistent rationale for this common wisdom by evaluating the strategic incentives to implement devaluations in a scenario where the incidence of exchange rate variations on prices is endogenous.

To obtain a comprehensive understanding of the impact of exchange rate variations on the market structures, we need to briefly review the role of exchange rates in affecting market competition. Imagine first a market in a foreign country in which international firms produce and compete with independent production units. This is typical of multinationals which are directly active in other countries where they sell their products. Under price competition, this case of local currency pricing with market power implies no pass-through of nominal exchange rate variations on prices. In this situation, a devaluation is not going to affect the equilibrium in the foreign market. All firms would choose the same prices in foreign currency after a devaluation, but the profits of the domestic firm would be artificially increased in the domestic currency. The same would happen under quantity competition, since production decisions abroad would be independent from the exchange rate again, but profits in domestic currency would be inflated by a devaluation. From a welfare point of view, the gains in profits from such a devaluation should be compared with the losses for the society in terms of higher prices of the imports. More importantly, in this situation there is not a strategic incentive to implement a competitive devaluation: this policy does not give a real strategic advantage to the domestic firm in the foreign market but just artificially increases its profits.

As we have seen in Chapter 4, a totally different situation occurs when all firms produce in their domestic country, bear production costs in domestic currency, choose their strategy taking into account the exchange rate and then export abroad (under price competition this corresponds to the case of producer currency pricing). Such a case is typical of SMEs which are active at a national level, often producing typical domestic products and exporting some of them abroad, but also of larger firms which are not directly active in the foreign market under consideration but sell their goods to distributors of this market. In this situation competitive devaluations are always desirable to provide a strategic advantage to domestic firms as long as entry in the international market is endogenous. The reason is that devaluations induce an aggressive behavior (lower prices or larger production) of the domestic firm in the international market, and this is the only way to shift positive profits at home when entry in those markets is endogenous. The domestic firm ends up with a larger market share and positive profits compared to the unsubsidized competitors. Finally, we need to remember that the competitive devaluation can be effective only in the short run, while inflation differentials tend to re-establish the situation pre-devaluation in the long run.

In a world where exchange rates fluctuate freely, our results need to be reinterpreted. In such a world, a depreciation of the domestic currency, due to any international or monetary reason, has a strategic impact on the exporting firms. Consider the likely case in which these firms export in foreign markets whose access is open to any international firm. As a consequence of the depreciation, the domestic firms are always led to reduce their prices in foreign currency, because the prices earned in domestic currency have been artificially increased and these firms can always earn more profits by reducing mark ups a bit to gain a larger market share. Some international firms will try to follow this price cut and the average mark ups in the industry will go down, with a consequent reduction in the number of firms able to remain in the market. Nevertheless, the domestic firms manage to earn larger profits in domestic currency thanks to the increase of their market shares.[50] The same identical result follows if firms compete in quantities: the domestic firms increase their production because the marginal revenues in domestic currency are higher, they gain market shares and earn more profits. More generally, depreciations provide a strategic advantage to the exporters under any market conditions as long as entry in the international markets is endogenous, and they always increase exports and market shares of the domestic firms in these markets. Of course, this has a positive effect on the trade balance and on the current account, at least in the short run, before inflation differentials neutralize the impact of the depreciation.

We can summarize our results as follows:

PRINCIPLE 9. IN THE SHORT RUN, A DEVALUATION OR A DEPRECIATION OF THE EXCHANGE RATE INCREASES MARKET SHARES AND PROFITS OF THE DOMESTIC FIRMS COMPETING IN INTERNATIONAL MARKETS WITH ENDOGENOUS ENTRY AND INCREASES ALWAYS DOMESTIC WELFARE.

The depreciation of the Dollar. An immediate application of our discussion on exchange rates concerns the effects of the depreciation of the Dollar between 2002 and 2008 (see Figure 6.8). This started in the aftermath

[50] The reader should keep in mind that in the absence of endogenous entry, the opposite result occurs: a depreciation leads the domestic firms to reduce their prices, which induces a price war in the foreign market. As long as entry is exogenous, the ultimate effect is simply to reduce profits for all the firms. Therefore, a depreciation would hurt exporters.

of the terrorist attack of September 11, 2001 and was strengthened by the expansionary monetary policy implemented first by Greenspan and later by Bernanke to contrast the downturn of the economy, especially during the first part of 2008, when the reduction of the U.S. interest rates compared to the European rates (see Figure 6.7), and a general tendency of the international central banks to diversify their monetary reserves contributed to weaken the Dollar in favor of the Euro.

The consequence of the depreciation of the Dollar was that U.S. firms started reducing their prices in foreign markets and gain market shares, sometimes inducing the exit of other firms from those markets. Meanwhile, European firms were forced to increase their prices in the U.S. and loose market shares, or to maintain similar prices and reduce their effective mark ups. In markets where firms price to market, price differentials between the same goods sold in Europe and the U.S. became substantial.[51]

At the aggregate level, the depreciation of the Dollar was slowly contributing to promote U.S. exports and reduce the huge American current account deficit, from a high of 6% of GDP in 2005 to 3.3% in the third quarter of 2008. However, the process of depreciation of the Dollar was inverted in the second half of 2008, when also the ECB started reducing the interest rates to face the crisis.

The current global scenario is characterized by a dichotomy between two groups of countries. On one side U.S. and U.K. (with few other European countries including Spain) running current account deficits that are driven by trade balance deficits an now also by huge public deficits. On the other side, countries as China, Germany, Japan and other energy-exporting countries running large surpluses, led by trade balance surpluses and now only partially reduced by larger public deficits (see Figure 6.9). The first group of countries has been characterized by low (or negative) saving rates of the private sector that only now are increasing, the second group by consistently high saving rates. This dichotomy has worked fine until recently, with net savers/exporters on one side financing net investors/importers on the other side, but such an unbalance may create sustainability problems now. Especially if deficit spending is stronger in the group of countries running current account deficits, as it appears to be the case.

To have an idea of the substantial difference between these groups, notice that in the last years U.S. household consumption was above 70 % of GDP,

[51] This is the case of luxury cars. In the summer of 2008, a new Porsche Carrera Convertible sold in the U.S. at about 94,000 $, could be bought in Europe (where it is produced) at about 106,000 €. At the exchange rate of 2002 the two prices would have been approximately equivalent, but at the moment of maximal depreciation of the Dollar, the differential was about 45,000 €! The same happened for other cars sent from the other side of the Atlantic: the price differential was 56,000 € for an Aston Martin DB9 and 40,000 € for a Maserati Gran Turismo. European car makers could have increased their U.S. prices, but they would have lost market shares: their best option was to sacrifice profits temporarily and wait for the depreciation of the Euro.

while Chinese consumption was about 35 % of GDP. As we noticed in Chapter 5, there are good reasons for which the largest country (U.S.) can specialize in technological investments and import final goods from the rest of the world while running a current account deficit. However, this cannot go on forever, in particular while another country (China) is growing at a much higher rate through excessive savings (probably beyond the border of dynamic inefficiency), and is expected to become the largest economy in the near future.

In front of the current situation, we need further international policy coordination (possibly with a bigger role for a reorganized IMF) or even rethinking the international monetary system. However, without substantial policy changes, the main correcting factor could be the exchange rate. As of now, America (as the U.K.) has implemented a massive monetary expansion, much stronger than in Europe. While the recession is now depressing inflation and in the short run prices are expected to be stable (or even declining in the U.S., where core inflation remains positive nevertheless), in the medium run this policy may create heavy inflationary pressures in America (and in the U.K.). This may lead to further depreciation of the Dollar (and the Pound), especially if China starts consuming more and/or lessening its wide Dollar exposure. This, in turn, may finally increase U.S. exports and reduce the global unbalance. In such a case, two scenarios are possible for the medium term, and they are equally likely as of now.

In the first (positive) scenario, we will arrive at inflationary pressures with an ongoing recovery of the real economy and of the stock market, which will boost both investment in business creation and consumption: this may be an easy way out from the recession because interest rates and taxes could be gradually increased to avoid excessive inflation and to keep public finances under control. However, notice that we may end up in a quite different world from the one we are used to: without U.S. imports driving foreign growth, but possibly with a pro-active role for Europe, which is now the largest integrated market in the world, and, mainly, with a new role for the emerging China, and especially for Chinese consumption and for Chinese investments in the Chinese economy (rather than in the American one, as in the last decade).

In the second (negative) scenario, the stock market will fail to recover enough (after a weak rebound) and investments and consumption will remain weak while inflationary pressures emerge, leaving monetary and fiscal authorities with two options. The first one will be to fight inflation back and increase taxes to keep deficits under control, so as to fall in another recessionary phase before recovering. The second option will be to inflate the economy and reduce both private and public debt by means of inflation. Also in this case, the growth model based on U.S. imports from the rest of the world would have to change drastically.

However, there is a third scenario for the medium term that we need to take into account. This is a scenario with an appreciated Dollar and a recover-

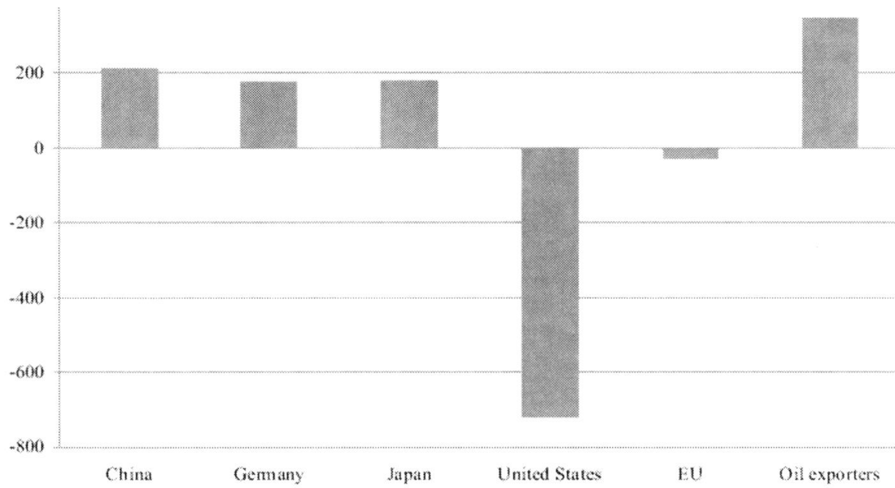

Fig. 6.9. World payment imbalances in Billions of U.S. $ (Average 2004-2007). Source: Di Noia and Micossi (2009)

ing stock market driving new investments in business creation, with an American demand-based fiscal policy that artificially sustains a weak consumption boom while increasing U.S. public and foreign debt, and with European and Chinese economies unable to drive global growth (respectively through innovations or expanding consumption rates). However, it is unlikely that such an unstable scenario could last for long.

6.10 Innovation and Competition Policy

The main message of the EMSs approach to the macroeconomy is that the structures of the markets and their determinants (of technological, behavioral, strategic and policy nature) are crucial for the evolution of the aggregate economy. For this reason industrial policy, including 1) innovation policy affecting competition *for* the markets and 2) antitrust policy affecting competition *in* the markets, must be taken in consideration when evaluating the general macroeconomic policy. This becomes more important for markets whose efficiency has a direct impact on other markets and whose technological progress is a main driver of the entire economy, namely high-tech markets.

We have already encountered a number of results concerning industrial policy. It is now time to put them together and draw the implications for the policies aimed at enhancing efficiency and growth in the global economy. The basic principle emerging from the EMSs approach can be stated as follows:

PRINCIPLE 10. THE OPTIMAL INDUSTRIAL POLICY HAS TO PROMOTE ENTRY IN THE COMPETITION IN AND FOR THE MARKETS, SUBSIDIZE R&D AND PROTECT IPRS.

In the following subsections we will briefly comment on innovation and competition policy issues, topics on which we have extensively discussed in Etro (2007,a, Ch. 5) and on which we return in this last part of the book.

6.10.1 Innovation policy and coordination

The general need to subsidize R&D and to protect IPRs departs from the ambiguous results of the traditional theory of Schumpeterian growth and derives from a result that we have emphasized repeatedly: in the presence of EMSs, firms tend to invest too little in innovation and they tend to be engaged in excessive duplication of their investments. This result is independent from the aggregate size of the investment in innovation, which could be above or below the efficient level: in either case, the organization of R&D could be made more efficient concentrating it in larger firms that invest more in innovation. R&D subsidies and a stronger protection of the IPRs can achieve the same outcome, but they do it in a different way and they should both be used.

A second implication of the analysis of EMSs in the competition for the market is that incumbent leaders tend to invest more than their rivals when they face a strong entry pressure. This leads them to innovate more frequently, to persist in their leadership position and to drive the growth process. Understanding the nature of this phenomenon is crucial to implement the proper industrial policy toward market leaders of dynamic sectors. Contrary to a populistic ideology, the EMSs approach has shown that the protection of IPRs is fundamental to promote sequential innovations, because it strengthens the incentives to invest by both the incumbent leaders and the followers. Therefore, a policy discriminating against the IPRs of persistent leaders can have deleterious consequences on the aggregate incentives to invest in R&D.[52]

R&D policy becomes even more important in a global perspective. When firms compete for international markets, their investments generate global growth, but each country tends to free ride in the implementation of the optimal R&D policy. This happens because each country subsidizes unilaterally

[52] Nevertheless, a major economist as Stiglitz (2006) keeps being in favor of firm specific policies, claiming that a solution to the "problem" of the persistence of the leadership of Microsoft in the software market "might involve limiting Microsoft's intellectual property protection for its operating system to, say, three years. That would provide strong incentives for it to provide innovations of the kind that users value and for which they would be willing to pay. If it failed to innovate, others could innovate off its old operating system - it would become a free platform, on top of which innovations in applications could be built". It can be extremely dangerous to neglect the aggregate consequences of this kind of policies.

its domestic firms with the purpose of shifting expected profits toward home and not with the purpose of promoting global growth. Alternatively, countries tend to protect more the IPRs of the domestic firms than those of the foreign ones, sometimes with a discriminatory use of antitrust policy. These tendencies lead to suboptimal investments in new technologies at the global level.

International coordination for the protection of IPRs and the coordination of fiscal policies to promote R&D are required to improve the allocation of resources at the global level. While some countries have been trying to coordinate the support of R&D activities at a supra-national level, heterogeneities in policy views and lack of binding commitments have undermined these efforts (in the case of the European Union, think of the Lisbon Agenda or the attempts to harmonize patent protection between the member states). Nevertheless, further coordination for the protection of IPRs could enhance the global incentives to invest in R&D in a substantial way. Additional delegation to an international organization may be difficult because there are still different views on innovation policy (with the U.S. much more in favor of the strengthening of IPRs protection than the E.U.). However, as in the case of other forms of coordination, it would be efficient to establish international standards for R&D promotion and subsidization, probably at the level of the TRIP agreements, leaving individual countries to do more.

6.10.2 Competition policy in the U.S. and the E.U.

The promotion of competition in the market and the reduction of entry barriers are crucial elements for the proper functioning of market economies and become particularly important in a period of crisis, when tendencies to relax competition and increase mark ups emerge naturally.

Antitrust policy plays a fundamental role in this sense, especially with its action to deter collusive behavior aimed at increasing mark ups and to stop mergers than can be detrimental to future competition.[53] On this front both the American and European antitrust authorities have adopted a similar and wise approach, focusing on markets characterized by barriers to entry and not on markets where endogenous entry forces can neutralize attempts to exercise market power. It is important to notice that the borderline between price-fixing and merger cases that do and do not require intervention is not an absolute one, but it depends on the macroeconomic conditions: in times of crisis, especially when dealing with failing firms, antitrust intervention should be reduced.[54]

[53] European coordination involves an additional aspect which is quite important in a period of crisis, the control of government intervention that limits competition in the European markets.

[54] Nevertheless, the crisis has made clear that there is a particular market, the financial market, where deregulation has been excessive. This market requires new rules at the global level, at least to coordinate general principles of micro-

The main differences between U.S. and E.U. industrial policy emerge in the general approach to market dominance and in the antitrust treatment of monopolization issues, which are extremely important not only for their impact on the effectiveness of competition, but also for their possible interference with aggressive competition (which is often borderline with abusive practices) and with innovation policy (which must protect some degree of market power to guarantee the proper incentives). The different approaches are well illustrated by the report issued by the U.S. Department of Justice in September 2008, "Competition and Monopoly: Single-Firm Conduct under Section 2 of the Sherman Act", and by the "Guidance on the Commission's Enforcement Priorities in Applying Article 82 to Abusive Exclusionary Conduct", issued by the European Commission three months later. These documents contain the general principles that guide the two authorities in deciding which cases to pursue and how to deal with specific types of conducts.

The American approach emerging from the Report is aimed at the defense of the competitive process both in principle and in practice, reflecting "a national commitment to the use of free markets to allocate resources efficiently and to spur the innovation that is the principal source of economic growth." The analysis of dominance pays a lot of attention on the limits imposed by endogenous entry, emphasizing the role of entry pressure in disciplining market leaders in spite of their large market shares. The Report provides an enlightening example which is in perfect accordance with the implications of the EMSs approach: "Suppose a large firm competes with a fringe of small rivals, all producing a homogenous product. In this situation, the large firm's market share is only one determinant of its market power over price ... if the fringe firms can readily and substantially increase production at their existing plants in response to a small increase in the large firm's price (that is if the fringe supply is highly elastic), a decision by the large firm to restrict output may have no effect on market prices."

More in general, the Report recognizes the poor correlation that can exist between market share and market power, especially in high-tech sectors: "in markets characterized by rapid technological change, for example, a high market share of current sales or production may be consistent with the presence of robust competition over time rather than a sign of monopoly power. In those situations, any power a firm may have may be both temporary and essential to the competitive process." As a consequence the U.S. Department of Justice adopts a non-intrusive role for antitrust policy in the competition in and for the markets. For instance, predatory pricing can be established only when recoupment is likely, that is only when entry is difficult once the market is monopolized. Moreover, the efficiency role of tying is recognized as a primary role (against a long-lasting hostility), especially for technological tying,

prudential regulation (concerning capital requirements, management incentives, and information disclosure) and a proper supervision of the international financial markets.

"an area where enforcement intervention poses a particular risk of harming consumers more than it helps them in the long run. Technological tying often efficiently gives consumers features they want and judicial control of product design risks chilling innovation." Finally, the Report marginalizes also the need for intervention in case of a refusal to supply, because "forcing a competitor with monopoly power to deal with rivals can undermine the incentives of either or both to innovate" and because "judges and enforcement agencies are ill-equipped to set and supervise the terms on which inputs, property rights, or resources are provided". In conclusion, the U.S. approach is based on the belief that competitive entry forces are the main constraints on the exercise of market power and when they are present antitrust intervention should be a marginal or residual necessity.[55]

The European approach is still more interventionist than the American one.[56] The cited Guidance of the European Commission states the adoption of an "effect-based" approach that is aimed at maximizing consumer welfare and protecting an effective competitive process, and not simply competitors. There is an important new aspect in the Guidance, the emphasis given to the role of entry in determining whether a dominant position exists or not. The key element in the Guidance definition of dominance is the extent to which the firm can behave independently of its competitors, customers and consumers, which relates to the degree of competitive constraints exerted on this firm by the supply of actual competitors, by the threat of expansion of competitors and potential entrants and by the bargaining power of customers. Therefore, entry plays a crucial role and dominance should be incompatible with the presence of a threat of endogenous entry. In particular, a leader "can be deterred from increasing prices if expansion or entry is likely, timely and sufficient", but it would be also important to recognize that the same entry can induce the leader to decrease its prices below those of the rivals, or to adopt other aggressive strategies, without any anti-competitive purpose, as the EMSs approach has made clear.

Beyond this, we have a strong concern on the way the positive premises of the Guidance are carried through its details. The defense of consumers is strongly emphasized in theory but not in practice: most of the focus of the Guidance is on the foreclosure of competitors and not on the relation between this and the harm to consumers, which is what should matter.

[55] The Report largely reflects the Chicago approach to antitrust that was prevailing during the Bush Administration. It is not entirely clear that the Obama Administration will adhere fully to the approach laid down in the Report. However, for a related point by a moderate leader of the so-called Harvard approach to antitrust, see Hovenkamp (2005).

[56] The rest of this section is based on the Opening Speech to the International Conference on the Recent Developments on Antitrust Policy and the Enforcement of Art. 82, organized by Intertic at the Autorità Garante della Concorrenza e del Mercato (Rome, March 6, 2009).

A related concern is about the nature of the foreclosure effects under the "effects-based" approach. The Guidance indicates that a key element of abuse is anti-competitive foreclosure, defined as "a situation where effective access of actual or potential competitors to supplies or markets is hampered or eliminated as a result of the conduct of the dominant undertaking" which is likely to profitably increase its prices with harm for the consumers. However, it is not entirely clear which facts are going to prove foreclosure and which not. For instance, consider a situation in which new competitors enter in the market and some competitors increase their market share to a significant extent: one would expect that this proves that the dominant company's practice is not abusive, but not even this can be taken for granted on the basis of the E.U. Guidance (we provide an example below).

Another issue is about the standard of undistorted competition. As regards pricing abuses, the European approach introduces the "as efficient competitor" test: "the Commission will normally intervene where the conduct concerned has already been or is capable of hampering competition from competitors which are considered to be as efficient as the dominant undertaking". However, the document introduces several exceptions to this principle (for instance, a dynamic view for which less efficient competitors may become as efficient in the future through network or learning effects), and the test does not apply to non-pricing abuses. This means that companies are left without a clear standard.

As a last issue we welcome the confirmation in the Guidance of an efficiency defense: a dominant firm may justify a conduct leading to foreclosure on the ground that efficiencies are sufficient to guarantee that consumers are not penalized. Now, while the consideration of efficiencies generated by a conduct is extremely important to re-direct antitrust policy toward the maximization of consumer welfare, in our view the Guidance appears to adopt a too vague approach and to make it hard, if not impossible, for dominant companies actually to avail themselves of the efficiency defense. The main reason is that their verification appears to be postponed after the establishment of an anti-competitive foreclosure that harms consumers, and not during the decision on whether the same foreclosure harms consumers. Moreover, there appears to be a bias against the possibility that efficiencies can occur: for instance, technological tying is not even mentioned as a source of efficiency in tying cases, but it is actually considered a source of greater risk of anticompetitive foreclosure (because more costly to reverse).

Notice that, to assert a successful efficiency defense under the European framework, dominant firms will be required to show that there are no other less anticompetitive alternatives to achieve the claimed efficiencies. Does the current rule mean that a defense must be rejected if the conduct creates more efficiency gains than other conducts but is more restrictive on competitors? In other words, is it the size of the efficiencies that matters or what matters is the amount of restrictions imposed on competition to obtain those efficiencies? Imagine the dominant company trying to manage these various imponderables: it is much easier just to forego the conduct and, possibly, deprive consumers of an important benefit. Is that what competition policy is supposed to do?

Last, it is not clear why the possibility of an efficiency defense (and with it the possibility to enhance consumer welfare) is to be off-limits for an entire class of companies, as the Guidance makes clear when it states that an "exclusionary conduct which maintains, creates or strengthens a market position approaching that of a monopoly can normally not be justified on the grounds that it also creates efficiency gains". It is positive that the Commission eliminated the reference to firms with a market share above seventy-five per cent which appeared in its 2005 document, but still, in our view, efficiencies should be assessed in the same manner in all cases, regardless of the defendant's market share.

Finally, the new guidelines do not seem to reduce the amount of uncertainty that is associated with the move toward the rule of reason approach. For instance, the potential conflicts between IPRs protection and antitrust policy remain entirely unsolved: while the U.S. have taken a clear position against the possibility of compulsory licensing of IPRs, the E.U. approach still contemplates this possibility under vague conditions. This kind of uncertainty can be a source of inefficiency and distorted behavior, especially when decision rules are imperfect and subject to errors.[57] More in general, antitrust uncertainty on exclusionary strategies may deter genuinely competitive or innovative strategies to be adopted by leading firms, and therefore it may exert negative consequences on consumer welfare. Leading scholars of competition policy have noticed that "the welfare cost of this lack of clarity and excessive caution must be enormous to the E.U. economy as a whole - something the E.U. can ill-afford given its lack of competitiveness relative to other international blocks and the stated objectives of the Lisbon Agenda in this regard."[58]

In conclusion, the E.U. competition policy remains largely linked to a naive version of the post-Chicago approach which is biased against market leaders and in favor of their competitors in a way that is largely unrelated to the real protection of consumers. The U.S. approach, closer to the principles of the Chicago school, has proved to be much more useful in promoting competition, business creation and innovation, especially in high-tech sectors.

6.10.3 Entry and competition in the browsers' market

Elsewhere we have discussed a symptomatic example of the E.U. approach to abuse of dominance issues, the Microsoft case on the tying of Windows

[57] The lack of legal certainty is particularly regrettable in a context of increasing punitive fines and important efforts by the Commission to increase the scope for private enforcement to complement public enforcement of E.U. competition law.

[58] O'Donoghue and Padilla (2006, xi).

MediaPlayer to the Windows operating system and the refusal to supply interoperability information protected by IPRs.[59] Since the Microsoft "saga" appears set to continue with an ongoing preliminary investigation by the European Commission on the bundling of the browser Internet Explorer (IE) with Windows, we return to this important issue for the evolution of the New Economy.[60]

For twelve years, Microsoft has distributed its operating system with IE and for eight of those twelve years, this has been done under a Consent Decree issued by the U.S. antitrust authorities. Alternative browsers can be easily installed on every PC and competition in the field is on the basis of quality and functionality, at least since the introduction of IE in the mid 90s led to a drop of the price to zero. In the recent years Mozilla's Firefox has seen considerable success, with the gap between IE and Firefox's respective market shares narrowing with every passing month (see Figure 6.10 for world market shares). Opera and Safari have consolidated their market positions, while Google's new Chrome quickly picked up a few percent of the global market following its launch in autumn of 2008. This tendency is even stronger in Europe, where the most recent data (from AT Internet Institute) show a large drop of IE's market share, from about 85 % a few years ago to 66 % in January 2008 and 58 % in January 2009, while Firefox has been growing up to 28 % in January 2008 and 33% in January 2009, Opera reached respectively 3.2 % and 4.1 % and Safari 2.1 % and 3 % (with Chrome at 1.5 % in January 2009). Notice that browsers are now an integral part of computer experience and they have promoted the rapid development of all the Internet markets, starting with online sales and online advertising. These kinds of markets represent the main engine of innovation, and in times of crisis they contract as well, though they remain crucial drivers of the economic recovery.

It is odd, to say the least, that the European Commission has decided at this moment to pursue an investigation of Microsoft for abuse of dominance in connection with the integration of IE into Windows.[61] It is an issue already raised and solved in the U.S. Clearly, the Commission is applying the judgment rendered by the Court of First Instance in the earlier European case. In that case, Microsoft was accused of excluding competition in the market for media players and was forced to commercialize a new operating system without its media player - which, by the way, was not bought by anybody, except for a few hundred collectors. Today, the issue emerges with IE. As with media functionality, a domain that has seen a flourishing of competitors' products

[59] See Etro (2007,a, Ch. 6). On September 17, 2007 the Court of First Instance concluded the Appeal of the case and essentially upheld the 2004 Commission decision.

[60] This section is based on an article for the Italian newspaper *Libero* ("La Commissione UE ci Riprova con Gates. A Che Pro?" by F. Etro, Libero Mercato, January 23, 2009, p.11).

[61] The original complaint was by the European competitor Opera, later backed by Firefox and Google.

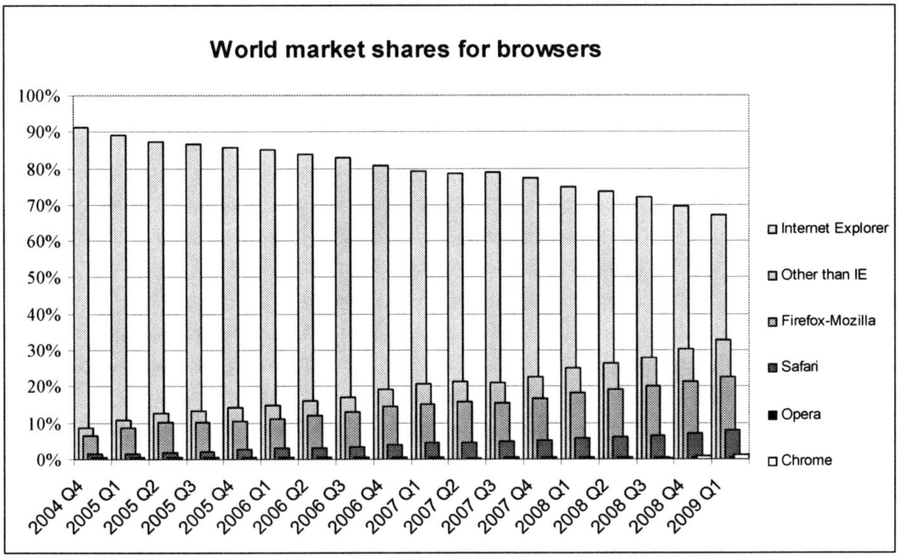

Fig. 6.10. Market structure in the browser market (2004-2009). Source: our calculations on Net Applications data

such as Apple's iTunes, despite the alleged anticompetitive conduct, the market for web browsers is marked by lively competition and a wide and easy diffusion (rather than foreclosure) of rival products. The market can be read as extremely competitive, with a leader in a primary market (Microsoft for operative systems) pressured by entry and innovation in a secondary market (browsers) to adopt aggressive strategies. These include tying of the two products to be sold at a very low price and heavy investments in R&D to preserve the leadership. The consequence has been a strong competitive and innovative pressure from other browser producers, with Firefox as the main alternative to IE, and important benefits accruing to consumers in terms of price, quality and product variety.

Moreover, there do not seem to be solid economic motivations in support of the Commission's thesis. It seems unlikely that Microsoft's strategy can have a predatory purpose because any increase in the price of IE is now unrealistic (meaning recoupment is impossible). Moreover, Microsoft mostly gains from the introduction and the diffusion of other browsers because this increases the quality of PCs and therefore the demand for Windows, its main product. Many users try different browsers before choosing their favorite one, and it is hard to imagine a more competitive scenario than this.[62] Finally,

[62] Notice that the new version of Windows allows users to turn off applications such as Media Player and IE, avoiding any limit to the exclusive use of competing applications.

there are clear (technological) efficiencies from the design of an operating system including a browser, which, as a matter of fact, can be substituted with another one in a few seconds and freely. In conclusion, there are no reasons for which the tying of Windows and IE could harm consumers, whose interest (not the interest of the competitors) should drive antitrust policy.

If the Commission is going to pursue this direction, most likely it will fine Microsoft and force it to commercialize a new operating system without IE (which, as it happened for the one without media player, will not be bought by anybody) or with the option to install other browsers (saving few seconds of free download). Apparently, such an outcome would not have any impact on consumers, but the uncertainty on the freedom of innovation and efficient product integration could reduce the incentives to invest in R&D for Microsoft, for all the software companies producing applications for Windows and IE, and for many other firms in similar situations, with harmful consequences for the future consumers. Moreover, a "must-carry" remedy would strengthen the dominance of Google as a search engine, since Mozilla and Opera are currently paid to pre-set Google as the default search engine, and original equipment manufacturers would be paid to do the same in the future, so as to limit competition in the market for online advertising even more.[63]

These results are not what we expect from policymaking aimed at promoting consumer welfare and growth, especially during a crisis that should suggest other priorities for policymaking.

6.11 Conclusions

In this chapter we have applied the main principles of the EMSs approach to discuss three broad topics: the evolution of global markets of the New Economy and their endogenous structures (as for cloud computing, online advertising or browsers), the evolution of macroeconomic phenomena (as business creation, long run growth, globalization and innovation), and the prescriptions for policymakers (on macroeconomic policy, trade policy, innovation policy and competition policy).

With this mix of applications we want to make a key point: there is no way to understand the macroeconomy that does not start from the structure of the markets that belong to it, especially the high-tech and global markets whose shocks, innovations and exchanges are at the basis of economic fluctuations, growth and trade.

[63] For this reason, Google is heavily supporting the investigation on Microsoft, while advertisers and content providers fear such a bonanza for Google.

7. Epilogue

I have a friend who's collecting unemployment insurance. This guy has never worked so hard in his life as he has to keep this thing going. He's down there every week, waiting on the lines and getting interviewed and making up all these lies about looking for jobs. If they had any idea of the effort and energy that he is expending to avoid work, I'm sure they'd give him a raise. **Jerry Seinfeld**

With this short chapter we conclude our *excursus* on the theory of the endogenous market structures (EMSs), started in Etro (2007,a) with an analysis of static deterministic partial equilibrium models and their implications for industrial, antitrust and innovation policy, and extended here to examine dynamic stochastic general equilibrium models and their implications for macroeconomic, trade and growth policy.

Our interest in the EMSs relies on at least three reasons. The first one is that markets characterized by few firms, some of which are incumbent leaders, and by fixed costs of entry and production that rationalize the same presence of few firms are the normality in the real world, not the exception. Therefore their analysis and the associated consideration of strategic interactions should be the default choice of economic research, not only in the study of single markets but also in the study of the macroeconomy. The failure of the perfectly competitive paradigm, based on price-taking behavior and constant returns to scale, in describing partial and general equilibrium phenomena, or to account for inefficiencies that emerge clearly during dramatic recessions as the current one, should not lead economists to reject entirely the Chicago approach. It should rather lead us to extend its spirit and methodology to take into account more realistic forms of technological structures and of competition based on rational choices of strategies and entry decisions.

The second related reason is that (contrary to what many observers claim) we believe that excessive reliance on rationality has not been a limit of modern economic theory, but quite the opposite: departing from perfect competition, most economic theories have neglected the rational choice of firms to enter in markets, resulting in a partial analysis of industrial organization and macroeconomic phenomena with exogenous entry. Therefore, the characterization of endogenous entry and its interplay with strategic interactions should be seen as a natural step in the long process of introduction of rationality (as a neutral hypothesis, not as an extreme one) in economic analysis, started with the introduction of microfounded demand functions and rational expectations.

The third reason has to do with the empirical evidence presented in this book and elsewhere, which strongly supports the necessity of taking into account EMSs in the analysis of market structures in micro- and macroeconomics.

First, microeconomic evidence suggests that market structures are affected by exogenous factors in ways that are incompatible with the implications of perfect competition, exogenous entry or monopolistic behavior, but that are consistent with the EMSs approach: positive shocks appear to attract entry, strengthen competition, reduce mark ups and increase average production of the firms. These phenomena, largely ignored by the standard theories, are unlikely to be neutral toward the aggregate behavior of the economy.

Second, the implications of the EMSs approach for the behavior of incumbent leaders in the innovative activity are supported by empirical research: leaders tend to invest more in R&D if and only if they are pressured by endogenous entry. This result, together with other results on the differential behavior of leaders in markets where entry can be regarded as endogenous, has critical consequences for competition and innovation policy.

Third, the macroeconomic evidence on entry and profits, that are procyclical, and on mark ups, that appear to be countercyclical, is in line with the results of the EMSs approach, while it remains unexplained within traditional theories. Moreover, simple dynamic stochastic general equilibrium models in the EMSs tradition are able to provide a better characterization of the cyclical behavior of the U.S. economy compared to standard neoclassical models in the Real Business Cycles tradition.

Accordingly, we consider our approach as complementary (and not alternative) to the neoclassical approach, because it draws on modern rational game theory to analyze the microeconomic structure of markets where firms maximize profits and entry is endogenous, and uses this methodology for positive and normative analysis. As a matter of fact, the antitrust implications of the EMSs approach are much in line with those of the old Chicago school of antitrust (which, however, did not employ game theoretical analysis), and the implications for macroeconomic policy are closer to those of the Chicago tradition rather than to those of the Keynesian tradition.

In conclusion, the study of markets characterized by strategic interactions and endogenous entry has provided new insights on:

1) the role of strategic choices at the firm level, which is radically different with or without endogenous entry for a number of applications (as R&D spending, signalling strategies, advertising and demand enhancing in-

vestments, choices on the financial structure, pricing for two-sided markets and markets with network effects);

2) the role of market leaders, that tend to be accommodating or aggressive depending on the entry conditions (with implications for a number of pricing and non-pricing strategies, for the incentives to invest in static and sequential innovations, and more);

3) competition policy issues, in particular on abuse of dominance (predatory pricing, tying, vertical restraints, price discrimination, refusals to supply) for which entry conditions are key, but also on other issues (mergers, price-fixing agreements and state aids);

4) the long run determinants of macroeconomic equilibria, which are closely related to the endogenous structure of the markets (endogenous mark ups, production and business creation depending on behavioral and technological factors, market size effects on the number of firms, their mark ups and their production, and through them on the aggregate variables);

5) the mechanism of propagation of macroeconomic shocks to the real economy, especially through the supply side, the business creation process, and the competition effect on mark ups and wages (which creates procyclical entry and profits and countercylical mark ups);

6) the effects of international trade that create three sources of gains (from competition through lower prices, from variety through more goods available, and from selection through endogenous entry of more efficient international firms) and new channels of propagation of shocks and policy (in particular through real exchange rate effects);

7) technology-driven growth, which is based on the protection of intellectual property rights (that promotes investment in innovation) and is led mainly by incumbent leaders under the pressure of innovative entrants (innovation by leaders), but can be characterized by inefficiencies (dynamic inefficiency associated with excessive savings, too much business creation and underinvestment in R&D by the single firms);

8) macroeconomic policies in terms of fiscal policy (which should be supply-based, with countercyclical indirect sales taxes and wage taxes, plus capital income taxes), monetary policy (which should be aimed at minimizing distortions on business creation and equity prices), and exchange rate policy for the open economy (which can promote domestic exporters through unilateral competitive devaluations, but only in the short run);

9) the optimal trade policy, which includes unilateral import tariffs and, most of all, it always requires positive export subsidies (and other export promoting policies) to give a strategic advantage to the domestic firms active in international markets (without actually reducing trade volumes or hurting the rest of the world);

10) the optimal growth policy, which requires R&D subsidies and strong IPRs protection (possibly coordinated at the international level) to optimize private investments in R&D.

We have compared our results with those of traditional models of perfect competition, which lead to indeterminate market structures, and of standard models of monopolistic or strategic behavior between a given number of firms, which lead to exogenous market structures. In many cases, the endogeneity of the market structure reaches new conclusions that qualify or invert old positive and normative results.

For these reasons, we hope that future research will adopt EMSs more often as a tool of analysis of microeconomic and macroeconomic issues.

8. References

Acemoglu, Daron, 1999, Changes in Unemployment and Wage Inequality: An Alternative Theory and Some Evidence, *The American Economic Review*, 89, 1259-78

Acemoglu, Daron, 2002, Directed Technical Change, *Review of Economic Studies*, 69, 781-809

Acemoglu, Daron, 2008, Innovation by Incumbents and Entrants, mimeo, M.I.T.

Acemoglu, Daron, 2009, *Introduction to Modern Economic Growth*, Princeton University Press

Acemoglu, Daron, Georgy Egorov and Konstantin Sonin, 2008, Dynamics and Stability of Constitutions, Coalitions, and Clubs, mimeo, MIT

Agénor, Pierre-Richard and Peter Montiel, 1999, *Development Macroeconomics*, Princeton University Press

Aghion, Philippe, Richard Blundell, Rachel Griffith, Peter Howitt and Susanne Prantl, 2009, The Effects of Entry on Incumbent Innovation and Productivity, *Review of Economics and Statistics*, 91, 1, 20-32

Aghion, Philippe and Peter Howitt, 1992, A Model of Growth Through Creative Destruction, *Econometrica*, 60, 2, 323-51

Aghion, Philippe and Peter Howitt, 2009, *The Economics of Growth*, MIT Press

Akerlof, George and Janet Yellen, 1985, Can Small Deviations from Rationality Make Significant Differences to Economic Equilibria?, *The American Economic Review*, 75, 4, 708-20

Akerlof, George and Robert Shiller, 2009, *Animal Spirits*, Princeton University Press

Alesina, Alberto, 1988, Credibility and Policy Convergence in a Two-party System with Rational Voters, *The American Economic Review*, 78, 796-806

Alesina, Alberto, Ignazio Angeloni and Federico Etro, 2001a, The Political Economy of International Unions, NBER WP 8645

Alesina, Alberto, Ignazio Angeloni and Federico Etro, 2001b, Institutional Rules for Federations, NBER WP 8646

Alesina, Alberto, Ignazio Angeloni and Federico Etro, 2005, International Unions, *The American Economic Review*, 95, 3, 602-15

Alesina, Alberto, Ignazio Angeloni and Ludger Schuknecht, 2005, What does the European Union do?, *Public Choice*, 123, 3-4, 275-319

Alesina, Alberto and Robert Barro, 2002, Currency Unions, *Quarterly Journal of Economics*, 117, 2, 409-36

Alesina, Alberto and Allan Drazen, 1991, Why Are Stabilizations Delayed?, *The American Economic Review*, 81, 5, 1170-88

Alesina, Alberto and Enrico Spolaore, 2003, *The Size of Nations*, MIT Press

Anderson, James and Eric van Wincoop, 2003, Gravity with Gravitas: A Solution to the Border Puzzle, *The American Economic Review*, 93, 1, 170-92

Anderson, Simon and Andrè de Palma, 2008, Price Dispersion and the Economics of Attention in the Information (Overload) Age, mimeo, University of Virginia, Special Lecture presented at the 2008 Intertic Conference

Arrow, Kenneth, Hollis Chenery, Bagiha Minhas and Robert Solow, 1961, Capital-Labor Substitution and Economic Efficiency, *Review of Economics and Statistics*, 43, 225-50

Arrow, Kenneth and Gerard Debreu, 1954, Existence of a Competitive Equilibrium for a Competitive Economy, *Econometrica*, 22, 3, 265-90

Atkeson, Andrew and Ariel Burnstein, 2008, Pricing to Market, Trade Costs, and International Relative Prices, *The American Economic Review*, 98, 5, 1998-2031

Axarloglou, Kostas, 2003, The Cyclicality of New Product Introductions, *Journal of Business*, 76, 1, 29-48

Azariadis, Costas, 1990, *Intertemporal Macroeconomics*, Blackwell Publishers

Azariadis, Costas and Allan Drazen, 1990, Threshold Externalities in Economic Development, *Quarterly Journal of Economics*, 105, 2, 501-26

Azariadis, Costas and Pietro Reichlin, 1996, Increasing Returns and Crowding out, *Journal of Economic Dynamics and Control*, 20, 5, 847-77

Azariadis, Costas and Bruce Smith, 1998, Financial Intermediation and Regime Switching in Business Cycles, *The American Economic Review*, 88, 3, 516-36

Backus, David, Patrick Kehoe, and Finn Kydland, 1992, International Real Business Cycles, *Journal of Political Economy*, 100, 745-75

Backus, David and John Driffill, 1985, Rational Expectations and Policy Credibility Following a Change in Regime, *Review of Economic Studies*, 52, 2, 211-21

Bagwell, Kyle and Robert Staiger, 2004, National Sovereignity in an Interdependent World, NBER wp N. 10249

Baldwin, Richard and Forslid, Rikard, 2000, Trade Liberalisation and Endogenous Growth: A q-theory approach, *Journal of International Economics*, 50, 2, 497-517

Baldwin, Richard and Paul Krugman, 1989, Persistent Trade Effects of Large Exchange Rate Shocks, *Quarterly Journal of Economics*, 104, 4, 635-54

Ball, Laurence and David Romer, 1990, Real Rigidities and the Non-Neutrality of Money, *Review of Economic Studies*, 57, 183-203

Ball, Laurence and David Romer, 1991, Sticky Prices as Coordination Failure, *The American Economic Review*, 81, 3, 539-52

Barro, Robert, 1974, Are Government Bonds Net Wealth?, *Journal of Political Economy*, 81, 6, 1095-117

Barro, Robert, 1979, On the Determination of Public Debt, *Journal of Political Economy*, 87, 940-71

Barro, Robert, 1981, Output Effects of Government Purchases, *Journal of Political Economy*, 89, 6, 1086-121

Barro, Robert, 1990, Government Spending in a Simple Model of Endogenous Growth, *Journal of Political Economy*, 98, 5, II, S103-105

Barro, Robert, 1999, Laibson Meets Ramsey in the Neoclassical Growth Model, *Quarterly Journal of Economics*, 114, 4, 1125-52

Barro, Robert, 2009, Rare Disasters, Asset Prices, and Welfare Costs, *The American Economic Review*, 99, 1, 243-64

Barro, Robert and David Gordon, 1983a, Rules, Discretion and Reputation in a Model of Monetary Policy, *Journal of Monetary Economics*, 12, 101-21

Barro, Robert and David Gordon, 1983b, A Positive Theory of Monetary Policy in a Natural Rate Model, *Journal of Political Economy*, 91, 4, 589-610

Barro, Robert and Robert King, 1984, Time Separable Preference and Intertemporal Substitution Effects, *Quarterly Journal of Economics*, 99, 817-39

Barro, Robert and Xavier Sala-i-Martin, 1992, Convergence, *Journal of Political Economy*, 100, 2, 223-51

Barro, Robert and Xavier Sala-i-Martin, 1995, *Economic Growth*, Mc Graw Hill (2nd Edition: 2004, MIT Press)

Barro, Robert and Silvana Tenreyro, 2006, Closed and Open Economy Models of Business Cycles with Marked up and Sticky Prices, *The Economic Journal*, 116, 511, 434-56

Basker, Emek, 2008, The Evolving Food Chain: Competitive Effects of Wal-Mart's Entry into the Supermarket Industry, mimeo, University of Missouri, presented at the 2008 Intertic Conference

Baxter, Marianne and Robert King, 1992, Productive Externality and Cyclical Volatility, WP 245, University of Rochester

Baxter, Marianne and Robert King, 1993, Fiscal Policy in General Equilibrium, *The American Economic Review*, 83, 3, 315-34

Benabou, Roland, 2000, Unequal Societies: Income Distribution and the Social Contract, *The American Economic Review*, 90, 1, 96-129

Bénassy, Jean-Pascal, 1996a, Monopolistic Competition, Increasing Returns to Specialization and Output Persistence, *Economic Letters*, 52, 187-91

Bénassy, Jean-Pascal, 1996b, Analytical Solutions to an RBC Model with Imperfect Competition, Increasing returns and Underemployment, *Recherches Economiques de Louvain*, 62, 3-4, 287-97

Benhabib, Jess, Stephanie Schmitt-Grohe and Martin Uribe, 2001, Monetary Policy and Multiple Equilibria, *The American Economic Review*, 91, 1, 167-86

Berger, Allen, Seth Bonime, Lawrence Goldberg and Lawrence White, 2004, The Dynamics of Market Entry: The Effects of Mergers and Acquisitions on Entry in the Banking Industry, *Journal of Business*, 77, 797-834

Bergin, Paul and Giancarlo Corsetti, 2008, The Extensive Margin and Monetary Policy, *Journal of Monetary Economics*, 55, 1222-37

Bernanke, Ben and Alan Blinder, 1988, Credit, Money, and Aggregate Demand, *The American Economic Review*, 78, 2, 435-39

Bernanke, Ben and Mark Gertler, 1989, Agency Costs, Net Worth, and Business Fluctuations, *The American Economic Review*, 79, 1, 14-31

Bernard, Andrew, Stephen Redding and Peter Schott, 2008, Multi-Product Firms and Product Switching, mimeo, Yale University

Berry, Steven, 1992, Estimation of a Model of Entry in the Airline Industry, *Econometrica*, 60, 4, 889-917

Berry, Steven and Peter Reiss, 2007, Empirical Models of Entry and Market Structure, Ch. 29 in *Handbook of Industrial Organization*, M. Armstrong and R. Porter Eds., Elsevier, 3, 1845-86

Berry, Steven and Joel Waldfogel, 1999, Free Entry and Social Inefficiency in Radio Broadcasting, *The RAND Journal of Economics*, 30, 3, 397-420

Bertola, Giuseppe, 1993, Factor Shares and Savings in Endogenous Growth, *The American Economic Review,* 83, 5, 118-49

Bertoletti, Paolo, Eijlen Fumagalli and Clara Poletti, 2008, Price-Increasing Monopolistic Competition? The Case of IES Preferences, IEFE WP 15, presented at the 2008 Intertic Conference

Betts, Caroline and Michael Devereux, 2000, Exchange Rate Dynamics in a Model of Pricing-to-Market, *Journal of International Economics*, 50, 1, 215-44

Bhagwati, Jagdish, 1988, *Protectionism*, MIT Press

Bhagwati, Jagdish, Arvin Panagariya and T. N. Srinivasan, 1998, *Lectures on International Trade*, MIT Press

Bhattacharjea, Aditya, 1995, Strategic Tariffs and Endogenous Market Structures: Trade and industrial Policies under Imperfect Competition, *Journal of Development Economics*, 47, 287-312

Bilbiie, Florin, Fabio Ghironi and Marc Melitz, 2007, Endogenous Entry, Product Variety, and Business Cycles, NBER WP 13646

Bilbiie, Florin, Fabio Ghironi and Marc Melitz, 2008a, Monopoly Power and Endogenous Variety: Distortions and Remedies, NBER WP 14383

Bilbiie, Florin, Fabio Ghironi and Marc Melitz, 2008b, Monetary Policy and Business Cycles with Endogenous Entry and Product Variety, in *NBER*

Macroeconomic Annual 2007, D. Acemoglu, K. Rogoff and M. Woodford Eds., 299-353

Bils, Mark, 1987, The Cyclical Behavior of Marginal Cost and Price, *The American Economic Review*, 77, 838-55

Blanchard, Olivier and Stanley Fisher, 1989, *Lectures on Macroeconomics*, MIT Press

Blanchard, Olivier and Nobuhiro Kiyotaki, 1987, Monopolistic Competition and the Effects of Aggregate Demand, *The American Economic Review*, 77, 4, 647-66

Blanchard, Olivier and Francesco Giavazzi, 2003, Macroeconomic Effects of Regulation and Deregulation in Goods and Labor Markets, *Quarterly Journal of Economics*, 3, 879-907

Blundell, Richard, Rachel Griffith and John Van Reenen, 1999, Market Share, Market Value and Innovation in a Panel of British Manufacturing Firms, *Review of Economic Studies*, 66, 529-54

Blundell, Richard, Rachel Griffith and Frank Windmeijer, 2002, Individual Effects and Dynamics in Count Data Models, *Journal of Econometrics*, 108, 1, 113-31

Bokan, Nikola and Andrew Hughes Hallett, 2008, The Impact of Tax and Market Distortions on the Phillips Curve and the Natural Rate of Unemployment, mimeo, George Mason University

Boldrin, Michele, 1992, Dynamic Externalities, Multiple Equilibria, and Growth, *Journal of Economic Theory*, 58, 2, 198-218

Boldrin, Michele, Lawrence Christiano and Jonas Fisher, 2001, Habit Persistence, Asset Returns, and the Business Cycle, *The American Economic Review*, 91, 1, 149-66

Boldrin, Michele and Luigi Montrucchio, 1986, On the Indeterminacy of Capital Accumulation Paths, *Journal of Economic Theory*, 40, 1, 26-39

Boldrin, Michele and Aldo Rustichini, 1994, Growth and Indeterminacy in Dynamic Models with Externalities, *Econometrica*, 62, 2, 323-42

Boone, Jan, Delia Ionaşcu and Krešimir Žigić, 2006, Trade Policy, Market Leaders and Endogenous Competition Intensity, mimeo, Intertic

Bonanno, Giacomo, 1990, General Equilibrium Theory with Imperfect Competition, *Journal of Economic Surveys*, 4, 4, 297-328

Bordignon, Massimo and Sandro Brusco, 2006, On Enhanced Cooperation, *Journal of Public Economics*, 90, 10-11, 2063-90

Bowen, Harry, Edward Leamer and Leo Sveikauskas, 1987, Multicountry, Multifactor Tests of the Factor Abundance Theory, *The American Economic Review*, 77, 5, 791-809

Brander, James, 1995, Strategic Trade Policy, Ch. 27 in *Handbook of International Economics*, 3, 1395-455, Elsevier

Brander, James and Paul Krugman, 1983, A Reciprocal Dumping Model of International Trade, *Journal of International Economics*, 15, 313-21

Brander, James and Barbara Spencer, 1985, Export Subsidies and International Market Share Rivalry, *Journal of International Economics*, 16, 83-100

Bresnahan, Timothy and Peter Reiss, 1987, Do Entry Conditions Vary across Markets?, *Brookings Papers on Economic Activity*, 3, 833-81

Bresnahan, Timothy and Peter Reiss, 1990, Entry in Monopoly Markets, *Review of Economic Studies*, 57, 531-53

Brito, Paulo, Luìs Costa and Huw Dixon, 2008, Two Saddles and a Source: on Endogenous Mark-up, Mulitple Equilibria, and Threshold Effects, mimeo, Cardiff Business School

Broda, Christian and David Weinstein, 2006, Globalization and the Gains from Variety, *Quarterly Journal of Economics*, 121, 2, 541-85

Broda, Christian and David Weinstein, 2009, Product Creation and Destruction: Evidence and Price Implications, mimeo, Columbia University, *The American Economic Review*, in press

Brou, Daniel and Michele Ruta, 2006, Special Interests and the Gains from Political Integration, *Economics & Politics*, 18, 2, 191-218

Bulow, Jeremy and Kenneth Rogoff, 1991, Sovereign Debt Repurchases: No Cure for Overhang, *Quarterly Journal of Economics*, 106, 4, 1219-35

Burda, Michael and Charles Wyplosz, 2005, *Macroeconomics: a European Text*, Oxford University Press

Caballero, Ricardo and Mohamad Hammour, 1998, The Macroeconomics of Specificity, *Journal of Political Economy*, 106, 4, 724-67

Calvo, Guillermo, 1978, On the Time Consistency of Optimal Policy in a Monetary Economy, *Econometrica*, 46, 6, 1411-28

Calvo, Guillermo, 1983, Staggered Prices in a Utility-Maximizing Framework, *Journal of Monetary Economics*, 12, 3, 383-98

Calvo, Guillermo, 1987, Balance of Payments Crises in a Cash-in-Advance. Economy, *Journal of Money, Credit, and Banking*, 19, 19-32

Calvo, Guillermo and Carmen Reinhart, 2002, Fear of Floating, *Quarterly Journal of Economics*, 117, 2, 379-408

Campbell, Jeffrey and Hugo Hopenhayn, 2005, Market Size Matters, *Journal of Industrial Economics*, 53, 1, 1-25

Campbell, John, 1994, Inspecting the Mechanism: An Analytical Approach to the Stochastic Growth Model, *Journal of Monetary Economics*, 33, 3, 463-506

Carmignani, Fabrizio, Emilio Colombo and Patrizio Tirelli, 2008, Exploring Different Views on Exchange Rate Regime Choice, *Journal of International Money and Finance*, 27, 7, 1177-97

Cavallari, Lilia, 2007, A Macroeconomic Model of Entry with Exporters and Multinationals, *The B.E. Journal of Macroeconomics*, 7, 1, 32

Cavallari, Lilia, 2008, Exports and Foreign Direct Investments in an Endogenous-Entry Model with Real and Nominal Uncertainty, mimeo, University of Rome III, presented at the 2008 Intertic Conference

Chamley, Christophe, 1986, Optimal Taxation of Capital Income in General Equilibrium with Infinite Lives, *Econometrica*, 54, 3, 607-22

Chang, Roberto and Andrés Velasco, 2001, A Model of Financial Crises in Emerging Markets, *Quarterly Journal of Economics*, 116, 489-517

Chatterjee, Satyajit and Russell Cooper, 1993, Entry and Exit, Product Variety and the Business Cycle, NBER WP 4562

Chatterjee, Satyajit, Russell Cooper and B. Ravikumar, 1993, Strategic Complementarity in Business Formation: Aggregate Fluctuations and Sunspots Equilibria, *Review of Economic Studies*, 60, 795-811

Cheikbossian, Guillaume, 2008, Rent-seeking, Spillovers and the Benefits of Decentralization, *Journal of Urban Economics*, 63, 1, 217-28

Chiang, Shin-Hwan and Ahmed Mahmud, 2008, Federations, Coalitions, and Risk Diversification, *Public Choice*, 137, 1, 403-26

Christiano, Lawrence, Martin Eichenbaum and Charles Evans, 1999, Monetary Policy Shocks: What have we learned and to what end?, Ch. 2 in *Handbook of Macroeconomics*, J. B. Taylor and M. Woodford Eds., Elsevier, 1, 65-148

Christiano, Lawrence, Martin Eichenbaum, and Charles L. Evans, 2005, Nominal Rigidities and the Dynamic Effects of a Shock to Monetary Policy, *Journal of Political Economy*, 113, 1, 1-45

Clarida, Richard, Jordi Galì and Mark Gertler, 1999, The Science of Monetary Policy: A New Keynesian Perspective, *Journal of Economic Literature*, 37, 4, 1661-707

Coase, Ronald, 1960, The Problem of Social Cost, *Journal of Law and Economics*, 3, 1-44

Cohen, Wesley and Steven Klepper, 1996, A Reprise of Size and R&D, *The Economic Journal*, 106, 925-51

Cohen, Andrew and Mark Manuszak, 2005, Endogenous Market Structure with Discrete Product Differentiation and Multiple Equilibria: An Empirical Analysis of Competition Between Banks and Thrifts, mimeo, Federal Reserve Board of Governors

Colciago, Andrea and Federico Etro, 2008, Real Business Cycles with Cournot Competition and Endogenous Entry, WP 135, University of Milan, Bicocca, Dept. of Economics

Cook, David, 2001, Time to Enter and Business Cycles, *Journal of Economic Dynamics & Control*, 25, 1241-61

Cooley, Thomas (Edited by), 1995, *Frontiers of Business Cycle Research*, Princeton University Press

Cooper, Russell, 1999, *Coordination Games*, Cambridge University Press

Cooper, Russell and Andrew John, 1988, Coordinating Coordination Failures in Keynesian Models, *Quarterly Journal of Economics*, 103, 441-63

Corsetti, Giancarlo and Paolo Pesenti, 2001, Welfare and Macroeconomic Interdependence, *Quarterly Journal of Economics*, 116, 2, 421-45

Creane, Anthony and Hideo Konishi, 2009, The Unilateral Incentives for Technology Transfers: Predation (and Deterrence) by Proxy, *International Journal of Industrial Organization*, in press

Creane, Anthony and Thomas Jeitschko, 2009, Endogenous Entry in Markets with Adverse Selection, mimeo, Michigan State University

Czarnitzki, Dirk and Federico Etro, 2009, Endogenous Market Structures: Aggregate Implications and Empirical Evidence, mimeo, Intertic

Czarnitzki, Dirk, Federico Etro and Kornelius Kraft, 2008, The Effect of Entry on R&D Investment of Leaders: Theory and Empirical Evidence, ZEW DP 08-78, presented at the 2008 Intertic Conference

Da Costa, Carlos and Ivan Werning, 2008, On the Optimality of the Friedman Rule with Heterogenous Agents and Nonlinear Income Taxation, *Journal of Political Economy*, 116, 1, 82-112

Dasgupta, Partha and Joseph Stiglitz, 1980, Industrial Structure and the Nature of Innovative Activity, *The Economic Journal*, 90, 266-93

Davis, Steven, John Haltiwanger and Scott Schuh, 1996, *Job Creation and Destruction*, MIT Press

Deardorff, Alan, 1980, The General Validity of the Law of Comparative Advantage, *Journal of Political Economy*, 88, 941-57

De Bondt, Raymond and Jan Vandekerckhove, 2008, Innovation Races with Reward Sharing, *Review of Business and Economics*, 53, 4, 354-74

Debrun, Xavier, Paul Masson and Catherine Pattillo, 2005, Monetary Union in West Africa: Who might Gain, Who might Lose, and Why?, *Canadian Journal of Economics*, 38, 2, 454-81

Dedola , Luca and Stefano Neri, 2007, What does a Technology Shock do? A VAR Analysis with Model Based sign restrictions, *Journal of Monetary Economics*, 54, 512-549

van De Klundert, Theo and Sjak Smulders, 1997, Growth, Competition and Welfare, *Scandinavian Journal of Economics*, 99, 1, 99-118

Denicolò, Vincenzo and Piercarlo Zanchettin, 2009, Leadership Cycles, mimeo, University of Bologna

De Santis, Roberto and Frank Stahler, 2001, Trade Policies for Exporting Industries under Free Entry, *German Economic Review*, 2, 4, 327-38

De Santis, Roberto and Frank Stahler, 2004, Endogenous Market Structures and the Gains from Foreign Direct Investment, *Journal of International Economics*, 64, 545-65

Devaney, Robert, 1989, *An Introduction to Chaotic Dynamical Systems*, Addison-Wesley

Devereux, Michael, Allen Head and Beverly Lapham, 1996, Aggregate Fluctuations with Increasing Returns to Specialization and Scale, *Journal of Economic Dynamics and Control*, 20, 627-56

Devereux, Michael and Khang Min Lee, 2001, Dynamic Gains from International Trade with Imperfect Competition and Market Power, *Review of Development Economics*, 5, 2, 239-55

Diamond, Peter, 1965, National Debt in a Neoclassical Growth Model, *The American Economic Review*, 55, 1126-50
Diamond, Peter, 1982, Aggregate Demand Management in Search Equilibrium, *Journal of Political Economy*, 90, 5, 881-94
Di Bartolomeo, Giovanni, Patrizio Tirelli and Nicola Acocella, 2008, Trend Inflation as a Workers Disciplining Device in a General Equilibrium Model, WP 142, University of Milan, Bicocca, Dept. of Economics
Di Noia, Carmine and Stefano Micossi, 2009, *Keep it Simple. Policy Responses to the Financial Crisis*, Assonime
Dixit, Avinash, 1984, International Trade Policy for Oligopolistic Industries, *The Economic Journal*, 94, 1-16
Dixit, Avinash, 2008, Economic Governance, "*Intertic Lecture*" presented at the 2008 Intertic Conference
Dixit, Avinash, Gene Grossman and Elhanan Helpman, 1997, Common Agency and Coordination: General Theory and Application to Government Policy Making, *Journal of Political Economy*, 105, 4, 752-69
Dixit, Avinash and Victor Norman, 1980, *Theory of International Trade*, Cambridge University Press
Dixit, Avinash and Robert Pyndick, 1994, *Investment under Uncertainty*, Princeton University Press
Dixit, Avinash and Joseph Stiglitz, 1977, Monopolistic Competition and Optimum Product Diversity, *The American Economic Review*, 67, 297-308
Doi, Noriyuki, 2008, Price Effects of Market Leadership Volatility in Japan, mimeo, Kwansei Gakuin University
Domar, Evsey, 1946, Capital Expansion, Rate of Growth, and Employment, *Econometrica*, 14, 137-47
Dornbusch, Rudiger, 1976, Expectations and Exchange Rate Dynamics, *Journal of Political Economy*, 84, 1161-76
Dornbusch, Rudiger, 1983, Real Interest Rate, Home Goods, and Optimal External Borrowing, *Journal of Political Economy*, 91, 141-53
Dornbusch, Rudiger, 1987, Exchange Rates and Prices, *The American Economic Review*, 77, 1, 93-106
Dornbusch, Rudiger, Stanley Fisher and Paul Samuelson, 1977, Comparative advantage, trade, and payments in a ricardian model with a continuum of goods, *The American Economic Review*, 67, 5, 823-39
Drazen, Allan, 1999, *Political Economy in Macroeconomics*, Princeton University Press
Dumav, Martin, 2008, R&D Competition and Persistence of Monopoly: the Strategic Role of Technological Complementarities, mimeo, University of Texas at Austin
Dur, Robert and Hein Roelfsema, 2005, Why does Centralization Fail to Internalize Policy Externalities?, *Public Choice*, 122, 3-4, 395-416
Eaton, Jonathan and Gene Grossman, 1986, Optimal Trade and Industrial Policy under Oligopoly, *Quarterly Journal of Economics*, 101, 386-406

Elkhoury, Marwan and Tommaso Mancini Griffoli, 2007, Monetary Policy with Endogenous Firm Entry, mimeo, Swiss National Bank

Ellickson, Bryan, 1993, *Competitive Equilibrium. Theory and Applications*, Cambridge University Press

Engel, Charles, 2000, Local-Currency Pricing and the Choice of Exchange-Rate Regime, *European Economic Review*, 44, 8, 1449-72

Erkal, Nisvan and Daniel Piccinin, 2007, Welfare-Reducing Mergers in Differentiated Oligopolies with Free Entry, mimeo, University of Melbourne, presented at the 2008 Intertic Conference

Etro, Federico, 2002, International Policy Coordination with Economic Unions, *Rivista Internationale di Scienze Sociali*, 110, 2, 187-211

Etro, Federico, 2003a, Globalization and Political Geography, CESifo W.P. 986

Etro, Federico, 2003b, Fiscal Sustainability: An Empirical Analysis for OECD countries, 2003 Memorandum for the Minister of Economy, Ministry of Economy, Rome

Etro, Federico, 2004a, Innovation by Leaders I: Patent Races and R&D Investment, *The Economic Journal*, 114, 4, 495, 281-303

Etro, Federico, 2004b, The Political Economy of Fiscal and Monetary Unions, *Giornale degli Economisti e Annali di Economia*, 63, 3-4, 289-328

Etro, Federico, 2006, Political Geography, *Public Choice*, 127, 3-4 (June), 329-51

Etro, Federico, 2007a, *Competition, Innovation, and Antitrust*, Springer

Etro, Federico, 2007b, Endogenous Market Structures and Macroeconomic Theory, *Review of Business and Economics*, 52, 4, 543-66

Etro, Federico, 2007c, Innovation by Leaders II: Schumpeterian Growth with Endogenous Persistence of Monopoly, *Rivista Internazionale di Scienze Sociali*, 2007, 115, 1, 81-108

Etro, Federico, 2007d, Infrastructures and Growth, WP 119, University of Milan, Bicocca, Dept. of Economics

Etro, Federico, 2007e, Public-Private Partnerships, WP 120, University of Milan, Bicocca, Dept. of Economics

Etro, Federico, 2007f, *The Endogenous Entry Approach to Antitrust*, Experts' Series on Intellectual Property and Competition, London, Stockholm Network Press

Etro, Federico, 2008a, Growth Leaders, *Journal of Macroeconomics*, 30, 1148-72

Etro, Federico, 2008b, Stackelberg Competition with Endogenous Entry, *The Economic Journal*, 118, 1670-97

Etro, Federico, 2008c, Market Leaders, Antitrust Policy and the Software Market, *Icfai Journal of Industrial Economics*, 5, 1, 7-30

Etro, Federico, 2009a, The Economic Impact of Cloud Computing on Business Creation, Employment and Output in the E.U. An application of the

Endogenous Market Structures approach to a GPT innovation, *Review of Business and Economics*, 54, 2, in press

Etro, Federico, 2009b, Dynamic Markets and the Abuse of a Dominant Position, Ch. 4 in *The Reform of EC Competition Law*, I. Lianos and I. Kokkoris Eds., Kluwer, in press

Etro, Federico, 2010, Endogenous Market Structures and Strategic Trade Policy, *International Economic Review*, forthcoming

Etro, Federico and Andrea Colciago, 2007, Endogenous Market Structure and the Business Cycle, WP 126, University of Milan, Bicocca, Dept. of Economics

Etro, Federico and Piero Giarda, 2007, Redistribution, Decentralization and Constitutional Rules, Ch. 10 in *The Economics of Transparency in Politics*, A. Breton, G. Galeotti, P. Salmon and R. Wintrobe Eds., Ashgate, 209-30

Facchini, Giovanni and Cecilia Testa, 2009, Who is Against a Common Market?, *Journal of the European Economic Association*, in press

Farmer, Roger, 1993, *Macroeconomics of Self-Fulfilling Prophecies*, MIT Press

Feenstra, Robert, 1986, Functional Equivalence between Liquidity Costs and the Utility of Money, *Journal of Monetary Economics*, 17, 2, 271-91

Feenstra, Robert, 2003, A Homothetic Utility Function for Monopolistic Competition Models, Without Constant Price Elasticity, *Economic Letters*, 78, 79-86

Feenstra, Robert, 2004, *Advanced International Trade. Theory and Evidence*, Princeton University Press

Feldstein, Martin and Charles Horioka, 1980, Domestic Saving and International Capital Flows, *The Economic Journal*, 90, 358, 314-29

Fernández, Borja Martinez, Iraj Hashi and Marc Jegers, 2008, The Implementation of the European Commission's Merger Regulation 2004: An Empirical Analysis, *Journal of Competition Law and Economics*, 4, 3, 791-809

Fischer, Stanley, 1977, Long-term Contracts, Rational Expectations, and the Optimal Money Supply Rule, *Journal of Political Economy*, 85, 1, 191-205

Fleming, Marcus, 1962, Domestic Financial Policies under Fixed and under Floating Exchange Rates, *International Monetary Fund Staff Papers*, 9, 369-79

Friedman, Milton, 1957, *A Theory of the Consumption Function*, Princeton University Press

Friedman, Milton, 1968, The Role of Monetary Policy, *The American Economic Review*, 58, 1, 1-17

Fuhrer, Jeffrey and George Moore, 1995, Inflation Persistence, *Quarterly Journal of Economics*, 110, 1, 127-60

Fumagalli, Chiara, Massimo Motta and Thomas Rønde, 2008, Exclusive Dealing: the Interaction between Foreclosure and Investment Promotion, mimeo, University of Copenhagen, presented at the 2008 Intertic Conference

Fujita, Masahisa, Paul Krugman and Anthony Venables, 1998, *The Spatial Economy. Cities, Regions, and International Trade*, Cambridge, The MIT Press

Fujita, Masahisa, 1989, *Urban Economic Theory. Land Use and City Size*, Cambridge University Press

Gabszewicz, Jean Jaskold, 2003, *Strategic Multilateral Exchange: General Equilibrium With Imperfect Competition*, Edward Elgar

Galì, Jordi, 2008, *Monetary Policy, Inflation, and the Business Cycle*, Princeton University Press

Galì, Jordi, Mark Gertler and David López-Salido, 2007a, Markups, Gaps, and the Welfare Costs of Business Fluctuations, *Review of Economics and Statistics*, 89, 1, 44-59

Galì, Jordi, David López-Salido and Javier Vallés, 2007b, Understanding the Effects of Government Spending on Consumption, *Journal of the European Economic Association*, 5, 1, 227-70

Galor, Oded, 1992, A Two-Sector Overlapping-Generations Model: A Characterization of the Dynamical System, *Econometrica*, 60, 351-86

Galor, Oded and Omer Moav, 2000, Ability-Biased Technological Change, Wage Inequality and Economic Growth, *Quarterly Journal of Economics*, 115, 469-98

Galor, Oded and Harl Ryder, 1989, Existence, Uniqueness and Stability of Equilibrium in an Overlapping-Generations Model with Productive Capital, *Journal of Economic Theory*, 49, 360-75

Galor, Oded and Joseph Zeira, 1993, Income Distribution and Macroeconomics, *Review of Economic Studies*, 60, 1, 35-52

Gautier, Axel, Kanisha Dam and Manipushpak Mitra, 2007, Efficient Access Pricing and Endogenous Market Structures, mimeo, Universitè de Liege, presented at the 2008 Intertic Conference

Geroski, Paul, 1995, What Do We Know about Entry?, *International Journal of Industrial Organization*, 13, 421-40

Ghironi, Fabio and Marc Melitz, 2005, International Trade and Macroeconomic Dynamics with Heterogenous Firms, *Quarterly Journal of Economics*, 120, 865-915

Ghironi, Fabio and Viktors Stebunovs, The Domestic and International Effects of Financial Deregulation, mimeo, Boston College

Goldberg, Arthur, 1991, *A Course in Econometrics*, Harvard University Press

Goldberg, Pinepoli, and Giovanni Maggi, 1999, Protection for Sale: An Empirical Investigation, *The American Economic Review*, 89, 5, 1135-55

Grandmont, Jean Michel, 1985, On Endogenous Competitive Business Cycles, *Econometrica*, 53, 995-1045

Greene, William, 2003, *Econometric Analysis*, 5th Ed., Prentice Hall

Greenwald, Bruce and Joseph Stiglitz, 1993, Financial Market Imperfections and Business Cycles, *Quarterly Journal of Economics*, 108, 1, 77-114

Greenspan, Alan, 2007, *The Age of Turbolence. Adventures in a New World*, Penguin Books

Gregorini, Filippo, 2009, Political Geography and Income Inequalities, WP 152, University of Milan, Bicocca, Dept. of Economics

Griliches, Zvi and Jacques Mairesse, 1984, Productivity and R&D at the firm level, in Z. Griliches (Eds), *R&D, Patents and Productivity*, The University of Chicago Press, Chicago, 152-75

Grossman, Gene and Elhanan Helpman, 1991, *Innovation in the Global Economy*, MIT Press

Grossman, Gene and Elhanan Helpman, 1994, Protection for Sale, *The American Economic Review*, 84, 4, 833-50

Grossman, Gene and Elhanan Helpman, 2002, *Special Interest Politics*, MIT Press

Grossman, Gene and Elhanan Helpman, 2005, Outsourcing in a Global Economy, *Review of Economic Studies*, 72, 1, 135-59

Grossman, Gene and Giovanni Maggi, 2000, Diversity and Trade, *The American Economic Review*, 90, 5, 1255-75

Hafer, Catherine and Dimitri Landa, 2007, Public Goods in Federal Systems, *Quarterly Journal of Political Science*, 2, 253-75

Hall, Bronwin, Zvi Griliches and Jerry Hausman, 1986, Patents and R&D: Is There a Lag?, *International Economic Review*, 27, 2, 265-83

Hall, Robert, 1978, Stochastic Implications of the Life-Cycle/Permanent Income Hypothesis: Theory and Evidence, *Journal of Political Economy*, 96, 971-87

Hall, Robert, 1986, Market Structure and Macroeconomic Fluctuations, *Brookings Papers on Economic Activity*, 2, 285-322

Hall, Robert, 1990, The Relationship between Price and Marginal Cost in U.S. Industry, *Journal of Political Economy*, 96, 921-47

Hansen, Gary, 1985, Indivisible Labor and the Business Cycle, *Journal of Monetary Economics*, 16, 3, 309-27

Harrod, Roy, 1948, *Toward a Dynamic Economics: Some Recent Developments of Economic Theory and Their Application to Policy*, Macmillan

Hart, Oliver, 1982, A Model of Imperfect Competition with Keynesian Features, *Quartely Journal of Economics*, 97, 109-38

Hartsad, Bard, 2007, Harmonization and Side Payments in Political Cooperation, *The American Economic Review*, 97, 3, 871-89

Hausman, Jerry, 1978, Specification Tests in Econometrics, *Econometrica*, 46, 6, 1273-91

Hausman, Jerry, 1983, Specification and Estimation of Simultaneous Equations Models, in *Handbook of Econometrics*, Z. Griliches and M.D. Intriligator Eds., North Holland

Hausman, Jerry, Bronwin Hall and Zvi Griliches, 1984, Econometric Models for Count Data with an Application to the Patents-R&D Relationship, *Econometrica*, 52, 909-38

Heckman, James, 1979, Sample Selection Bias as a Specification Error, *Econometrica*, 47, 1, 153-61

Heckscher, Eli, 1929, A Plea for a Theory in Economic History, *The Economic Journal*, 525-534

Hellmann, Thomas, Kevin Murdock and Joseph Stiglitz, 2000, Liberalization, Moral Hazard in Banking, and Prudential Regulation: Are Capital Requirements Enough?, *The American Economic Review*, 90, 1, 147-65

Helpman, Elhanan, 2006, Trade, FDI, and the Organization of Firms, *Journal of Economic Literature*, 44, 589-630

Helpman, Elhanan and Paul Krugman, 1985, *Market Structure and Foreign Trade. Increasing Returns, Imperfect Competition and the International Economy*, MIT Press

Helpman, Elhanan and Paul Krugman, 1989, *Trade Policy and Market Structure*, MIT Press

Helpman, Elhanan, Marc Melitz and Yona Rubinstein, 2008, Estimating Trade Flows: Trading Partners and Trading Volumes, *Quarterly Journal of Economics*, 123, 2, 441-87

Hicks, John, 1937, Mr. Keynes and the Classics. A Suggested Interpretation, *Econometrica*, 5, 2, 147-59

Horstmann, Ignatius and James Markusen, 1986, Up the Average Cost Curve: Inefficient Entry and the New Protectionism, *Journal of International Economics*, 20, 225-47

Horstmann, Ignatius and James Markusen, 1992, Endogenous Market Structures in International Trade (Natura Facit Saltum), *Journal of International Economics*, 32, 109-29

Hovenkamp, Herbert, 2005, *The Antitrust Enterprise. Principle and Execution*, Harvard University Press

Hughes, Danny, 2007, Innovation, Firm Efficiency, and Market Structure, mimeo, University of South Alabama

Impullitti, Giammario, 2007, Foreign Technological Competition and the U.S. R&D Subsidies Response in the 1970s and 1980s: a Quantitative Welfare Analysis, mimeo, EUI

Ino, Hiroaki and Toshihiro Matsumura, 2007, How Many Firms Should Be Leaders? Beneficial Concentration Revisited, mimeo, Institute of Social Science, mimeo, University of Tokyo, presented at the 2008 Intertic Conference

Irwin, Douglas, 1991, Mercantilism as Strategic Trade Policy: The Anglo-Dutch Rivalry dor the East India Trade, *Journal of Political Economy*, 99, 6, 1296-314

Ishida, Junichiro, Toshihiro Matsumura and Noriaki Matsushima, 2008, When Market Competition Benefits Firms, mimeo, Osaka University

Jaimovich, Nir, 2007, Firm Dynamics and Markup Variations: Implications for Multiple Equilibria and Endogenous Economic Fluctuations, *Journal of Economic Theory*, 137, 1, 300-25

Jaimovich, Nir and Max Floetotto, 2008, Firm Dynamics, Markup Variations, and the Business Cycle, *Journal of Monetary Economics*, 55, 7, 1238-52

Johnson, Harry, 1953-54, Optimum Tariffs and Retaliation, *Review of Economic Studies*, 21, 2, 142-53

Jones, Charles, 1995, R&D-Based Models of Economic Growth, *Journal of Political Economy*, 103, 4, 759-84

Jones, Larry and Rodolfo Manuelli, 1990, A Convex Model of Equilibrium Growth: Theory and Policy Implications, *Journal of Political Economy*, 98, 5, 1008-38

Jones, Ronald, 1965, The Structure of Simple General Equilibrium Models, *Journal of Political Economy*, 73, 557-72

Kaldor, Nicholas, 1940, A Note on Tariffs and the Terms of Trade, *Economica*, 7, 377-80

Kaplan, Steven and Luigi Zingales, 1997, Do Investment-Cash Flow Sensitivities Provide Useful Measures of Financing Constraints?, *Quarterly Journal of Economics*, 112, 1, 169-215

Kato, Masatoshi and Yuji Honjo, 2007, The Persistence of Market Leadership: Evidence from Japan, mimeo, Hitotsubashi University

Katsoulacos, Yannis, 2008, Optimal Legal Standards for Refusals to License IP: An Welfare Based Analysis, mimeo, Athens University, presented at the 2008 Intertic Conference

Katsoulacos, Yannis and Anastasios Xepapadeas, 1995, Endogenous Policy under Oligopoly with Endogenous Market Structure, *Scandinavian Journal of Economics*, 97, 3, 411-20

Kleibergen, Frank and Richard Paap, 2006, Generalized Reduced Rank Tests Using the Singular Value Decomposition, *Journal of Econometrics*, 133, 97-126

Keynes, John Maynard, 1936, *The General Theory of Employment, Interest and Money*, Macmillan Cambridge University Press

Kim, Jinill, 2004, What Determines Aggregate Returns to Scale?, *Journal of Economic Dynamics & Control*, 28, 1577-94

King, Robert, Charles Plosser and Sergio Rebelo, 1988,a,b, Production, Growth and Business Cycles: I. The Basic Neoclassical Model. II. New Directions, *Journal of Monetary Economics*, 21, 2-3, 195-232, 309-341

King, Robert and Sergio Rebelo, 2000, Resuscitating Real Business Cycles, Ch. 14 in *Handbook of Macroeconomics*, J. B. Taylor & M. Woodford Eds., Elsevier, 1, 927-1007

Kiyotaki, Nobuhiro and John Moore, 1997, Credit Cycles, *Journal of Political Economy*, 105, 2, 211-48

Klette, T.J., 1994, Strategic Trade policy for Exporting Industries: More General Results in the Oligopolistic Case, *Oxford Economic Papers*, 46, 296-310

Kortum, Samuel, 1993, Equilibrium R&D and the Patent-R&D Ratio: U.S. Evidence, *The American Economic Review, Papers & Proceedings*, 83, 2, 450-7

Kothenburger, Marko, 2008, Revisiting the "Decentralization Theorem". On the Role of Externalities, *Journal of Urban Economics*, 64, 1, 116-22

Kováč, Eugen, Viatcheslav Vinogradov and Krešimir Žigić, 2009, Technological Leadership and the Persistence of Monopoly under Endogenous Entry: Static *versus* Dynamic Analysis, mimeo, CERGE-EI

Kremer, Michael, 1993, The O-Ring Theory of Economic Development, *Quarterly Journal of Economics*, 108, 3, 551-75

Krugman, Paul, 1979a, Increasing Returns, Monopolistic Competition and International Trade, *Journal of International Economics*, 9, 469-79

Krugman, Paul, 1979b, A Model of Balance-of-Payments Crises, *Journal of Money, Credit and Banking*, 11, 3, 311-25

Krugman, Paul, 1980, Scale Economies, Product Differentiation, and the Pattern of Trade, *The American Economic Review*, 70, 950-9

Krugman, Paul, 1981, Intraindustry Specialization and the Gains from Trade, *Journal of Political Economy*, 89, 959-74

Krugman, Paul, 1987, The Narrowing Moving Band, the Dutch Disease and the Economic Consequences of Mrs. Thatcher, *Journal of Development Economics*, 27, 41-55

Krugman, Paul, 1991a, Increasing Returns and Economic Geography, *Journal of Political Economy*, 99, 3, 483-99

Krugman, Paul, 1991b, Target Zones and Exchange Rate Dynamics, *Quarterly Journal of Economics*, 106, 3, 669-82

Krugman, Paul, 2000, Technology, Trade and Factor Prices, *Journal of International Economics*, 50, 1, 51-71

Krugman, Paul, 2008, *The Return of Depression Economics and the Crisis of 2008*, Penguin Books

Kydland, Finn and Edward Prescott, 1977, Rules Rather than Discretion: The Inconsistency of Optimal Plans, *Journal of Political Economy*, 85, 473-91

Kydland, Finn and Edward Prescott, 1982, Time to Build and Aggregate Fluctuations, *Econometrica*, 50, 6, 1345-70

Lahiri, Amartya and Carlos Vegh, 2003, Delaying the Inevitable: Interest Rate Defense and Balance of Payments Crises, *Journal of Political Economy*, 111, 2, 404-24

Lahiri, Sajal and Yoshiyasu Onu, 1995, The Role of Free Entry in an Oligopolistic Heckscher-Ohlin Model, *International Economic Review*, 36, 3, 609-24

Laibson, David, 1997, Golden Eggs and Hyperbolic Discounting, *Quarterly Journal of Economics*, 62, 443-77

Le, Thanh, 2008, A Dual Economy Model of Endogenous Growth with R&D and Market Structure, *Journal of Evolutionary Economics*, in press

Leamer, Edward, 1980, The Leontief Paradox, Reconsidered, *Journal of Political Economy*, 3, 495-503

Leamer, Edward, 1987, Paths of Development in the Three-Good n-Factor General Equilibrium Model, *Journal of Political Economy*, 95, 5, 961-99

Leamer, Edward, 1999, Effort, Wages and the International Division of Labor, *Journal of Political Economy*, 107, 6, 1, 1127-63

Leamer, Edward, 2000, What's the Use of Factor Contents?, *Journal of International Economics*, 50, 1, 17-50

Leontief, Wassily, 1941, *The Structure of American Economy, 1919-1939: An empirical application of equilibrium analysis*, Oxford University Press

Lerner, Abba, 1934, The Diagrammatic Representation of Demand Conditions in International Trade, *Economica*, 1, 319-34

Lerner, Abba, 1936, The Symmetry between Import and Export Taxes, *Economica*, 3, 11, 306-13

Lerner, Abba, 1952, Factor Prices and International Trade, *Economica*, 19

Levitt, Steven and Stephen Dubner, 2005, *Freakonomics: A Rogue Economist Explores the Hidden Side of Everything*, William Morrow

Lewis, Vivien, 2007a, Business Cycle Evidence on Firm Entry, *Macroeconomic Dynamics*, forthcoming

Lewis, Vivien, 2007b, Optimal Fiscal and Monetary Policy Responses to Firm Entry, mimeo, Deutsche Bundesbank

Ljungqvist, Lars and Thomas Sargent, 2004, *Recursive Macroeconomic Theory*, 2nd Ed., MIT Press

Lockwood, Ben, 2008, Voting, Lobbying, and the Decentralization Theorem, *Economics & Politics*, in press

Long, John and Charles Plosser, 1983, Real Business Cycles, *Journal of Political Economy*, 91, 1, 39-69

van Long, Ngo, Horst Raff and Frank Stähler, 2009, Innovation and Trade with Heterogeneous Firms, mimeo, McGill University

Lorz, Oliver and Gerald Willmann, 2005, On the Endogenous Allocation of Decision Power in Federal Structures, *Journal of Urban Economics*, 57 (2), 242-57

Lossani, Marco, Piergiovanna Natale and Patrizio Tirelli, 2000, Macroeconomics and Politics Revisited: Do Central Banks Matter?, *Economics and Politics*, 12, 1, 53-67

Loury, Glenn, 1979, Market Structure and Innovation, *Quarterly Journal of Economics*, 1, 429-36

Lucas, Robert, 1972, Expectations and the Neutrality of Money, *Journal of Economic Theory*, 4, 103-24

Lucas Robert, 1978, Asset Prices in an Exchange Economy, *Econometrica*, 46, 1429-46

Lucas Robert, 1980, Equilibrium in a Pure Currency Economy, *Economic Inquiry*, 18, 2, 203-20

Lucas Robert, 1988, On the Mechanincs of Development, *Journal of Monetary Economics*, 222, 3-42

Lucas, Robert, 1990, Why Doesn't Capital Flow from Rich to Poor Countries, *The American Economic Review*, 80, 92-6

Lucas, Robert, 2000, Inflation and Welfare, *Econometrica*, 68, 247-74

Lucas, Robert and Leonard Rapping, 1969, Real Wages, Employment, and Inflation, *Journal of Political Economy*, 77, 5, 721-54

Lucas, Robert and Nancy Stockey, 1983, Optimal Fiscal and Monetary Policy in an Economy without Capital, *Journal of Monetary Economics*, 12, 1, 55-93

Lyons, Bruce, Catherine Matraves and Peter Moffatt, 2001, Industrial Concentration and Market Integration in the European Union, *Economica*, 68, 1-26

Maci, Ilir and Krešimir Žigić, 2008, Competition Policy and Market Leaders, mimeo, Intertic, presented at the 2008 Intertic Conference

Malcomson, James, 1981, Unemployment and the Efficiency Wage Hypothesis, *The Economic Journal*, 91, 364, 848-66

Mankiw, Gregory, 1985, Small Menu Costs and Large Business Cycles: A Macroeconomic Model, *Quarterly Journal of Economics*, 100, 2, 529-38

Mankiw, Gregory and Ricardo Reis, 2002, Sticky Information Versus Sticky Prices: A Proposal To Replace The New Keynesian Phillips Curve, *Quarterly Journal of Economics*, 117, 4, 1295-328

Mankiw, Gregory, David Romer and David Weil, 1992, A Contribution to the Empirics of Economic Growth, *Quarterly Journal of Economics*, 107, 2, 407-37

Mankiw, Gregory and Mark Taylor, 2006, *Microeconomics*, Thomson Learning

Mankiw, Gregory and Michael Whinston, 1986, Free Entry and Social Inefficiency, *The RAND Journal of Economics*, 17, 1, 48-58

Manuszak, Mark, 2002, Endogenous Market Structure and Competition in the 19th Century American Brewing Industry, *International Journal of Industrial Organization*, 20, 673-932

Manuszak, Mark and Charles Moul, 2008, Prices and Endogenous Market Structure in Office Supply Superstores, *Journal of Industrial Economics*, 56, 94-112

Marchesi, Silvia and Laura Sabani, 2007, IMF Concern for Reputation and Conditional Lending Failure: Theory and Empirics, *Journal of Development Economics*, 84, 640-66

Markowitz, Harry, 1959, *Portfolio Selection: Efficient Diversification of Investments*, John Wiley & Sons

Markusen, James, 1981, Trade and Gains from Trade with Increasing Returns and Imperfect Competition, *Journal of International Economics*, 11, 531-51

Markusen, James and Anthony Venables, 1988, Trade Policy with Increasing Returns and Imperfect Competition, *Journal of International Economics*, 24, 299-316

Marx, Karl, 1867, *Capital*, Progress Publishers

MacDonald, Ian and Robert Solow, 1981, Wage Bargaining and Employment, *The American Economic Review*, 71, 896-908

Mas Colell, Andreu, Michael Whinston and Jerry Green, 1995, *Microeconomic Theory*, Oxford University Press

Maskin, Eric and Jean Tirole, 1988, A Theory of Dynamic Oligopoly, I: Overview and Quantity Competition with Large Fixed Costs. II: Price Competition, Kinked Demand Curves, and Edgeworth Cycles, *Econometrica*, 56, 3, 549-99

Matsuyama, Kiminori, 1999, Growing Through Cycles, *Econometrica*, 67, 335-47

Mazzeo, Michael, 2002, Competitive Outcomes in Product-differentiated Oligopoly, *Review of Economics and Statistics*, 84, 4, 716-28

McCracken, Scott and Frank Stähler, 2007, Economic Integration and the Choice of Commodity Tax Base with Endogenous Market Structures, mimeo, University of Otago

Meade, James, 1952, *A Geometry of International Trade*, Allen and Unwin

Mehra, Rajnish and Edward Prescott, 1985, The Equity Premium: A Puzzle, *Journal of Monetary Economics*, 15, 145-61

Melitz, Marc, 2003, The Impact of Trade on Intra-Industry Reallocations and Aggregate Industry Productivity, *Econometrica*, 71, 6, 1695-725

Melitz, Marc and Giancarlo Ottaviano, 2008, Market Size, Trade, and Productivity, *Review of Economic Studies*, 75, 295-316

Mendola, Mariapia, 2008, Migration and Technological Change in Rural Households: Complements or Substitutes?, *Journal of Development Economics*, 85, 1-2, 150-75

Merton, Robert, 1973, An Intertemporal Capital Asset Pricing Model, *Econometrica*, 41, 867-87

Miller, Merton and Franco Modigliani, 1961, Dividend Policy, Growth, and the Valuation of Shares, *Journal of Business*, 34, 411-33

Minniti, Antonio, 2008, Growth, Inter-industry and Intra-industry Competition and Welfare, *Japanese Economic Review*, in press

Mitra, Devashish, 1999, Endogenous Lobby Formation and Endogenous Protection: A Long-Run Model of Trade Policy Determination, *The American Economic Review*, 89, 5, 1116-34

Modigliani, Franco and Richard Brumberg, 1954, Utility Analysis and the Consumption Function: An Interpretation of Cross-Section Data, in K. Kurihara, Ed., *Post Keynesian Economics*, Rutgers University Press

Monacelli, Tommaso and Roberto Perotti, 2008, Fiscal Policy, Wealth Effects, and Markups, NBER WP 14584

Mortensen, Dale and Christopher Pissarides, 1994, Job Creation and Job Destruction in the Theory of Unemployment, *Review of Economic Studies*, 61, 3, 397-415

Mukherjee, Arijit, 2008, Endogenous Market Structures in a Vertical Structure, mimeo, University of Nottingham, presented at the 2008 Intertic Conference

Mulligan, Casey and Kevin Tsui, 2009, Political Entry, Public Policies and the Economy, mimeo, University of Chicago

Mundell, Robert, 1968, *International Economics*, Macmillan

Mundlak, Yair, 1961, Empirical Production Function Free of Management Bias, *Journal of Farm Economics*, 43, 1, 44-56

Murphy, Kevin, Andrei Shleifer and Robert Vishny, 1989, Industrialization and the Big Push, *Journal of Political Economy*, 97, 5, 1003-26

Mussa, Michael, 1974, Tariffs and the Distribution of Income: The Importance of Factor Specificity, Substitutability, and Intensity in the Short and Long Run, *Journal of Political Economy*, 82, 1191-204

Muth, John, 1961, Rational Expectations and the Theory of Price Movements, *Econometrica*, 39, 315-34

Nordhaus, William, 1969, *Invention, Growth, and Welfare*, MIT Press

Nordhaus, William, 1975, The Political Business Cycle, *Review of Economic Studies*, 42, 169-90

Obstfeld, Maurice, 1994, Risk-Taking, Global Diversification, and Growth, *The American Economic Review*, 84, 5, 1310-29

Obstfeld, Maurice, 1996, Models of Currency Crises with Self-fulfilling Features, *European Economic Review*, 40, 3-5, 1037-47

Obstfeld, Maurice and Kenneth, Rogoff, 1995, Exchange Rate Dynamics Redux, *Journal of Political Economy*, 103, 3, 624-60

Obstfeld, Maurice and Kenneth Rogoff, 1996, *Foundations of International Macroeconomics*, MIT Press

Obstfeld, Maurice and Kenneth, Rogoff, 2000, New Directions for Stochastic Open Economy Models, *Journal of International Economics*, 50, 1, 117-53

O'Donoghue, Robert and Jorge Padilla, 2006, *The Law and Economics of Article 82 EC*, Hart Publishing

Ohlin, Bertil, 1933, *Interregional and International Trade*, Harvard University Press

Palfrey, Thomas and Svetlana Pevnitstaya, 2008, Endogenous Entry and Self-selection in Private Value Auctions: An Experimental Study, *Journal of Economic Behavior and Organization*, 66, 3-4, 731-47

Pasinetti, Luigi, 1962, The Rate of Profit and Income Distribution in Relation to the Rate of Economic Growth, *Review of Economic Studies*, 29, 267-79

Peretto, Pietro, 1996, Sunk Costs, Market Structure and Growth, *International Economic Review*, 37, 895-923

Peretto, Pietro, 1999, Cost Reduction, Entry, and the Interdependence of Market Structure and Economic Growth, *Journal of Monetary Economics*, 43, 173-96

Peretto, Pietro, 2003, Endogenous Market Structure and the Growth and Welfare Effects of Economic Integration, *Journal of International Economics*, 60, 177-201

Peretto, Pietro, 2008a, A Schumpeterian Analysis of Deficit-Financed Dividend Tax Cuts, Nimeo, Duke University, presented at the 2008 Intertic Conference

Peretto, Pietro, 2008b, Is the "Curse of Natural Resources" Really a Curse?, mimeo, Duke University

Peretto, Pietro and Michelle Connolly, 2005, The Manhattan Metaphor, mimeo, Duke University

Persson, Torsten and Lars Svensson, 1989, Why a Stubborn Conservative would Run a Deficit: Policy with Time–consistent Preferences, *Quarterly Journal of Economics*, 104, 325-45

Persson, Torsten and Tabellini, Guido, 1990, *Macroeconomic Policy, Credibility, and Politics*, Harwood Academic Publishers GmbH

Persson, Torsten and Tabellini, Guido, 2000, *Political Economics: Explaining Economics Policy*, MIT Press

Phelps, Edmund, 1961, The Golden Rule of Accumulation: A Fable for Growhtmen, *The American Economic Review*, 51, 4, 638-43

Phelps, Edmund, 1973, Inflation in the Theory of Public Finance, *Scandinavian Journal of Economics*, 75, 67-82

Phillips, Alban, 1958, The Relation Between Unemployment and the Rate of Change of Money Wage Rates in the United Kingdom, 1861-1957, *Economica*, 25, 283-99

Pissarides, Christopher, 2000, *Equilibrium Unemployment Theory*, MIT Press

Polo, Michele and Giovanni Immordino, 2008, Judicial Errors and Innovative Activity, mimeo, Bocconi University, presented at the 2008 Intertic Conference

Prescott, Edward, 1986, Theory Ahead of Business Cycle Measurement, *Quarterly Review, Federal Reserve Bank of Minneapolis*, Fall, 9-22

Ramsey, Frank, 1927, A Contribution to the Theory of Taxation, *The Economic Journal*, 37, 145, 47-61

Ramsey, Frank, 1928, A Mathematical Theory of Saving, *The Economic Journal*, 38, 152, 543-59

Ravn, Morten, Stephanie Schmitt-Grohe and Martin Uribe, 2006, Deep Habits, *Review of Economic Studies*, 73, 1, 195-218

Rebelo, Sergio, 1991, Long-Run Policy Analysis and Long-Run Growth, *Journal of Political Economy*, 99, 3, 500-521

Ricardo, David, 1817, *On the Principles of Political Economy and Taxation*, John Murray

Rivera-Batiz, Luis and Paul Romer, 1991, Economic Integration and Endogenous Growth, *Quarterly Journal of Economics*, 106, 2, 531-55

Rodrik, Dani, 1989, Optimal Trade Taxes for a large Country with Non-Atomistic Firms, *Journal of International Economics*, 26, 157-67

Rogerson, Richard, 1988, Indivisible Labor, Lotteries and Equilibrium, *Journal of Monetary Economics*, 21, 1, 3-16

Rogoff, Kenneth, 1985, The Optimal Degree of Commitment to an Intermediate Monetary Target, *Quarterly Journal of Economics*, 100, 4, 1169-89

Romer David, 1996, *Advanced Macroeconomics*, McGraw Hill

Romer, Paul, 1986, Increasing Returns and Long-run Growth, *Journal of Political Economy*, 94, 5, 1002-37

Romer, Paul, 1990, Endogenous Technological Change, *Journal of Political Economy*, 98, 5, S71-102

Rotemberg, Julio, 1982a, Sticky Prices in the United States, *Journal of Political Economy*, 90, 6, 1187-211

Rotemberg, Julio, 1982b, Monopolistic Price Adjustment and Aggregate Output, *Review of Economic Studies*, 49, 4, 517-31

Rotemberg, Julio, 2008, Entry, Aggregat Output and Monetary Policy, in *NBER Macroeconomic Annual 2007*, D. Acemoglu, K. Rogoff and M. Woodford Eds.

Rotemberg, Julio and Garth Saloner, 1986, A Supergame-Theoretic Model of Price Wars during Booms, *The American Economic Review*, 76, 3, 390-407

Rotemberg, Julio and Michael Woodford, 1992, Oligopolistic Pricing and the Effects of Aggregate Demand on Economic Activity, *Journal of Political Economy*, 100, 6, 1153-207

Rotemberg, Julio and Michael Woodford, 1997, An Optimization-based Econometric Framework for the Evaluation of Monetary Policy, in *NBER Macroeconomic Annual*, 2, 297-346

Rotemberg, Julio and Michael Woodford, 2000, The Cyclical Behavior of Prices and Costs, Ch. 16 in *Handbook of Macroeconomics*, J. B. Taylor and M. Woodford Eds., Elsevier, 1, 1051-135

Rybczynski, T. M., 1955, Factor Endowment and Relative Commodity Prices, *Economica*, 22, 336-41

Sachs, Jeffrey, 1981, The Current Account and Macroeconomic Adjustment in the 1970s, *Brookings Papers on Economic Activities*, 1, 201-68

Salop, Steven, 1979, A Model of the Natural Rate of Unemployment, *The American Economic Review*, 69, 117-25

Samuelson, Paul, 1948, International Trade and the Equalisation of Factor Prices, *The Economic Journal*, 58, 163-84

Samuelson, Paul, 1954,The Pure Theory of Public Expenditure, *Review of Economics and Statistics*, 36, 4, 387-89

Samuelson, Paul, 1958, An Exact Consumption-Loan Model of Interest with or without the Social Contrivance of Money, *Journal of Political Economy*, 66, 6, 467-82

Samuelson, Paul, 1975, Optimum Social Security in a Life-Cycle Growth Model, *International Economic Review*, 16, 3, 539-44

Sargan, John, 1958, The Estimation of Economic Relationships using Instrumental Variables, *Econometrica*, 26, 393-415

Sargent, Thomas, 1987a, *Macroeconomic Theory*, Academic Press

Sargent, Thomas, 1987b, *Dynamic Macroeconomic Theory*, Harvard University Press

Sargent, Thomas and Neil Wallace, 1975, "Rational" Expectations, the Optimal Monetary Instrument, and the Optimal Money Supply Rule, *Journal of Political Economy*, 83, 2, 241-54

Sargent, Thomas and Neil Wallace, 1981, Some Unpleasant Monetarist Arithmetic, *Federal Reserve Bank of Minneapolis Quarterly Review*, Fall, 1-17

Schumpeter, Joseph, 1942, *Capitalism, Socialism and Democracy*, Harper & Row Publishers

Segerstrom, Paul, 2007, Intel Economics, *International Economic Review*, 48, 1, 247-80

Segerstrom, Paul, T. Anant and Elias Dinopoulos, 1990, A Schumpeterian Model of the Product Life Cycle, *The American Economic Review*, 80, 1077-92

Siegfried, John and Laurie Evans, 1994, Empirical Studies of Entry and Exit: A Survey of the Evidence, *Review of Industrial Organization*, 9, 121-55

Shapiro, Carl and Joseph Stiglitz, 1984, Equilibrium Unemployment as a Worker Discipline Device, *The American Economic Review*, 74, 3, 433-44

Sharpe, William, 1964, Capital Asset Prices - A Theory of Market Equilibrium Under Conditions of Risk, *Journal of Finance*, 19, 3, 425-42

Shiller, Robert, 2005, *Irrational Exuberance*, 2nd Ed., Princeton University Press

Sidrauski, Michael, 1967, Rational Choices and Patterns of Growth in a Monetary Economy, *The American Economic Review*, 57, 534-44

Smith, Adam, 1776, *An Inquiry into the Nature and Causes of the Wealth of Nations*, Strahan and Cadell

Smith, Richard and Richard Blundell, 1986, An Exogeneity Test for a Simultaneous Equation Tobit Model with an Application to Labor Supply, *Econometrica*, 54, 3, 679-85

Solow Robert, 1956, A Contribution to the Theory of Economic Growth, *Quarterly Journal of Economics*, 70, 65-94

Solow Robert, 1957, Technological Change and the Aggregate Production Function, *The Review of Economics and Statistics*, 39, 312-20

Solow Robert, 1979, Another Possible Source of Wage Stickiness, *Journal of Macroeconomics*, 1, 79-82

Spence, Michael, 1976, Product Selection, Fixed Costs and Monopolistic Competition, *Review of Economic Studies*, 43, 217-35

Spencer, Barbara and James Brander, 1983, International R&D Rivalry and Industrial Strategy, *Review of Economic Studies*, 50, 707-22

Staiger, Douglas and James Stock, 1997, Instrumental Variables Regression with Weak Instruments, *Econometrica*, 65, 3, 557-86

Stiglitz, Joseph, 2006, *Making Globalization Work*, W.W. Norton & Co.

Stock, James and Motohiro Jogo, 2005, Testing for Weak Instruments in Linear IV Regression, in *Identification and Inference for Econometric Models: Essays in Honor of Thomas Rothenberg*, D.W.K. Andrews and J.H. Stock Eds., Cambridge University Press, 80-108

Stockey, Nancy and Robert Lucas with Edward Prescott, 1989, *Recursive Methods in Economic Dynamics*, Harvard University Press

Stolper Wolfgang and Paul Samuelson, 1941, Protection and Real Wages, *Review of Economic Studies*, 9, 58-73

Summers, Lawrence, 1988, Relative Wages, Efficiency Wages, and Keynesian Unemployment, *The American Economic Review*, 78, 2, 383-88

Sutton, John, 1991, *Sunk Costs and Market Structure*, MIT Press

Sutton, John, 1998, *Technology and Market Structure*, MIT Press

Sutton, John, 2007a, Quality, Trade and the Moving Window: the Globalisation Process, *The Economic Journal*, 117, F469-98

Sutton, John, 2007b, Market Share Dynamics and the "Persistence of Leadership" Debate, *The American Economic Review*, 97, 1, 222-41

Sutton, John, 2007c, Market Structure: Theory and Evidence, Ch. 35 in *Handbook of Industrial Organization*, M. Armstrong & R. Porter Eds, Elsevier, 3, 2301-68

Sutton, John, 2008, Endogenous Market Structures and Globalisation, "*Stackelberg Lecture*" presented at the 2008 Intertic Conference

Svensson, Lars, 1985, Money and Asset Prices in a Cash-in-Advance Economy, *Journal of Political Economy*, 93, 5, 919-44

Svensson, Lars, 1997, Optimal Inflation Targets, 'Conservative' Central banks, and Linear Inflation Contracts, *The American Economic Review*, 87, 98-114

Swan, Trevor, 1956, Economic Growth and Capital Accumulation, *Economic Record*, 344-61

Symeonidis, George, 2000, Price and Nonprice Competition with Endogenous Market Structure, *Journal of Economics & Management Strategy*, 9, 1, 53-83

Tabellini, Guido and Alberto Alesina, 1990, Voting on the Budget Deficit, *The American Economic Review*, 80, 37-49

Taylor, John, 1980, Aggregate Dynamics and Staggered Contracts, *Journal of Political Economy*, 88, 1, 1-23

Taylor, John, 1993, Discretion versus policy rules in practice, *Carnegie-Rochester Conference Series on Public Policy*, 39, 195-214

Taylor, John, 2009, The Financial Crisis and the Policy Responses: An Empirical Analysis of What went Wrong, NBER WP 14631

Tesoriere, Antonio, 2008a, Endogenous Timing with Infinitely Many Firms, *International Journal of Industrial Organization*, 26, 1381-88

Tesoriere, Antonio, 2008b, Stackelberg Equilibrium with Many Leaders, U-shaped Average Costs, and Endogenous Number of Followers, mimeo, University of Palermo

Tirelli, Patrizio, 1993, *Monetary and Fiscal Policy, the Exchange Rate and Foreign Wealth*, Macmillan Press

Tobin, James, 1958, Estimation of Relationships for Limited Dependent Variables, *Econometrica*, 19, 26, 24-36

Tobin, James, 1965, Money and Economic Growth, *Econometrica*, 33, 4, 671-84

Tobin, James, 1969, A General Equilibrium Approach To Monetary Theory, *Journal of Money, Credit and Banking*, 1, 1, 15-29

Tornell, Aaron and Andes Velasco, 1992, The Tragedy of the Commons and Economic Growth: Why Does Capital Flow from Poor to Rich Countries?, *Journal of Political Economy*, 100,6, 1208-31

Toshiki, Tamai, 2006, Endogenous Market Structure and Fiscal Policy in an Endogenous Growth Model with Public Capital, *Economics Bulletin*, 15, 11, 1-10

Trefler, Daniel, 1995, The Case of the Missing Trade and Other Mysteries, *The American Economic Review*, 85, 5, 1029-46

Turnovsky, Stephen, 1997, *International Macroeconomic Dynamics*, MIT Press

Turnovsky, Stephen, 2000, *Methods of Macroeconomc Dynamics*, MIT Press

Uzawa, Hirofumi, 1964, Optimal Growth in a Two-Sector Model of Capital Accumulation, *Review of Economic Studies*, 31, 1-24

Yang, Xiaokai and Ben Heijdra, 1993, Monopolistic Competition and Optimum Product Diversity: Comment, *The American Economic Review*, 83, 1, 295-301

Venables, Anthony, 1986, Trade and Trade Policy with Impefect Competition: the Case of Identical Products and Free Entry, *Journal of International Economics*, 19, 1-19

Ventura, Jaume, 1997, Growth and Interdependence, *Quarterly Journal of Economics*, 112, 1, 57-84

Vives, Xavier, 2008, Innovation and Competitive Pressure, *Journal of Industrial Economics*, 56, 3, 419-69

Wälde, Klaus and Weiss, Pia, 2007, International Competition, Downsizing and Wage Inequality, *Journal of International Economics*, 73, 2, 396-406

Walsh, Carl, 1995, Optimal Contracts for Central Bankers, *The American Economic Review*, 85, 1, 150-67

Walsh, Carl, 1998, *Monetary Theory and Policy*, Cambridge, The MIT Press

Weinschenk, Philipp, 2008, Persistence of Monopolies and Research Specialization, mimeo, Bonn Graduate School of Economics

Weiss, Andrew, 1980, Job Queues and Layoffs in Labor Markets with Flexible Wages, *Journal of Political Economy*, 88, 3, 526-38

Wen, Yi, 2006, Demand Shocks and Economic Fluctuations, *Economics Letters*, March 2006, 90, 3, 373-83

Wickens, Michael, 2008, *Macroeconomic Theory. A Dynamic General Equilibrium Approach*, Princeton University Press

Wilson, Bart and Reynolds, Stanley, 2005, Market Power and Price Movements over the Business Cycle, *Journal of Industrial Economics*, 52, 2, 145-74

Wong, Kar-Yiu, 1995, *International Trade in Goods and Factor Mobility*, MIT Press

Woodford, Michael, 2003, *Interest and Prices*, Princeton University Press

Wooldridge, Jeffrey, 2002, *Econometric Analysis of Cross Section and Panel Data*, MIT Press

Wrede, Matthias, 2004, Small States, Large Unitary States and Federations, *Public Choice*, 119, 1-2, 219-40

Wu, Yangru and Junxi Zhang, 2000, Endogenous Markups and the Effects of Income Taxation: Theory and Evidence from OECD Countries, *Journal of Public Economics*, 77, 383-406

Wu, Yangru and Junxi Zhang, 2001, The Effects of Inflation on the Number of Firms and Firm Size, *Journal of Money, Credit and Banking*, 33, 2, 251-71

Yang, Xiaokai and Ben Heijdra, 1993, Monopolistic Competition and Optimum Product Diversity: Comment, *The American Economic Review*, 83, 1, 295-301

Young, Alwyn, 1991, Learning by Doing and the Dynamic Effects of International Trade, *Quarterly Journal of Economics*, 106, 368-405

Zeira, Joseph, 2004, Innovations, Patent Races and Endogenous Growth, mimeo, Harvard University

Žigić, Krešimir, 2008, Stackelberg Leadership with Product Differentiation and Endogenous Entry: Some Comparative Statics and Limiting Results, mimeo, Intertic

Index

Abuse of dominance, 301–305
Acemoglu, Daron, 30, 34, 40, 163, 197, 213, 214, 216, 222, 223, 226, 228, 230, 258
Adaptive expectations, 5, 48, 52
Adjustment cost of capital accumulation, 29, 40, 44
Adverse selection and endogenous entry, 93
Agénor, Pierre-Richard, 29
Aghion, Philippe, 24, 47, 55, 92, 143, 197, 212, 222, 223, 225, 226, 228, 234, 240, 271
Aghion-Howitt model, 222, 223, 225, 234
Airbus, 291
Akerlof, George, 47, 252
Alesina, Alberto, 9, 23, 30, 242, 269
Altavista, 274
Amazon, 264, 265
American Recovery and Reinvestment Act of 2009, 281
Angeloni, Ignazio, 30, 269
Animal spirits, 254
Animoto, 264
Antitrust policy and endogenous market structures, 59, 73, 211, 212, 273, 298–305, 308
Antitrust policy and the Microsoft case, 305
Arrow effect, 198–200, 223, 226, 227
Arrow, Kenneth, 15, 17, 198, 200, 226, 281
Article 82 E.C., 301–305
As efficient competitor test, 303
Asset pricing, 5, 110
Asset-backed commercial paper, 255
Atkeson, Andrew, 107, 148, 168
Auctions and endogenous entry, 93
Automatic stabilizers, 283
Azariadis, Costas, 18, 19, 24
Backus, David, 23, 44, 168
Bagwell, Kyle, 36
Balance of payments, 27, 32, 167
Baldwin, Richard, 164, 240, 283
Ball, Laurence, 140
Bank of America, 254, 288
Bank of China, 254
Bank of England, 288, 290
Barro, Robert, 6–8, 18, 23, 24, 40, 43, 50, 112, 126, 133, 136, 214, 215, 217, 219, 222, 226, 234, 235, 237, 242, 280
Baumol-Allais-Tobin approach, 9
Baxter, Marianne, 43, 45, 128
Bear Stearns, 254
Bellman equation, 38, 229
Benassy, Jean-Pascal, 97
Benhabib, Jess, 49
Bergin, Paul, 105, 137
Berlusconi, Silvio, 279
Bernanke, Ben, 141, 144, 255, 288, 296
Bernard, Andrew, 104
Berry, Steven, 54, 146, 149
Bertrand competition, 63, 69, 80, 109, 118, 160, 181, 190, 215
Betts, Caroline, 188
Bhagwati, Jagdish, 145, 173
Bhattacharjea, Aditya, 148, 170
Big push, 85
Bilbiie, Florin, 97, 105, 129, 130, 133–139, 239
Bils, Mark, 95
Blanchard, Olivier, 40, 47, 88, 95, 141, 142, 285
Blanchard-Giavazzi model, 141, 142
Blinder, Alan, 144
Blundell, Richard, 198, 209, 211
Boldrin, Michele, 24, 38, 44, 282
Bonanno, Giacomo, 79
Boone, Jan, 173
Brander, James, 54, 91, 145, 147, 173–175, 184, 186, 245, 291

Bresnahan, Timothy, 54, 146, 149, 150
Broda, Christian, 163, 260–262
Browsers, 305–308
Brumberg, Richard, 4
Buchanan, James, 282
Bundling, 302, 304, 305, 308
Burnstein, Ariel, 107, 148, 168
Bush, George W., 281, 302
Buy American, 281

Caballero, Ricardo, 141, 255
Calibration, 41, 120
Calvo, Guillermo, 23, 32, 47, 137, 237, 281
Campbell, Jeffrey, 54, 146, 154, 155
Campbell, John, 40
Capital account, 27, 32
Capital Asset Pricing Model, 6
Capital income tax, 45, 234, 235, 279, 280
Case-Shiller index, 279
Cash-in-advance constraint, 9, 137, 239
Cato Institute, 281
Cavallari, Lilia, 169
Centre for European Economic Research (ZEW), 201
CES production function, 17, 26, 40, 213
Chamley, Christophe, 45
Chaos, 38
Chicago approach to antitrust policy, 300–305
China Construction Bank, 254
Cholesky decomposition, 104
Christiano, Lawrence, 44, 49
Chrome, 305, 307, 308
Cisco, 265
Citigroup, 254, 288
Clarida, Richard, 50
Clinton, Bill, 251
Cloud computing, 77, 260, 263–267, 308
Cloud platforms, 263
Cobb-Douglas production function, 11, 12, 14, 18, 24, 25, 28, 33, 40, 131, 213
Cohen, Wesley, 4, 198, 220
Colciago, Andrea, 96, 100, 105, 127
Commerzbank, 255
Community Innovation Survey, 201
Comparative advantage, 35, 244
Competition effect, 96, 122, 159, 250, 268, 269
Competitive devaluations, 187–193, 294–296
Constant Returns to Scale, 11, 13, 131, 213, 220, 222
Consumption beta, 6
Consumption-output anomaly, 28, 168
Cook, David, 98
Cooley, Thomas, 44
Cooper, Russell, 55, 56, 97, 140
Corsetti, Giancarlo, 105, 137, 187
Cost minimization, 11
Countercyclical fiscal policy, 8, 45, 135, 136, 280, 281
Cournot competition, 60, 67, 76, 79, 81, 96, 108, 115, 158, 170, 177, 185, 189, 215
Creane, Anthony, 54, 93
Credit constraints, 143
Credit easing, 288
Credit market imperfections, 142
Credit rationing, 141, 255, 278
Current account, 27, 32, 167, 244, 296–298
Current account sustainability, 27
Czarnitzki, Dirk, 54, 150–153, 198, 200, 206, 272

Da Costa, Carlos, 11
Dasgupta, Partha, 53, 74
Dasgupta-Stiglitz model, 53, 54, 74
De Santis, Roberto, 147, 173
Deardoff, Alan, 36
Debreu, Gerard, 15
Debt overhang, 28
Deep habits, 97
Dell, 265
Denicolò, Vincenzo, 226
Department of Justice, U.S.A., 301, 302
Deregulation, 142, 168
Deutsche Bank, 251
Devereux, Michael, 90, 97, 98, 137, 148, 188
Diamond, Peter, 16, 17, 37, 56
Directed technological change, 213, 214, 222
Display advertising, 275
Divine coincidence, 49
Dixit, Avinash, 13, 36, 54, 60, 64, 66, 69, 71, 95, 97, 109, 145, 149–153, 157, 162, 173, 175, 186, 258, 271
Dixon, Huw, 219
Doi, Noriyuki, 198
DoJ Report on Competition and Monopoly, 301
Domar, Evsey, 17

Dominance, 301, 303
Dornbusch, Rudiger, 27, 31, 35, 188, 190, 294
DoubleClick, 275, 276
Draghi, Mario, 269
Drazen, Allan, 9, 24
Driffill, John, 23
Dynamic inefficiency, 87, 114, 116, 117, 119, 133, 199, 233, 277, 297
Dynamic programming, 38, 228

Eaton, Jonathan, 145, 173, 175, 182, 291
Economic geography, 156
Effects-based approach, 303
Efficiency defense, 304
Efficiency wages, 141
Efficient wage bargaining model, 141
Eichenbaum, Martin, 49
Elastic cloud computing, 264
Elkhoury, Marwan, 137
Empirical evidence on the EMSs approach, 54, 98, 103–105, 149–152, 154–156, 200–205, 207–209, 211, 212, 260–262
EMSs and growth, 91, 212, 240, 243
EMSs and trade, 90, 156
EMSs in a two-period exchange economy, 75
EMSs in general equilibrium, 78
EMSs in partial equilibrium, 60
EMSs in steady state, 81, 87, 114, 224
EMSs with isoelastic utility, 65
EMSs: definition, 52
Endogeneity test, 209
Endogenous growth, 24, 40, 47, 91, 197
Engel, Charles, 188
Equity premium puzzle, 6
Equity price stabilization, 139
Escape competition effect, 227
Etro, Federico, 30, 54–57, 59, 71, 75, 96, 105, 127, 147, 148, 150, 153, 173, 175–183, 186, 198, 200, 206, 220, 225, 227, 228, 234, 235, 242, 245, 269, 272
Euler equation: definition, 3
European Central Bank, 287–290, 297
European Commission, 301–305
European Patent Office, 203
European Union, 30, 246, 256, 266, 283, 289, 291, 300–305
Evans, Charles, 49
Exchange rate overshooting, 31
Exchange rate pass-through, 192
Exclusionary abuses in antitrust policy, 301–305
Exhange rate policy, 30–32, 169, 187–190, 294
Exogenous market structures, 46, 170, 174

Facebook, 264, 265
Factor content of trade, 35
Factor price equalization, 33, 35, 40, 161
Factor-biased technological progress, 34
Fannie Mae and Freddie Mac, 254
Farmer, Roger, 219
FDIC, 255
Federal mandates, 30
Federal Reserve, 254, 255, 287–290, 298
Federal Reserve Economic Database, 100
Feenstra, Robert, 9, 35, 65, 157
Feldstein, Martin, 28
FIAT, 257
Financial crisis of 2008, 251, 252, 254, 256, 258, 281, 283, 288, 289
Financial Stability Forum, 269
Firefox, 305, 307, 308
Fiscal gap, 21
Fiscal theory of the price level, 21
Fisher equation, 10, 31, 45, 49
Fisher, Stanley, 35, 40, 47
Fixed Effects, 152
Fixed exchange rate, 31, 169, 187–193, 294
Flexible exchange rate, 30, 31, 169, 187, 294
Foreign direct investment, 27, 147, 165
Fortis, 255
Friedman rule, 10, 23, 46, 137, 236
Friedman, Milton, 4, 137, 236, 238, 285
Frish elasticity, 5, 42
Fujita, Masahisa, 156

Gabszewicz, Jean Jaskold, 79
Gains from trade, 36, 90, 145, 158, 161, 165, 267–269
Galì, Jordi, 49, 50, 95
Galor, Oded, 18, 33, 143, 163
Geithner, Tim, 255
General Motors, 257
General Purpose Technology, 77, 89, 224, 250, 260, 263–267, 272
German Federal Statistical Office, 153
German Monopolies Commission, 152

German Patent Office, 203
Geroski, Paul, 207
Gertler, Mark, 50, 141
Ghironi, Fabio, 55, 83, 91, 96, 97, 105, 129, 130, 133–139, 143, 146, 148, 163, 164, 166–169, 195, 239
Ghironi-Melitz model, 163–168
Giavazzi, Francesco, 141, 142, 283
Global EMSs, 240, 243
Globalization backlash, 242, 243, 292, 293
Golden rule, 20, 39, 114
Goldman Sachs, 251, 254, 288
Google, 264, 265, 267, 273–276, 305, 307, 308
Gordon, David, 23, 50
Grandmont, Jean Michel, 22
Gravity equation, 166
Great Depression of 1929, 252, 255, 282, 293
Green, Jerry, 15, 79
Greene, William, 152, 204, 206
Greenspan, Alan, 288, 296
Griffith, Rachel, 198
Grossman, Gene, 35, 145, 165, 173–175, 182, 184–186, 212, 235, 240, 291
Growth convergence, 18
Guidance on the Commission's Enforcement Priorities in Applying Art. 82, 301

Habits, 44, 97
Hall, Robert, 5, 95
Hansen J test, 209, 211
Hansen, Gary, 44
Harrod, Roy, 17
Harrod-Balassa-Samuelson effect, 164
Hausman, Jerry, 209
Head, Allen, 97, 98, 137
Heckman, James, 149
Heckscher, Eli, 33, 145
Heckscher-Ohlin theorem, 35, 161
Hedge funds, 254
Heijdra, Ben, 109
Helpman, Elhanan, 36, 145, 157, 165, 169, 173, 184–186, 212, 235, 240, 267
Herfindahl-Hirschman Index, 151–153, 202, 204, 212
Heteroscedasticity, 204
Hewlett-Packard, 265
Hicks, John, 15
Hodrick-Prescott filter, 44, 100, 103, 104
Home bias, 28
Hopenhayn, Hugo, 54, 146, 154, 155
Horioka, Charles, 28
Horstmann, Ignatius, 147, 148
Hovenkamp, Herbert, 302
Howitt, Peter, 24, 47, 55, 92, 143, 197, 212, 222, 223, 225, 226, 228, 234, 240, 271
HSBC, 254
Human capital, 24, 270, 271

I welfare Theorem, 15, 20, 36, 38, 41
IBM, 265
Iceberg costs, 161, 164, 241
IDC, 264
II Welfare Theorem, 15
IMF, 187, 283, 297
Imperfect collusion, 74, 110, 123
Impulse response functions: demand shocks, 43, 125
Impulse response functions: supply shocks, 42, 121
Income distribution, 13, 86, 269, 270
Income inequality, 34
Industrial & Commercial Bank of China, 254
Industrial districts in Italy, 278
IndyMac Bank, 254
Infant industry argument, 184
Inflation persistence, 48, 139
Inflation targets, 287
Inflation tax, 23
Instrumental variables, 207–211
International business cycles, 44, 168, 195
Internet Explorer, 305–308
Intertemporal approach to the current account, 27
Intertemporal substitution, 5, 42, 77, 78, 89, 96, 122, 167
INTERTIC Lecture, 258, 271
Investment banking, 254
Involuntary unemployment, 141
IPRs protection, 58, 193, 194, 197, 223, 230, 232, 233, 246, 271, 279, 287, 298, 300
Irrational exuberance, 251
IS-LM model, 15

Jaimovich, Nir, 98, 109, 218
Job creation & destruction, 98, 142, 162
John, Andrew, 56
Johnson, Harry, 36

Jones, Charles, 222
Jones, Ronald, 34
JP Morgan Chase, 254, 288

Kaldor, Nicholas, 36
Katsoulacos, Yannis, 73
Kehoe, Patrick, 44, 168
Keynes, John Maynard, 8, 15, 31, 85, 280
Keynesian multiplier, 15
King, Mervin, 288
King, Robert, 40, 42–45, 120, 126, 128, 132
Kiyotaki, Nobuhiro, 47, 88, 95, 141, 285
Klein, Lawrence, 281
Klepper, Steven, 4, 198, 220
Kortum, Samuel, 198, 220
Kovac, Eugen, 71, 72
Kraft, Kornelius, 198, 200, 206
Kroes, Neelie, 273
Krugman, Paul, 32, 34, 36, 47, 54, 90, 91, 145, 147, 149, 153, 156–158, 160, 162, 163, 167, 169, 173, 195, 240, 255, 267, 268, 282
Kydland, Finn, 9, 23, 40, 44, 88, 168, 249

Labor market imperfections, 140, 162, 270
Labor unions, 141, 142, 278, 283
Laffer curve, 22, 23
Lahiri, Amartya, 32
Lahiri, Sajal, 147, 161
Laibson, David, 40
Lapham, Beverly, 97, 98, 137
Law of one price, 30, 188
Leamer, Edward, 34, 35
Lee, Khang Min, 90, 98, 148
Lehman Brothers, 251, 254
Leontief, Wassily, 17
Lerner, Abba, 33, 36, 145, 179
Lewis, Vivien, 105, 136, 137, 139
Life-cycle Hypothesis, 4
Lisbon agenda, 246, 300, 305
Lloyds, 254, 255
Lobbying, 184–186
Local currency pricing, 188, 295
Lost decade of Japan, 290
Loury, Glenn, 198
Lucas paradox, 244
Lucas, Robert, 5, 6, 9, 11, 22, 24, 29, 38, 45, 48, 88, 228, 244, 289

M.E.T.I., 291
Maastricht criteria, 283
MacDonald, Ian, 141
Maggi, Giovanni, 35, 185
Mancini Griffoli, Tommaso, 137
Mankiw, Gregory, 2, 24, 47, 48, 68, 285
Mannheim Innovation Panel, 201, 272
Manuszak, Mark, 54, 146, 149, 150
Marchionne, Sergio, 257
Market leaders, 71, 110, 124, 200, 226, 271
Market shares in the browsers' market, 306
Market structure indeterminacy, 13
Markowitz, Harry, 6
Markusen, James, 147, 148, 161
Marshall-Lerner condition, 187
Marx, Karl, 1
Mas Colell, Andreu, 15, 79
Maskin, Eric, 75
Matching grants, 30
Matching models, 142
Matsumura, Toshihiro, 54
Matsuyama, Kiminori, 216
Meade, James, 36
Mehra, Rajnish, 6
Melitz, Marc, 55, 83, 91, 96, 97, 105, 129, 130, 133–139, 146–148, 163–169, 195, 239
Merchantilism, 147
Merril Lynch, 254
Merrill Lynch, 251
Merton, Robert, 6
Method of undetermined coefficients, 228
Microsoft, 265, 273–276, 300, 305, 306, 308
Migration, 29
Miller, Merton, 112
Milton, Friedman, 10
Minimum wage, 141
Modigliani, Franco, 4, 112
Modigliani-Miller Theorem, 112
Monacelli, Tommaso, 105, 127
Monetary base, 9
Money in the utility function, 9, 137, 236
Monopolization, 301–305
Montiel, Peter, 29
Montrucchio, Luigi, 38
Moore, John, 141
Moral hazard, 254
Moral hazard in the banking sector, 143, 254

Morgan Stanley, 251, 254, 288
Motta, Massimo, 73
Mozilla, 305, 307, 308
Multinationals, 147, 165, 295
Mundell, Robert, 31, 187
Mundell-Fleming model, 31, 32, 187, 188
Mussa, Michael, 35
Muth, John, 5

NACE 3-digit level, 152, 208
NACE 4-digit level, 153
NAIRU, 48
National Bureau of Economic Research, 101
National Economic Council, 255
Neoclassical growth model, 16
Neoclassical theory of trade policy, 36, 290, 291
Neoclassical trade theory, 33
Network effects, 266, 274, 275
New Deal, 252
New Economy, 51, 213, 249–251, 263, 271, 273, 275, 276, 308
New Keynesian macroeconomics, 37, 47, 95, 96, 137–140, 237, 238, 250, 284, 285, 289
New Keynesian Phillips curve, 48, 49, 139, 140
New open economy macroeconomics, 169, 187, 188, 294
Nominal exchange rate, 30–32, 164
Non-tradable goods, 28, 31, 36, 40, 164, 167, 269
Nordhaus, William, 23
Norman, Victor, 36
Northern Rock, 254
Numeraire, 14

Obama, Barack, 255, 281, 282, 302
Obstfeld, Maurice, 27, 40, 169, 187, 294
Ohlin, Bertil, 33, 145
Okun's Law, 48
OLS, 203
On-premises platforms, 263
Online advertising, 272–276, 308
Ono, Yoshitasu, 147, 161
Opera, 305, 307, 308
Optimal export subsidy, 91, 148, 177–180, 182, 183, 186, 290–292
Optimal export tax, 36, 145, 173, 181, 291, 292
Optimal fiscal policy, 6, 8, 20, 29, 45, 133, 235, 239, 279
Optimal import tariff, 36, 169–171, 290
Optimal market structures, 64, 133
Optimal monetary policy, 10, 23, 30, 46, 49, 138, 236, 239, 285, 287
Optimal R&D subsidy, 194, 233, 235, 244, 245, 298
Optimal trade policy, 36, 169, 172, 290
Oslo manual, 201
Ottaviano, Giancarlo, 147
Output gap, 49, 285
Outsorcing, 165
Overlapping-generations model, 16, 17, 37
Overnight interbank rate, 289

Padilla, Jorge, 305
Panel data, 150, 152, 153, 202
Pasinetti, Luigi, 17
Patent races, 58, 199, 220
Patent stock, 202–205
Peretto model, 217, 218
Peretto, Pietro, 55, 92, 161, 198, 218, 235
Perfect capital mobility, 28
Perotti, Roberto, 105, 127
Perpetual inventory method, 202, 203
Persson, Torsten, 9, 23
Pesenti, Paolo, 187
Pharmaceutical sector, 257
Phelps, Edmund, 20, 24, 277
Phillips curve, 48
Phillips, Alban, 48
Plosser, Charles, 40
Political business cycles, 23
Political economy of export promotion, 184
Political economy of fiscal policy, 9
Political economy of international unions, 29, 30, 32, 36
Political economy of monetary policy, 23
Political economy of trade policy, 36, 184
Political geography, 242, 243
Polo, Michele, 73
post-Chicago approach to antitrust policy, 300–305
Precautionary savings, 5
Predatory pricing, 302
Prescott, Edward, 6, 9, 23, 38, 40, 44, 88, 228, 249, 282
Price stickiness, 15, 31, 47, 49, 95, 137–140, 237–239, 285, 287

Pricing to market, 168, 188, 294, 296
Producer currency pricing, 189, 295
Production subsidy, 135, 170–172
Public capital, 24
Purchasing power parity, 30, 148, 164, 167, 195, 269, 284

Quantitative easing, 288, 289

R&D intensity, 200
Ramsey model, 37
Ramsey rule, 8
Ramsey, Frank, 8, 37
Random walk hypothesis, 5
Rapping, Leonard, 5, 88
Rare disasters, 6
Rational expectations, 1, 5, 22, 48, 52
Rational preferences, 3
Ravn, Morten, 97
Real Business Cycles, 37, 40–43, 128–130, 250
Real estate prices, 251, 252
Real exchange rate, 31, 164, 166
Rebelo, Sergio, 24, 40, 42, 44, 120, 128, 132
Redding, Stephen, 104
Refusal to supply, 302
Regulation and access pricing, 94
Reichlin, Pietro, 24
Reis, Ricardo, 48
Reiss, Peter, 54, 146, 149, 150
Relative scale effects, 242, 244
Ricardian equivalence, 7, 41, 112, 133, 143
Ricardian model, 35
Ricardo, David, 1, 7, 35
Ricardo-Viner model, 35
Right to Manage model, 142
Risk premium, 5
Rivera-Batiz, Luis, 240
Rodrik, Dani, 173, 269
Rogerson, Richard, 44
Rogoff, Kenneth, 23, 27, 28, 169, 187, 294
Romer model, 216
Romer, David, 24, 140
Romer, Paul, 24, 47, 55, 91, 92, 197–199, 212, 216, 240, 241, 271
Roosevelt, Franklin, 252
Rotemberg, Julio, 44, 47–49, 75, 95, 97, 101, 137, 139, 237, 285
Royal Bank of Scotland, 255
Rubinstein, Yona, 165
Rule of thumb consumers, 143
Rustichini, Aldo, 24
Rybczynski theorem, 34

Sachs, Jeffrey, 27
Saddle-path stability, 38, 113
Safari, 305, 307, 308
Sala-i-Martin, Xavier, 18, 214, 215, 217, 219, 222, 226, 234, 235
Sale subsidy, 135
Saloner, Garth, 75
Samuelson rule, 7, 8, 21, 29, 44, 134
Samuelson, Paul, 7, 20, 22, 33, 35, 134, 164, 281
Sargan test, 209, 211
Sargent, Thomas, 24, 40, 48, 228, 289
Schmitt-Grohe, Stephanie, 49, 97
Schott, Peter, 104
Schumpeter, Joseph, 197
Search advertising, 274
Second moments, 43, 128
Segerstrom, Paul, 212, 226, 228
Seigniorage, 22
Selection effect, 148, 165–168
Self-fulfilling financial crisis, 187
September 11, 2001, 251, 288
Shadow banking system, 254
Shapiro, Carl, 141
Sharpe, William, 6
Shiller, Robert, 251, 252
Sidrauski, Michael, 9, 236
SMEs, 189, 266, 267, 278, 285, 295
Smith, Adam, 1, 147, 172
Smith, Richard, 209, 211
Smith-Blundell test, 209
Smoot-Hawley Act, 293
Software as a Service, 263
Software market, 262–267, 272, 273, 275, 305–308
Software plus Services, 265
Solow model, 16, 81, 83
Solow, Robert, 16, 17, 82, 85, 91, 93, 141, 258, 281
Sovereign risk, 28
Special interest groups, 184
Spencer, Barbara, 54, 91, 145, 173–175, 184, 186, 245, 291
Spolaore, Enrico, 242
Stackelberg competition, 71, 110, 123
Stackelberg competition with endogenous entry, 71, 73, 227
Stagflation, 238
Staggered price adjustments, 47

Stahler, Frank, 147, 167, 173
Staiger, Douglas, 210
Staiger, Robert, 36
Stebunovs, Viktors, 143, 168
Stiglitz, Joseph, 53, 54, 60, 64, 66, 69, 71, 74, 95, 97, 109, 141, 143, 145, 149–153, 157, 162, 282, 300
Stock market bubbles, 111, 143, 251
Stock options, 254
Stock, James, 210
Stockey, Nancy, 38, 45, 228
Stolper-Samuelson theorem, 34
Subprime crisis, 252, 256, 281, 289
Subsidiarity, 30
Summers, Lawrence, 141, 255
Superneutrality, 236
Supply-based fiscal policy, 279–282, 284, 285
Sutton, John, 53, 163, 198, 207, 208
Svensson, Lars, 9, 50
Swan, Trevor, 16

Tabellini, Guido, 9, 23, 257, 283
Tax smoothing, 8, 136, 280
Taxation and endogenous entry, 94
Taylor principle, 45, 49, 139
Taylor rule, 50, 285, 287, 289
Taylor, John, 45, 47, 49, 50, 285, 289
Taylor, Mark, 2
Tenreyro, Silvana, 237
Term Asset-Backed Loan Facility, 288
Terms of trade, 36, 91, 145, 169, 173, 175, 176, 190, 291
Theory of permanent income, 4
Time inconsistency, 9, 23, 50, 237
Tirelli, Patrizio, 23, 31, 32, 141
Tirole, Jean, 75
Tobin's q, 84
Tobin, James, 23, 84, 203, 238
Tobit model, 203
Toxic assets, 254, 255
Toyota, 257
Trade balance, 27, 166, 187, 244, 296, 297
Transport costs, 161, 164, 184, 241
Trefler, Daniel, 35
Trichet, Jean-Claude, 288
TRIP agreements, 301
Turnovsky, Stephen, 40
Twin-deficits argument, 29
Two-period model and consumption, 3
Two-period model and current account, 27
Two-period model and EMSs, 75
Two-period model and fiscal policy, 6
Two-period model and labor supply, 4
Two-period model and money, 9
Two-SLS regression, 211
Tying, 302, 304, 305, 308

UBS, 251
Uncovered interest parity, 31
Unemployment, 15, 140, 142, 162, 257, 270, 279
Unpleasant monetaristic arithmetic, 24
Urban economics, 156
Uribe, Martin, 49, 97
Uzawa, Hirofumi, 24

Van Reenen, John, 198
VAR evidence, 41, 103–105, 139
Vegh, Carlos, 32
Venables, Anthony, 148, 156
Ventura, Jeaume, 40
Vinogradov, Viatcheslav, 71, 72
Vives, Xavier, 53, 74
Vox, 283

Wachovia, 254
Wage bargaining, 141
Wage subsidy, 136
Wallace, Neil, 24, 48, 289
Walras' law, 14, 113, 166
Walrasian equilibrium, 14, 15, 79
Walsh, Carl, 23
Weak and strong equilibrium bias, 213
Weil, David, 24
Weinstein, David, 163, 260–262
Wells Fargo, 254
Wen, Li, 128
Werning, Ivan, 11
Whinston, Michael, 15, 68, 79
Windows, 305–308
Windows Azure, 265
Windows MediaPlayer, 305
Wong, Kar-Yiu, 29
Woodford, Michael, 44, 49, 50, 75, 95, 97, 101, 137, 285
World Trade Organization, 269
WTO, 173, 291, 293, 294
Wu, Yangru, 98, 135, 136

Yahoo!, 265, 273–276
Yang, Xiaokai, 109
Yellen, Janet, 47

Zeira, Joseph, 143
Zhang, Junxi, 98, 135, 136
Zigic, Kresimir, 54, 71, 72
Zigic, Kreso, 173
Zingales, Luigi, 141

Printed in the United States
151198LV00001B/83/P

9 783540 874263